Plant
Biotechnology

Plant Biotechnology

S Umesha

CRC Press
Taylor & Francis Group
Boca Raton London New York

CRC Press is an imprint of the
Taylor & Francis Group, an **informa** business

The Energy and Resources Institute

CRC Press
Taylor & Francis Group
6000 Broken Sound Parkway NW, Suite 300
Boca Raton, FL 33487-2742

First issued in paperback 2023

ISBN 13: 978-1-03-265383-9 (pbk)
ISBN 13: 978-0-367-17504-7 (hbk)
ISBN 13: 978-0-429-05726-7 (ebk)

DOI: 10.1201/9780429057267

Print edition not for sale in South Asia (India, Sri Lanka, Nepal, Bangladesh, Pakistan or
Bhutan)

Publisher's Note
The publisher has gone to great lengths to ensure the quality of this reprint but points out that
some imperfections in the original copies may be apparent.

Library of Congress Cataloging in Publication Data
A catalog record has been requested

Visit the Taylor & Francis Web site at
http://www.taylorandfrancis.com

and the CRC Press Web site at
http://www.crcpress.com

Foreword

Plant biotechnology is a science that adapts plants for specific purpose by cross-breeding, extending their growing seasons, adjusting height, colour and texture and several other mechanisms. Some plant biotechnologies exist for the purpose of creating reliable and sustainable food sources for populations around the world, while other technologies using plants for scientific and medical research, and find ways to make crops resistant to droughts, diseases and other maladies.

Plant Biotechnology covers an extensive spectrum of different aspects of plant biotechnology based on the leading research articles in the field. The first chapter lays the foundation and the following chapters create the required structure according to the needs of the individual. Subsequently, somatic embryogenesis synthetic seeds, plant transformation technology, plant genetic engineering, applications of plant biotechnology and other important issues have been covered in detail. Many times, it becomes a problem to operate the acquired equipment. This book would serve as an excellent source for all undergraduate, postgraduate and PhD students in the field of plant biotechnology.

The author, Dr S. Umesha, my student and colleague at the Department of Studies in Biotechnology, University of Mysore, has done a very commendable job in filling the wide gap between the scientific workers and the fresh students. I look forward to many more such scientific accomplishments from Dr S. Umesha.

S.R.b
24/3/17

Prof. S. R. Niranjana
Vice Chancellor
Gulbarga University
Karnataka, India

Preface

The purpose of this book is to provide the basic and advanced theoretical information about the various aspects of plant biotechnology. Chapters 1 and 2 discuss all aspects of plant biotechnology effectively in detail, including tissue culture, nutrient medium, micronutrients, macronutrients, solidifying agents/supporting systems, growth regulators, sterilization of the medium, somatic embryogenesis, synthetic seeds, applications of tissue culture, production of haploids, diploids in genomics, germplasm conservations, production of secondary metabolites, industrial production of useful chemicals from plant tissue culture, industrial production of food additives and greenhouse technology.

Chapter 3 discusses plant transformation technology, hairy root culture, functions of *vir* proteins in T-DNA integration, gene expression reprogramming in response to *Agrobacterium*, development of binary vector systems, uses of 35S promoter, reporter genes, biolistic gene transfer, transgene stability, and gene silencing.

Chapters 4 and 5 analyse the plant genetic engineering for productivity and performance, resistance to herbicides, insect resistance, resistance to abiotic stresses, and molecular marker aided breeding, molecular markers, types of markers, biochemical markers, RFLP, linkage analysis, RAPD markers, microsatellites, SCAR, SNPs, AFLP, QTL, and mapping functions.

In Chapter 6, applications of plant biotechnology, industrial enzymes, genetic engineering, applications in food and food ingredients, dairy applications, baking applications, brewing and juicing applications, household and personal care applications, edible vaccines, plantibodies are dealt with. Chapter 7 examines propagation of economically important plants, biofertilizers, biopesticides, and vermiculture.

Chapter 8 explains different aspects of important issues in plant biotechnology, commercial status and public acceptance, biosafety guidelines, gene flow, and IPR.

I would like to thank many scientists who provided illustrations from their works and those who have helped in completing this mammoth task. I would like to express my deep sense of gratitude to Dr (Mrs.) Kaunain Roohie, for her keen interest in collecting the literature and helping me in preparation of the manuscript. Special thanks are due to the TERI Press for publishing the work with utmost interest.

I would appreciate suggestions/constructive criticisms from teachers and students for the improvement of this book subsequently.

Contents

Tissue Culture

1.1 INTRODUCTION TO CELL AND TISSUE CULTURE

Totipotency is one of the most wonderful phenomena of plant cells. An entire organism can be produced by this inherent capacity of cells when a suitable nutrient medium is present. In plant tissue culture, plant cells, tissues, or segments are grown for generating large numbers of new cells, tissues, or plants. Plant tissue culture is also used to study development of plant cells, and it provides mechanisms of genetic engineering and produces valuable chemicals found in plant cells. For understanding many applied aspects of plant science, it is essential to study the techniques of plant tissue culture. These techniques have been used in the past to study totipotency and determine the roles of hormones in cytodifferentiation and organogenesis. At present, an insight into plant molecular biology and gene regulation can be obtained from tissue culture and genetic engineering of plants. Plant tissue culture techniques also play a central role in innovative areas of applied plant science, such as plant biotechnology and agriculture.

Cloning is a very familiar concept for people who study the reproducibility of plants. Cloning is the process used to produce exact and multiple copies of a parent plant. Cuttings and grafts can be used for cloning. Cloning can also occur naturally through agents such as the propagules dispersed by spider plants. Much smaller pieces of tissue are used by plant tissue culture in the cloning process. Plant tissue culturists have many possibilities in this field and can do more than simply cloning plants. They can place new plants into culture or introduce new methods of working with plant cultures.

At the beginning of an experiment, a plant tissue culturist should make purposeful ideas about its necessity. There are different types of *in vitro* tissue culture approaches. The term *in vitro* refers to growing pieces of plants separately in artificial conditions. For instance, plant products can be harvested from select plants cultured or cloned as suspended cells. Moreover, tissue culture procedures are required for the management of genetically modified cells used to grow transgenic whole plants. These methods are also necessary for forming somatic haploid embryos to generate homozygous plants. Hence, tissue culture techniques will remain prominent in academic and applied plant sciences.

1.2 TISSUE CULTURE MEDIA (COMPOSITION AND PREPARATION)

The composition of the culture medium is an important factor that governs the growth and morphogenesis of plant tissues in the culture. Cultured plant cells have the same basic nutrient requirements as the whole plant. Some or all of the following components are present in the culture media of plant tissues and cells: micronutrients, macronutrients, amino acids or other nitrogen supplements, vitamins, sugar, solidifying agents or support systems, other undefined organic supplements, and growth regulators. Common media formulations used in majority of cell and tissue culture work include those described by White (1963); Murashige and Skoog (1962); Gamborg, Miller, and Ojima (1968); Schenk and Hildebrandt (1972); Nitsch and Nitsch (1969); and Lloyd and McCown (1980). The media high in macronutrients include the Murashige and Skoog medium, Schenk and Hildebrandt medium, and Gamborg B5 medium.

1.2.1 Macronutrients

Macronutrients consist of six major elements that are required for the growth of plant tissue. They include nitrogen (N), phosphorus (P), potassium (K), calcium (Ca), magnesium (Mg), and sulphur (S). The optimum concentration of each nutrient required for maximum growth rate depends on the species.

For proper plant cell growth, the culture media must contain at least 25–60 mM of inorganic nitrogen. Although plant cells may be grown on nitrates alone, better results can be obtained if nitrate and ammonium are used as nitrogen sources. For their cell growth, some species require ammonium or another source of reduced nitrogen. Nitrates are usually supplied in the range of 20–25 mM, while ammonium concentrations typically lie in the range 2–20 mM. However, ammonium concentrations in excess of 8 mM may be deleterious to cell growth in certain species. Cells can also be grown on a medium having ammonium as the only source of nitrogen if one or more of the tricarboxylic acid (TCA) cycles (e.g., citrate, succinate, or malate) are included in the medium at approximately 10 mM concentrations. When nitrate and ammonium are used together in the culture medium as source of nitrogen, the ammonium ions are consumed more rapidly and before the nitrate ions.

Potassium is an important element for cell growth in most plants. It is present in the form of nitrates or chlorides in most media at concentrations of 20–30 mM. When all other requirements for cell growth are met, the optimum concentrations of phosphorous, magnesium, sulphur, and calcium range from 1 to 3 mM. These nutrients may be required in higher concentrations when other nutrients are deficient.

1.2.2   Micronutrients

Iron (Fe), manganese (Mn), zinc (Zn), boron (B), copper (Cu), and molybdenum (Mo) constitute the micronutrients essential for plant cell and tissue growth. Culture media are mostly prepared from the chelated forms of iron and zinc. Of all the micronutrients, iron is the most essential. Iron citrate and tartrate may be used in the culture media. But these compounds sparsely dissolve and are mostly precipitated after the media are formed. To resolve this problem, Murashige and Skoog (1962) used an ethylenediaminetetraacetic acid (EDTA)–iron chelate. Cobalt (Co) and iodine (I) may be added in certain media, but it is not established whether they can promote cell growth. Although sodium (Na) and chlorine (Cl) are not essential for cell growth, they are often added in some media. Usually copper and cobalt, iron and molybdenum, iodine, zinc, manganese, and boron are added to culture media at the concentrations of 0.1, 1, 5, 5–30, 20–90, and 25–100 μM, respectively.

1.2.3   Carbohydrates

Scientists like to use sucrose as the carbohydrate in the plant cell culture media. In some cases, instead of sucrose, glucose or fructose can also be used. While glucose is as effective as sucrose, fructose is not so effective. Some other carbohydrates have also been tested, such as starch, lactose, rafinose, maltose, and galactose. The concentration of sucrose in the culture media generally ranges between 2% and 3%. Autoclaved fructose can be dangerous for cell growth.

The culture medium should be supplied with carbohydrates because only a few isolated plants cell lines are completely autotrophic, that is, capable of meeting their own carbohydrate needs by assimilating carbon dioxide during photosynthesis.

1.2.4   Vitamins

The vitamins required for development and growth are normally produced in plants. Vitamins catalyse various metabolic processes in plants. In plant cells and tissues grown *in vitro*, some vitamins may behave as limiting factors for cell growth. Vitamins frequently used in cell and tissue culture media are thiamine (B1), nicotinic acid, pyridoxine (B6), and myo-inositol. All cells normally require thiamine at concentrations of 0.1–10 mg/L for growth. Although nicotinic acid and pyridoxine are often added to the culture media, they are not necessary for cell growth in many species. Usually nicotinic acid and pyridoxine are used at the concentrations of 0.1–5.0 mg/L and 0.1–10 mg/L, respectively. Many vitamin stock solutions contain myo-inositol. Even though myo-inositol is a carbohydrate and not a vitamin, it is reported to stimulate growth in certain cell cultures. Its presence is not essential in the culture medium. However, it stimulates cell growth in most species when it is present in small quantities. Myo-inositol is mostly used in cell and tissue culture media at concentrations in the range 50–5000 mg/L.

Some of the cell culture media also contain other vitamins, such as ascorbic acid, biotin, folic acid, *p*-aminobenzoic acid, pantothenic acid, riboflavin, and vitamin E (tocopherol). Required in negligible amounts in plant cell cultures, these vitamins are generally not regarded as growth-limiting factors. The culture medium generally requires these vitamins only when the thiamine concentration is below the desired level or when it is preferable to cultivate cells at very low population density.

1.2.5   Amino Acids or other Nitrogen Supplements

Although cultured cells are normally capable of synthesizing all the required amino acids, certain amino acids or amino acid mixtures may be added to stimulate further cell growth. It is particularly important to use amino acids for forming cell cultures and protoplast cultures. The nitrogen present in amino acids is immediately available to plant cells, and it can be absorbed more rapidly than inorganic nitrogen.

Some of the common sources of organic nitrogen used in the culture media include amino acid mixtures (e.g., casein hydrolysate), adenine, L-asparagine, and L-glutamine. Generally, casein hydrolysate is used in the concentration range 0.05%–0.1%. Care must be taken when amino acids are added alone because they may hinder cell growth. The amino acids commonly added in the culture media to augment cell growth include glycine at 2 mg/L, glutamine up to 2 mg/L, asparagine at 100 mg/L, L-arginine and cysteine at 10 mg/L, and L-tyrosine at 100 mg/L. Morphogenesis in cell cultures is stimulated by the addition of tyrosine, but it should be used only in an agar medium. On supplementing the culture medium with adenine sulphate, the cell growth and shoot formation can be greatly enhanced.

1.2.6   Undefined Organic Supplements

When a wide variety of organic extracts are added to culture media, favourable tissue responses are obtained. Some tested supplements include coconut milk, ground banana, malt extracts, orange juice, protein hydrolysates, tomato juice, and yeast extracts. However, the use of undefined organic supplements should be the last resort, and only coconut milk and protein hydrolysates are used to any extent today. The concentration of protein (casein) hydrolysates generally added to the culture media is 0.05%–0.1%, while that of coconut milk is 5%–20% (v/v).

Activated charcoal (AC) may have beneficial effects on the culture media. These effects can be because of one of the following factors: absorption of growth regulators from the culture medium, absorption of inhibitory compounds, and darkening of the medium. Growth is inhibited in the presence of AC mostly because phytohormones get absorbed by AC. 1-Naphthaleneacetic acid (NAA), 6-benzylaminopurine (BAP), 6-γ-γ-dimethylallylaminopurine, indole-3-acetic acid (IAA), and kinetin all bind to AC. Cell growth is stimulated by AC because it binds to toxic phenolic compounds produced during culture. Before being added to the culture medium at the concentration of 0.5%–3.0%, AC is mostly acid washed.

1.2.7   Solidifying Agents or Support Systems

The most common gelling agent used for preparing semisolid and solid tissue culture media is agar. There are several advantages of using agar over other gelling agents. It can easily melt and solidify. When agar is mixed with water, it forms a gel that melts at approximately 60–100°C and solidifies at approximately 45°C. This makes agar gels stable at all feasible incubation temperatures. Besides, constituents of the media do not react with agar gels and plant enzymes do not digest them. The firmness of agar gel depends on the concentration and brand of agar used and the pH of the culture medium. The concentrations of agar commonly used in culture media lie in the range 0.5%–1%. At this range, agar forms a firm gel at various pH values typical of plant cell culture media.

Gelrite is another gelling agent commonly used for research and commercial purposes. To form a clear gel, this synthetic product must be used at a concentration of 1.25–2.5 g/L, which aids in detecting contamination. There are alternative methods of support systems such as filter paper bridges, filter paper wicks, perforated cellophane, polyester fleece, and polyurethane foam. Whether agar or other supporting agents will be more suitable for the growth of explants depends on plant species.

1.2.8   Growth Regulators

In plant tissue culture, there are four important classes of growth regulators: (i) auxin, (ii) cytokinin (CK), (iii) gibberellin (GA3), (iv) and abscisic acid (ABA). Skoog and Miller (1957) reported that the type and extent of organogenesis in plant cell cultures depend on the ratio of auxin to CK. Usually, auxin and CK are added to the culture media to promote morphogenesis. However, the ratio of hormones required for root and shoot induction is not the same universally. There exists considerable variability among genera, species, and even cultivars regarding the amount and type of auxin and CK required for morphogenesis.

Auxins commonly used in plant tissue culture media include 1H-indole-3-acetic acid (IAA), 1H-indole-3-butyric acid (IBA), 2,4-dichlorophenoxyacetic acid (2,4-D), and NAA. The only naturally occurring auxin found in plant tissues is IAA. There are other synthetic auxins used in plant tissue culture. They include 4-chlorophenoxyacetic acid or *p*-chlorophenoxyacetic acid (4-CPA, PCPA), 2,4,5-trichlorophenoxyacetic acid (2,4,5-T), 3,6-dichloro-2-methoxybenzoic acid (Dicamba), and 4-amino-3,5,6-trichloropicolinic acid (Picloram).

Auxins differ in their physiological activity and also to the extent that they are bound to cells, metabolized, or move through tissues. It has been found that naturally occurring IAAs have less physiological activity than synthetic auxins. On the basis of stem curvature assays, it has been found that 2,4-D has 8–12 times, 2,4,5-T has 4 times, PCPA and Picloram have 2–4 times, and NAA has 2 times the activity of IAA. Exposure to high levels (or for a prolonged period) of 2,4-D, 2,4,5-T, PCPA, and Picloram, particularly 2,4-D, results in suppressed morphogenic activity though these toxins are often used to promote rapid proliferation of cells. In a culture medium, auxin is usually added for callus induction and cell growth, somatic embryogenesis, root and shoot initiation, and stimulation of growth from shoot apices and shoot tip cultures.

Cytokinins often used in culture media, include 6-benzylaminopurine or 6-benzyladenine (BAP, BA), 6-γ-γ-dimethylaminopurine (2iP), N-(2-furanylmethyl)-1H-puring-6-amine (kinetin), and 6-(4-hydroxy-3-methyl-trans-2-butenylamino)purine (zeatin). Zeatin and 2iP are naturally occurring CKs, whereas BA and kinetin are synthetically produced. Adenine is another naturally occurring compound having a base structure similar to that of CK. Adenine has shown CK-like activity in some cases. For morphogenesis, many plant tissues require a specific CK, but some tissues are independent of CK and do not require it for organogenesis.

Cell division is initiated when CKs are added to a culture medium. Cytokinins also induce shoot formation and axillary shoot proliferation and inhibit root formation. The ratio and concentration of auxin and CK present in the tissue culture medium largely determines the type of morphogenesis in the culture. When the ratio of auxin to CK is high, root initiation of plantlets, embryogenesis, and callus initiation generally occur. On the contrary, adventitious and axillary shoot proliferation occurs at a low auxin to CK ratio. Therefore, the ratio and concentrations of auxin and CK are equally important.

Two other growth regulators occasionally used in culture media are GA3 and ABA. Growth of plant tissue cultures can be induced in the absence of GA3 or ABA, but these hormones might be required by certain species for enhanced growth. When GA3 is added to culture media, it promotes growth in low-density cell cultures, stimulates callus growth, and elongates stunted or dwarfed plantlets. When

ABA is added to culture media, it inhibits or stimulates callus growth (depending on the species), enhances shoot or bud proliferation, and impedes later stages of embryo development.

1.3 STOCK SOLUTIONS

Use of stock solutions reduces the repetition of procedures for preparing the culture media, thus minimizing the chance of human or experimental error. For tissue culture work, accurately weighing the media components (e.g., micronutrients and hormones), required only in milligram or microgram amounts in the final formulation, every time is very difficult. Therefore, the standard procedure involves preparing a concentrated stock solution of these components and subsequently diluting it in the final media. Moreover, concentrated solutions of some materials are more stable and can be stored for longer periods than their dilute solutions.

A stock solution is prepared by weighing the required amount of compound and putting it in a clean flask. Based on the solubility of the compound, stock solutions of $10x$ or $100x$ are generally prepared. To dissolve the compound placed in the flask, a small amount of water, ethyl alcohol, 1 N NaOH, or 1 N HCl is used. Then double-distilled water is slowly added to the flask with agitation. The flask is filled to the required volume of double-distilled water. Then the flask is labelled with the name of the reagent, preparation and expiry dates, and the name of the person who prepared it. Substances such as IAA should be prepared and stored in amber-coloured bottles to avoid photodecomposition.

1.3.1 Macronutrients

Stock solutions of the macronutrients are usually prepared of 10 times concentration of the final strength. To prevent precipitation, a separate stock solution of calcium salts may be required. Stock solutions of macronutrients can be safely stored for several weeks at 2–4°C in a refrigerator.

1.3.2 Micronutrients

Stock solutions of micronutrients prepared are 100 times their final strength. Until finally used, micronutrient stocks should be stored in a refrigerator (2–4°C) or a freezer (–20°C). These solutions can be stored in a refrigerator without any appreciable deterioration for up to 1 year. Stock solutions of iron must be prepared and stored separately from other micronutrients in an amber-coloured storage bottle.

1.3.3 Vitamins

Stock solutions of vitamins are prepared at $10x$ or $100x$ and stored in a freezer at –20°C. The solutions should be prepared each time if a refrigerator or a freezer is not available. Vitamin stock solutions can be safely stored for up to 2–3 months in a refrigerator (4°C) and should be thrown away after that.

1.3.4 Growth Regulators

Auxins, such as NAA, and 2,4-D, are considered stable. They can be stored at 4°C for several months. IAA should be stored at –20°C. Stock solutions of auxin are generally prepared at 100–1000 times the final working concentrations. NAA and 2,4-D solutions can be stored in a refrigerator for several months or for an indefinite period of time at –20°C. Normally, a small volume of 95% ethyl alcohol or KOH is used to dissolve IAA and 2,4-D, and the volume is made up to the desired level by adding double-distilled water. NAA can also be dissolved in a small amount of 1 N NaOH or KOH, which can be used to dissolve 2,4-D and IAA as well.

Cytokinin is regarded as stable and can be stored at –20°C. Stock solutions of CK are mostly prepared at concentrations of $10x$ or $100x$. Cytokinins are frequently reported to be difficult to dissolve and sometimes a few drops of 1 N HCl or 1 N NaOH are added to facilitate their dissolution.

1.3.5 Storage of Stock Solutions

Most of the conditions required for storing various stock solutions are already discussed. However, there are additional considerations. Many laboratories find it convenient to divide stock solutions into small aliquots adequate to prepare 1 L of medium. These aliquots are then put in small vials or plastic bags and stored in a freezer. Thus, a large volume of frozen stock does not require to be thawed every time the medium is prepared. Some find it satisfactory and quick to thaw frozen stock solutions by heating in a microwave oven.

1.4   STERILIZATION AND AGENTS OF STERILIZATION USED IN TISSUE CULTURE LABORATORIES

Sterilization involves the destruction of all living matter. Explants are disinfected, sinks are cleaned or disinfected, but media and transfer tools are sterilized. In disinfection, chemical agents are used to kill pathogenic microbes, but the material to which the chemical is applied is not necessarily sterilized. Sanitation is the process of considerably reducing the population of microbes in the air and on laboratory objects to acceptable limits. This includes cleaning surfaces by using water, soap, or detergent.

1.4.1   Methods of Sterilization

1.4.1.1   Sterilization by heat

In the presence of moist heat, organisms are destroyed when their enzymes and proteins get coagulated and denatured. The presence of water accelerates the process. Spores are killed when exposed to moist heat at 121°C for 10–30 min. When dry heat is used, organisms are killed because of the oxidative destruction of cell constituents. Spores are killed when they are exposed to dry heat at 160°C for 1 h.

1.4.1.1.1   Sterilization by dry heat

Dry heat sterilization can be used on many objects in the absence of water. Put the items to be sterilized in an oven at 160–170°C for 2–3 h. The oxidation of cell constituents and denaturation of proteins apparently result in the death of microbes. Even if moist heat is more effective than dry heat, there are certain advantages of dry heat over moist heat. Dry heat does not corrode glassware and metallic instruments unlike moist heat. Also, powders, oils, and similar items can be sterilized with dry heat. In most laboratories, glass Petri dishes and pipettes are sterilized with dry heat. Dry heat sterilization is, however, slow and unsuitable for materials such as plastics and rubber items, which are heat sensitive. In such cases, the following methods are used:

- **Red hot**   The object is heated in a flame until it turns red (e.g., inoculating wires, tip of forceps, spatulas, and so on).

- **Flaming**   The object is burnt in methylated spirit or exposed to gas flame (e.g., scalpels, needles, and basins). This method does not produce high temperature for sterilization.
- **Hot-air oven**   The oven has a thermostat to regulate temperature and a fan to circulate hot air. Generally, it is heated electrically at 160°C for 1 h. Glasswares, forceps, scalpels, scissors, throat swabs, syringes, and materials such as fats, oils, and powders are sterilized by this method.
- **Infrared radiation**   Objects such as glass syringes are exposed to infrared rays directed from an element electrically heated at 180°C. Surgical instruments are sterilized in a vacuum chamber at temperatures above 200°C. Nitrogen is circulated during the process of cooling to prevent oxidation.
- **Susceptibility to dry heat**   Dry heat at 100°C for 60 min kills vegetative bacteria. Fungal spores are killed at 115°C and bacterial spores at 120–160°C, both within 60 min.

1.4.1.1.2   Sterilization by moist heat

Sterilization by moist heat can be carried out at a temperature below 100°C, at 100°C (boiling water or free steam), and above 100°C, using saturated steam under increased pressure (e.g., autoclaves). For the pasteurization process, moist heat is used at a temperature below 100°C. Pasteurization of milk is carried out at 63–66°C for 30 min. The flash method is done at 72°C for 20 s. Both methods are able to destroy all non-spore forming, milk-borne microorganisms such as *Mycobacterium tuberculosis*, *Brucella*, and *Salmonella*. However, spores are not killed by the use of these methods.

The following are the applications of temperature at 100°C:

- **Boiling at 100°C**   All non-spore forming microorganisms are killed within 5–10 min when boiled at 100°C. It is generally used to disinfect blades and syringes.
- **Steaming at 100°C**   To destroy bacterial endospores, moist heat sterilization is carried out at temperatures above 121°C. Saturated steam under pressure is used for this purpose. An autoclave is used for steam sterilization, which is a device similar to a fancy pressure cooker. The invention of autoclave by Charles Chamberland in 1879 gave tremendous impetus to the study of microbiology. Steam from

the boiled water is released through the jacket into the autoclave chamber. The air already present in the chamber is forced out by the saturated steam until the chamber is filled with saturated steam, and then the outlets are closed. The chamber continues to be filled with hot, saturated steam until the desired temperature and pressure are reached, which are usually 121°C and 15 lb pressure, respectively. Under these conditions, saturated steam kills within 10–12 min all vegetative cells and endospores in a small volume of liquid. To provide a margin of safety, the treatment is carried out for about 15 min. In the case of a large volume of liquid, the sterilization time has to be longer because the centre of the liquid takes more time to reach 121°C; for example, 5 L of liquid may require 70 min for sterilization.

1.4.1.2 Sterilization by filtration

Filtration is an efficient method to reduce the population of microbes from heat-sensitive solutions (e.g., toxins, serum, and antibodies). It is sometimes used to sterilize solutions. The filter simply removes microorganisms instead of directly destroying them. A filter of pore size less than 0.75 μm can remove bacteria. Filters with much smaller pore sizes are required to exclude viruses. The filter's quality can be judged based on its efficiency of removing *Serratia marcescens* from the broth culture. Examples of some filters are as follows:

- Asbestos and asbestos–paper disks (Seitz)
- Sintered glass filters
- Microfilters
- Depth filters
- Cellulose membrane filters
- Gradocol membrane (made of cellulose nitrate)
- Cellulose acetate filter

1.4.1.3 Sterilization by radiation

Ultraviolet (UV) radiation is quite lethal at around 260 nm, but it cannot effectively penetrate glass, dirt, films, water, and other substances. This disadvantage limits the use of UV radiation as a sterilizing agent to only a few situations. UV lamps are generally placed on the ceiling of rooms or in biological safety cabinets to sterilize the air and any exposed surfaces. On being exposed to UV lamps, pathogens and other microorganisms present in a thin layer of water are destroyed. Many vegetative forms of microorganisms present in the air are killed by UV rays, but their susceptibility varies. These rays are utilized to sterilize surgical areas, instruments, hospitals, schools, and warehouses and also to sterilize plastic materials.

1.4.1.4 *Sterilization by chemical disinfectants*

Chemical disinfectants can be categorized into the following two types:

(i) **Strong disinfectants** Strong disinfectants are very effective but toxic to human beings; therefore, these should be applied to articles other than skin or tissues. Examples of strong disinfectants include formalin solution.

(ii) **Mild disinfectants** Mild disinfectants are non-toxic and can be applied to soft tissues and skin. Examples of mild disinfectants include ethyl alcohol, iodine, and soap.

The mode of action of most disinfectants is denaturing or altering proteins or lipids present in the cytoplasmic membrane of bacteria. Some antiseptics alter the energy-yielding systems within the cell, such as formation of adenosine triphosphate (ATP). Other disinfectants impede some steps of the metabolic pathways of the cell.

1.5   INITIATION AND MAINTENANCE OF CALLUS AND CELL SUSPENSION CULTURE

Usually, sterile pieces of a plant are used for developing cultures. These pieces are called explants and consist of parts of different organs, such as leaves and roots, or potentially available specific cell types, such as pollen or endosperm. The efficiency of culture initiation is affected by many features of explants. Generally, tissue at an early stage of development, that is, younger, more rapidly growing tissue, is most effective.

1.5.1 Callus

When cultured in an appropriate medium, usually with both auxin and CK, explants produce an unorganized, growing, and dividing mass of cells called callus. It is normally composed of unspecialized parenchyma cells. It is assumed that any plant tissue can act as an explant if it is provided with appropriate growing conditions. This proliferation can be maintained more or less indefinitely in the culture if the callus is periodically subcultured in a fresh medium. During the formation of callus, some degree of dedifferentiation (that is, the changes that occur during development and specialization are reversed to some extent) occurs in both morphology and metabolism. Most plant cultures cannot photosynthesize due to dedifferentiation, which is a major concern. This has serious consequences for the callus culture because there is a possibility that the metabolic profile will not match with that of the donor plant. As a result, it becomes necessary to add to the culture medium other components, such as vitamins and a carbon source, along with usual mineral nutrients. Callus culture is generally carried out in the dark because light can cause differentiation of the callus. In the long term, auxin and CK may not be required by the culture. This process is known as habituation and is common in callus cultures of some plant species such as sugar beet.

Callus cultures are extremely important in the field of plant biotechnology. By regulating the auxin to CK ratio in the medium, shoots, roots, or somatic, embryos can be developed and from which entire plants can be subsequently produced. Cell suspensions can also be initiated by callus cultures, which can be used in plant transformation studies in a number of ways.

1.5.2   Cell Suspension Cultures

Callus cultures are of two types: compact and friable. The cells in compact callus cultures are densely aggregated, whereas the cells in friable callus cultures are loosely associated with each other, which makes the callus soft and breakable. An inoculum is provided by friable callus to form cell suspension cultures. Explants from particular cell types or some plant species do not form friable callus, and thus it becomes difficult to initiate cell suspension cultures. It is sometimes possible to improve the friability of the callus by manipulating the medium components, by repeated subculturing, or by using a semisolid medium (a medium in which the concentration of the gelling agent is low).

On placing friable callus in a liquid medium (usually of the same composition as the solid medium) with agitation, single cells and small clumps of cells are released into the medium. These cells continue to grow and divide under suitable conditions, ultimately producing a cell suspension culture. A considerably large inoculum must be used while initiating cell suspensions to quickly build up the number of released cells. However, the use of extremely large inoculum should be avoided as toxic materials released from damaged or stressed cells can accumulate to lethal levels. Large cell clumps can be removed during subculture of the cell suspension.

It is relatively simple to maintain cell suspensions as batch cultures in conical flasks. By repeated subculturing into fresh medium, they are continually cultured. This results in the dilution of the suspension and the initiation of another batch growth cycle. The degree of dilution during subculturing should be empirically determined for each subculture. A very large degree of dilution will result in a greatly extended lag period or death of the transferred cells in extreme cases. After subculture, the cells divide and the biomass of the culture increases in a characteristic manner. This happens until nutrients in the medium are depleted or toxic by-products accumulate to inhibitory levels; this is called the stationary phase. Cells die when they are left in the stationary phase for a very long period, and thus the culture is lost. Hence, cells should be transferred as soon as they enter the stationary phase. Therefore, determining the batch growth cycle parameters for each cell suspension culture is important.

1.6 NUTRITIONAL REQUIREMENT FOR *IN-VITRO* CULTURE

1.6.1 Selection of a Suitable Medium

To devise a suitable medium, various combinations of different concentrations of major components such as plant growth regulators, salt, and sucrose may be tried. A suitable medium can be prepared by using various combinations of low, medium, and high concentration of components such as minerals, auxins, CK, and organic nutrients. The exact conditions required to initiate and sustain plant cells in a culture or to regenerate intact plants from cultured cells are different for each plant species. Each variety of species will have a particular set of requirements for culture. Despite advanced knowledge of plant tissue culture during the 20th century, these conditions can be identified for each variety only through experimentation.

1.7   CONCEPT OF CELLULAR TOTIPOTENCY

For understanding plant cell culture and regeneration, studying two concepts—plasticity and totipotency—is central. The sessile nature and long lifespan of plants have helped acquire a greater ability to endure extreme conditions and predation than animals. Most of the processes associated with plant growth and development adapt according to environmental conditions. Because of this plasticity, plants can alter their metabolism, growth, and development to suit their environment. In plant tissue culture and regeneration, the essential aspects of adaptations are the capacity of plants to initiate cell division from almost any tissue, regenerate lost organs, or undergo different developmental pathways in response to particular stimuli. Plant cells and tissues generally exhibit a high degree of plasticity when cultured *in vitro*. This allows one type of tissue or organ to be developed from another type, consequently regenerating the whole plant. The regeneration is based on the notion that all plant cells can express the total genetic potential of the parent plant if they are given the correct stimuli. This phenomenon of maintaining genetic potential is called totipotency. The most compelling evidence of totipotency can be found from plant cell culture and regeneration. However, in practice, it can be extremely difficult to determine the appropriate culture conditions and stimuli needed to manifest totipotency, and it is still a largely empirical process.

1.7.1   Totipotency

Plant cells can be distinguished from animal cells by their ability to regenerate fertile plants from any somatic tissue. The gene transfer technology, which is used to produce genetically modified/engineered plants of superior quality, is based on this property. Following the path of somatic embryogenesis, many plant cells can reorganize as embryos in a manner comparable to that observed after fertilization. Otherwise, somatic cells can form meristematic centres and regenerate mature plants via organogenic pathway under appropriate conditions. This ability to regenerate is not limited to diploid sporophytic cells; haploid plants can also be regenerated from microspores. Sometimes, the absence of high regenerating ability from single cells in species of agricultural and commercial interest, for example, corn, soya bean, and beans, has hindered the production of superior varieties through genetic engineering (Goldberg 1988).

1.8   SINGLE CELL CLONES AND ORGANOGENESIS

Single cell culture is the process by which free cells are isolated from plant organs or cell suspensions and grown *in vitro* as single cells, thus generating a clone of identical cells.

1.8.1   Isolation of Single Cells from Plant Organs

Generally, leaf tissue is isolated because it has homogeneous population of cells. For this purpose, the following two methods are employed:

(i) **Mechanical** Leaves are chopped into small pieces and macerated in mortar and pestle in a grinding buffer. The homogenate produced is filtered using a muslin cloth and then centrifuged to finally pellet down the cells.

(ii) **Enzymatic** Leaves are chopped into moderate pieces after removing the lower epidermis. These pieces are incubated in macerozyme or pectinase, which results in degradation of the middle lamella and cell wall of parenchymatous tissue. To prevent any damage to the cell wall because of enzymatic action, a suitable osmoticum (e.g., 0.3 M mannitol) is added to the culture.

1.8.2   Isolation of Single Cells from Cell Suspension

Cell isolation is carried out from suspension cultures made from friable calli by filtering and harvesting the cells through centrifugation. Isolated single cells cannot be divided in normal tissue culture media; hence, they are cultured on nurse tissue where well-grown callus cultures diffuse their exudates through a filter paper. The single cells derive their nutrition from the exudates present on the filter paper and are thus called nurse tissues. This method of culturing single cells is known as filter paper–raft nurse tissue technique. Several methods exist for culturing single cell. They include microdrop, microchamber, Bergmann's cell plating technique, and thin-layer liquid medium. Bergmann's cell plating technique is the most widely used method, in which free cells are suspended in a liquid medium. In a water bath, culture medium with agar (1%) is cooled and the temperature is maintained at 35°C. Equal quantities of liquid and agar medium are mixed and rapidly spread on a Petri dish. Using an inverted microscope, the cells that remain embedded in the soft agar medium are observed. When cell colonies develop, they are isolated and separately cultured. As light is detrimental to cell proliferation, single cells must be cultured in dark.

1.8.2.1   *Synchronization of suspension cultures*

Cell suspension is mostly asynchronous, with different cells having different sizes, shapes, DNA and nuclear contents, and they are also in different stages of cell cycle (G1, S, G2, and M). This asynchronous property is not suitable for cell metabolism studies. Therefore, it is essential to obtain synchrony in suspension cultures, which can be achieved by the following methods:

- **Starvation**   Cells are starved of a nutrient (e.g., a growth regulator) which is necessary for cell division. This results in the arrest of cell growth during G1 or G2 stages. When the nutrient is supplied, all arrested cells undergo synchronous division.

- **Inhibition**   Cell growth can be arrested at the G1 stage using a biochemical inhibitor of DNA synthesis such as 5-aminouracil. Removal of the inhibitor leads to synchronous division of cells.

- **Mitotic arrest**   Colchicine is commonly used to arrest cells at metaphase, but it is only for a short duration, as longer colchicine treatment might induce a normal mitosis and chromosome stickiness.

- **Cell viability test**   Cell suspension culture is carried out with the aim of achieving rapid growth rates and uniform cells, with all cells being viable.

The following approaches can determine the viability of cells:

- **Phase contrast microscopy**   A phase contrast microscope can be used to observe live cells having a well-defined healthy nucleus and streaming cytoplasm.

- **Reduction of tetrazolium salts**   When cell masses are stained with 1%–2% solution of 2,3,5-triphenyltetrazolium chloride (TTC), the live cells reduce TTC to red-coloured formazon. The formazone is then extracted and measured spectrophotometrically for quantifying viability. However, this approach is not employed for single cells.

- **Fluorescein diacetate**   Esterases within the cytoplasm of living cells cleave fluorescein diacetate (FDA) to release the fluorescent compound fluorescein. These fluoresces under UV ray and live cells appear green.

- **Evan's blue staining**   This is the only dye that is taken up by dead cells. Therefore, this procedure is usually used as a complement to FDA.

1.9 ORGANOGENESIS

Somatic embryogenesis relies on plant regeneration through a method similar to zygotic embryo germination. However, organogenesis is the development of plant organs from either an explant or a callus culture. Plant regeneration via organogenesis can be achieved through three methods. The first two methods rely on adventitious organs arising either from a callus culture or from an explant. The third method involves axillary bud formation and growth, which can also be employed to regenerate whole plants from some types of tissue culture.

The inherent plasticity of plant tissues is the basis of organogenesis, which is regulated by varying the components of the medium. Particularly, the developmental pathway that the regenerating tissues will take is determined by the auxin to CK ratio of the medium. Usually, shoot formation can be initiated by increasing the CK to auxin ratio in the culture medium. These shoots can then be rooted relatively simply. A classic example is the organogenesis from tobacco pith callus, which shows how varying plant growth regulator regimes can be used to alter the pattern of regeneration from plant tissue cultures. The callus proliferates when it is cultured in a medium containing both auxin and CK. If the ratio between auxin and cytokinin is increased, adventitious roots develop from the callus by organogenesis; on the other hand, if the ratio between auxin and cytokinin is low, adventitious shoots are generated. Shoots can be generated directly if the explants are cultured in a medium containing only CK. Tobacco leaf pieces can easily regenerate tobacco plants. Tobacco leaves are sliced into pieces of approximately 1 cm^2 using a sterile scalpel (avoiding large leaf veins and any damaged areas). The leaf pieces are then placed (right-side up) into gelled MS medium supplemented with 1 mg/L BA (CK) and 0.1 mg/L NAA (auxin). Callus develops from explants during the next few weeks, particularly around the cut surfaces. After three to five weeks, shoots develop either from explants or from callus derived from explants. When these shoots reach the height of about 1 cm, they are cut at the base and placed on to a solid MS medium without the addition of any plant growth regulators. The shoots will develop roots and plantlets in this medium, which can be subsequently transferred to soil.

1.10   ARTIFICIAL SEEDS

New vistas have been opened in agriculture by plant propagation using artificial or synthetic seeds developed from somatic and non-zygotic embryos. Artificial seeds make a promising method for the generation of transgenic plants, polyploids with elite traits, non-seed-producing plants. Plant lines face difficulty in seed propagation. The technique is clonal in nature, and thus it eliminates the arduous selection procedure of the conventional recombination breeding. Therefore, this technique will provide Indian farmers with a cost-effective approach. Abundant supply of the desired plant species can be ensured through the development of micropropagation techniques. In some crop species, and also in some seedless varieties of crop plants such as grapes and watermelon, seed propagation has been found to be unsuccessful mainly because of heterozygosity of seeds, presence of reduced endosperm, minute seed size, and the requirement of mycorrhizal fungi association for germination (e.g., orchids). Some of these species can be generated by vegetative means. However, *in vivo* vegetative propagation methods are expensive and time-consuming.

Artificial seed production technology can provide an alternative, effective, and efficient method of propagation for many commercially important agronomic and horticultural crops. It is considered a powerful technique for the mass propagation of elite plant species of a high commercial value. In artificial seed technology, tissue culture derived somatic embryos encased in a protective coating are produced. Artificial seeds are also referred to as synthetic seeds. But the term 'synthetic seed' should not be confused with commercial seeds of a synthetic cultivar, which is defined as an advanced generation of an open pollinated population consisting of a group of selected inbred clones or hybrids. The process of generating artificial or synthetic seeds is shown in Box 1.1.

Synthetic seed production is a low-cost, high-volume technology. New avenues for clonal propagation have been opened in many commercial crop species by the high-volume propagation potential of somatic embryos and the formation of synthetic seeds for low-cost delivery. As somatic embryos are bipolar structures, they have both apical and basal meristematic regions that can form shoots and roots, respectively. A plant generated from a somatic embryo is often termed 'embling'.

1.10.1 Basic Requirement for Production of Artificial Seeds

Recently, synthetic seeds of a few species have been produced by encapsulating somatic embryos. For applying synthetic seed technology in micropropagation, the generation of high-quality and vigorous somatic embryos is a prerequisite that can produce plants at the same rate as natural seeds. A major limitation in the development of synthetic seeds is the inability to recover such embryos. In synthetic seed technology, a large number of high-quality somatic embryos with synchronous maturation have to be produced at low cost. The overall quality of the somatic embryo is crucial for achieving high conversion frequencies. Encapsulation and coating systems are essential for delivery of somatic embryos, and they are not the limiting factors in the development of synthetic seeds. Currently, the absence of developmental synchrony in embryogenic systems hinders multi-step procedures from guiding somatic embryos through maturation. The principal barrier to the widespread commercialization of synthetic seeds is the lack of synchrony in somatic embryos. Synchronized embryonic development is needed for the effective production of synthetic seeds.

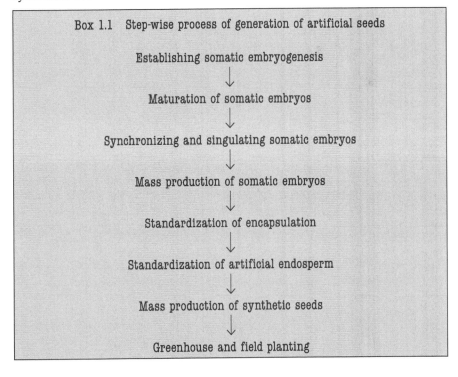

Box 1.1 Step-wise process of generation of artificial seeds

Establishing somatic embryogenesis
↓
Maturation of somatic embryos
↓
Synchronizing and singulating somatic embryos
↓
Mass production of somatic embryos
↓
Standardization of encapsulation
↓
Standardization of artificial endosperm
↓
Mass production of synthetic seeds
↓
Greenhouse and field planting

1.10.2   Types of Gelling Agents Used for Encapsulation

Several gels have been tested for synthetic seed production, including agar, alginate, polyco 2133, carboxymethyl cellulose, carrageenan, gelrite, guar gum, sodium pectate, and tragacanth gum. Out of these, alginate encapsulation has been found to be the most practicable and suitable technique. Artificial seeds are used for specific purposes, such as multiplication of non-seed-producing and ornamental plants (currently propagated by cuttings), or for the generation of polyploid plants with elite traits. These artificial seeds can also be used for the propagation of male or female sterile plants for hybrid seed production. Cryopreserved artificial seeds can be employed for preservation of germplasm, particularly in recalcitrant species (such as mango, cocoa, and coconut), because these seeds do not undergo desiccation. Moreover, transgenic plants that need separate growth facilities to retain original genotypes can also be preserved using somatic embryos.

Somatic embryogenesis is a potential tool in the genetic engineering of plants. A single gene can be potentially introduced into a somatic cell. The progeny of plants regenerated by somatic embryos from a single transgenic cell will not be chimeric. Genetic recombination does not take place in the process involving multiplication of elite plants selected from plant breeding programmes via somatic embryos. Therefore, this process does not permit continued selection inherent in conventional plant breeding and saves a considerable amount of time and other resources. Artificial seeds produced in tissue culture do not carry pathogens. This provides the advantage of transporting pathogen-free propagules across international borders, thus removing the requirement for bulk transportation of plants, quarantine, and spread of diseases.

1.11   MICROPROPAGATION

Micropropagation or clonal propagation is the *in vitro* vegetative propagation of plants by tissue culture to generate genetically identical copies of a cultivar. Plants that are sexually propagated (through seed generation) show a high amount of heterogeneity because their seed progenies are not true to their type. However, asexual reproduction (by multiplication of vegetative parts) results in the generation of genetically similar copies of the parent plant. Thus, micropropagation technique allows perpetuation of the parental characters of cultivars among the plants produced. Micropropagation is useful for propagation in the following cases:

(i) Sexually sterile species (e.g., triploids and aneuploids) that cannot be perpetuated by seeds

(ii) Seedless plants (e.g., banana)

(iii) Cross-bred perennials in which heterozygosity is to be maintained

(iv) Mutant lines (e.g., auxotrophs) that cannot be propagated *in vivo*

(v) Disease-free planting materials of fruit trees and ornamentals

Micropropagation involves the following major stages:

- **Stage 0** It involves selection and maintenance of stock plants for culture initiation (3 months).
- **Stage I** It involves preparation of the explant followed by its establishment on a suitable culture medium (3–24 months; usually shoot tips and axillary buds are used).
- **Stage II** It involves regeneration and multiplication of shoots or somatic embryos using a defined culture medium (10–36 months).
- **Stage III** It involves rooting of regenerated shoots/somatic embryo *in vitro* (1–6 weeks).
- **Stage IV** It involves transfer of plantlets under greenhouse environment to sterilized soil for hardening.

It should be noted that Stage III can be skipped for rooting of Stage II regenerated shoots *in vivo*. The advantages of micropropagation over the conventional propagation methods are as follows:

- Constitution of genotype is maintained because there is lesser variation in somatic embryos.
- Small-sized propagules facilitate easier transport and storage and they possess the ability to grow in soil-less medium.
- There is control over growing conditions because the production of planting material is completely under artificial control *in vitro*.
- As shoot multiplication has short cycle, reduced growth cycle and rapid multiplication are observed in micropropagation. Also, each cycle results in an exponential rise in the number of shoots.

- Selective multiplication can be carried out for auxotrophs and aneuploids. Sexual selection can be applied to dioecious species.
- Pathogen-free plants can be raised and maintained through meristem culture, which is the only available method.

The disadvantages of micropropagation over the conventional propagation methods are as follows:

- The process involves high cost.
- Any somaclonal variation likely to occur during multiplication might go unnoticed.
- Many tree species such as mango do not respond to *in vitro* growth.

1.11.1   Application of Micropropagation

Secondary metabolites are compounds or biochemicals that are not directly involved in primary metabolic processes such as respiration and photosynthesis. They include a variety of compounds, such as alkaloids, terpenoids, and phenylpropanoids, having various biological activities, which include antimicrobial, antibiotic, insecticidal, pharmacological, and pharmaceutical. Hence, micropropagation allows their commercial-scale production from cell cultures. For example, shikonin and its derivatives used in dyes and pharmaceuticals are produced from the cultured cells of *Lithospermum erythrorhizon*. Moreover, cultured cells of many plant species produce novel biochemicals, which are otherwise difficult to detect in whole plants.

Production of synthetic seeds   Synthetic seed is a bead of gel containing somatic embryo or shoot bud, growth regulators, nutrients, fungicides, and pesticides, which are essential for the development of a complete plantlet. These are superior propagules because they do not require hardening agents and can be directly sown in the field.

Raising somaclonal variants   The genetic variability occurring in somatic cells of plants produced *in vitro* by tissue culture is termed somaclonal variation. If these variants show traits of economic importance, they are raised and maintained by micropropagation.

Production of disease-free plants   Various fungi, viruses, and bacteria infect most horticultural fruits and ornamental crops. Micropropagation provides a time-saving technique for the generation of pathogen-free plants. In case of viral diseases, either the apical meristems of infected plants are free of viruses or they carry a very low amount of viruses. Hence, disease-free plants can be produced by culturing meristem tips.

1.11.2   Micropropagation Methods

It has been demonstrated from the capacity of a mature cell to dedifferentiate into callus tissue and the method of cloning an isolated single cell *in vitro* that somatic cells can generate a whole plant under specific conditions. This ability of a cell to divide and develop into a multicellular plant is termed the cellular totipotency. After dedifferentiation, the cell has to undergo redifferentiation or regeneration to express totipotency. Redifferentiation is the ability of dedifferentiated cells to form a plant or plant organs. This may occur through organogenesis or embryogenesis.

In organogenesis, meristematic cells present in callus redifferentiate into shoot buds, which are monopolar structures. These buds give rise to leaf primordial and apical meristem. They have procambial strands which are connected with pre-existing vascular tissue of the explant or callus. Many factors stimulate the shoot bud differentiation in plants, which varies from one plant species to another. Usually, it is promoted by CK and inhibited by auxin.

Skoog and Miller (1957) demonstrated that the nature of organogenesis in tobacco pith tissue largely depends on the relative CK to auxin ratio. In tobacco, bud formation is initiated by the high level of CK, whereas high concentration of auxin promotes root formation. However, some studies conducted on other plant species do not adhere to this concept of auxin to CK ratio. In most cereals, organogenesis occurs in callus tissue when the callus is subcultured in a medium in which 2,4-D is replaced by IAA or NAA. Generally, GA3 has an inhibitory effect on shoot buds, whereas ABA enhances shoot regeneration in many species. Different plant species respond differently to various growth regulators because the requirement of exogenous growth regulators depends on their endogenous levels, which varies across plant species and plant materials. The size and source of the explant also affect organogenesis. The likelihood of shoot bud formation increases when the size of the explant (containing parenchyma, cambium, and vascular tissue) increases. Moreover, shoot regeneration is affected by the genotype of explants because explants collected from different plant varieties of the same species exhibit different frequencies of shoot bud differentiation. It has also been found that light has an inhibitory effect on shoot buds. Effects of specific wavelength on organogenesis have also been observed. For instance, blue light induces shoot formation and red light influences root formation in tobacco. The required optimum temperature might vary across plant species.

Somatic embryogenesis is a process which leads to redifferentiation of meristematic cells into non-zygotic somatic embryos, which can germinate to form whole plants. Somatic embryos are bipolar structures having a radical and a plumule, whereas monopolar shoot buds have only the plumular end in organogenesis. The meristematic cells break any cytoplasmic or vascular links with other cells while developing into somatic embryos and become isolated. Contrary to shoot buds, the somatic embryos can thus be easily separated from explants. Somatic embryogenesis includes the following three steps, which are absent in organogenesis:

1. **Induction** Induction is the initiative phase in which callus cells are induced to differentiate into groups of meristematic cells called embryogenic clumps (ECs). These clumps develop into initial stages of somatic embryo, that is, globular stage.

2. **Maturation** In the maturation phase, somatic embryos evolve into mature embryos by differentiating from globular to heart-shaped and torpedo to cotyledonary stages. The mature embryo acquires hardiness by undergoing biochemical changes.

3. **Conversion** Embryos germinate to produce seedlings.

The following factors influence somatic embryogenesis:

- **Growth regulators** The presence of auxin (generally 2,4-D) in the medium is necessary in the induction phase. The dedifferentiation of explants to form ECs is induced by 2,4-D. On removing auxin or reducing its concentration, ECs convert to somatic embryos. Once induced, cells do not need plant growth regulators. However, some doses of CK at maturation and conversion can produce better plants. On culturing somatic embryos in a high-sucrose medium, maturation is reached. Moreover, adding ABA helps in hardening because of the water loss, which is essential for embryo maturation. Ethylene impedes both somatic embryogenesis and organogenesis. Hence, plant regeneration is promoted by adding silver nitrate to the medium as an inhibitor of ethylene.

- **Nitrogen source** Nitrogen in the NH_4^+ form is essential for induction of somatic embryogenesis, and nitrogen in the NO_3^- form is needed during maturation phase.
- **Other factors** Similar to shoot bud differentiation, somatic embryogenesis is influenced by the explant genotype. In cereals, both somatic embryo induction and maturation are promoted by the use of maltose as a carbohydrate source.

1.12 SOMATIC EMBRYOGENESIS

Somatic (asexual) embryogenesis is a process in which embryo-like structures are formed from somatic tissues transformation work. The somatic embryos can develop into particularly favoured whole plants in a way similar to zygotic embryos. Somatic embryos can be directly or indirectly produced. In direct somatic embryogenesis, the embryo is directly produced from a cell or a group of cells without the formation of an intermediate callus phase. Although direct somatic embryogenesis is common for some tissues (generally reproductive tissues, such as nucellus, styles, or pollen), it is relatively rare in comparison to indirect somatic embryogenesis. Figure 1.1 shows an example of direct somatic embryogenesis.

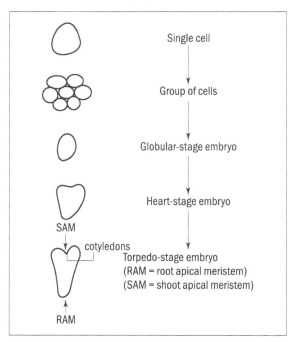

Fig. 1.1 *Steps involved in direct somatic embryogenesis*

A callus can be produced from explants from a wide range of carrot tissues by putting the explants on a solid medium (e.g., MS medium) containing 2,4-D (1 mg/L). A cell suspension can be produced using this callus by placing it in an agitated liquid MS medium containing 2,4-D (1 mg/L). The cell suspension can be maintained by subculturing repeatedly into a medium containing 2,4-D. Embryos are formed when the medium containing 2,4-D is replaced with a fresh medium containing ABA (0.025 mg/L). Young trifoliate leaves are used as the explant (Figure 1.2). The leaves are cut into small pieces after being plucked from the plant. These pieces are then washed in a medium that does not contain plant growth regulators and then placed in a liquid medium (B5) supplemented with 2,4-D (4 mg/L), kinetin (0.2 mg/L), adenine (1 mg/L), and glutathione (10 mg/L). The cultures are kept in an agitated liquid medium for around 10–15 days. On washing the explants and replacing the old medium with a B5 medium supplemented with maltose and polyethylene glycol (PEG), somatic embryos are produced. The somatic embryos can be matured on a solid medium containing ABA.

In both cases, even though the generation of somatic embryos from alfalfa requires the use of more complex media, the process of embryo production involves two steps. The initial medium containing 2,4-D is replaced by a medium that does not contain 2,4-D.

Callus is first produced from the explant in indirect somatic embryogenesis. Then embryos are produced from the callus or from a cell suspension formed from that callus. A classic example of indirect embryogenesis (embryo initiation) is the somatic embryogenesis from carrot. In this process, a high concentration of 2,4-D is used. In the next stage, embryos are produced in a medium with no, or very low concentration of 2,4-D. It has been found in many systems that somatic embryogenesis is improved when a source of reduced nitrogen, such as specific amino acids or casein hydrolysate, is supplied.

1.13 TRANSFER AND ESTABLISHMENT OF WHOLE PLANTS IN SOIL

When micropropagated plants are transferred from *in vitro* conditions to a greenhouse or field environment, a considerable number of them fail to survive. Compared to *in vitro* conditions, field and greenhouse environments have significantly lower relative humidity, higher

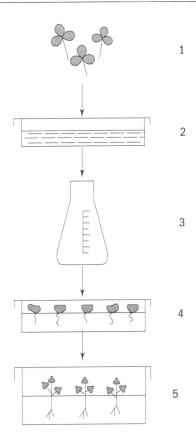

Fig. 1.2 *Direct somatic embryogenesis*

light level, and a septic environment, which are more stressful to micropropagated plants. However, the benefits of any micropropagation system can be completely reaped only when plantlets from tissue culture vessels are successfully transferred to *ex vitro* conditions. Most species produced *in vitro* need an acclimatization process so that a considerable number of plants survive and develop when they are transferred to soil.

Considerable efforts have been made to improve upon the conditions of *in vitro* stages of micropropagation. However, the process of acclimatization of micropropagated plants to the soil environment has not been fully studied. As a result, in the micropropagation of many plants, the transplantation stage continues to be a major complication. Plantlets or shoots grown *in vitro* are exposed to a unique microenvironment that provides minimal stress and optimum conditions for propagation of plants. In the micropropagation, plantlets are produced in culture vessels under low light levels and aseptic conditions, in an atmosphere with a high level of humidity, and in a nutrient medium that contains ample sugar that allows heterotrophic growth. These conditions result in a culture-induced phenotype which cannot survive in a greenhouse or field when placed directly in those environmental conditions. The physiological and anatomical characteristics of micropropagated plantlets make it imperative that they should be gradually accustomed to the environment of the greenhouse or field. Although specific details of acclimatization may vary, certain generalizations can be made.

1.13.1   Sugar in the Medium

The most common source of carbon used in plant cell, tissue, and organ culture media is sucrose. Studies have found that sucrose may also affect secondary metabolism in cell and organ culture (Fowler 1983). In these studies, it was assumed that larger persistent leaves packed with larger amounts of storage compounds contribute more after transplantation. Hence, the nutrient function of persistent leaves will likely be maximized by the increase in the concentration of sugar in the medium (Gorut and Millam 1885). This strategy is regarded as appropriate to increase evapotranspiration losses in transplants. However, it might prove helpful for only some plants. Wainwright and Scrace (1989) reported that maximum values for shoot height and fresh and dry weights of *Potentilla fruticosa* and *Ficus lyrata* can be obtained *in vivo* when previously conditioned with 2% or 4% sucrose. Plantlet establishment decreased in the absence of sucrose. Sucrose concentration of 40 g/L before transferring watercress microcutting to *in vivo* conditions maximized the dry weight of established plantlets. Preconditioning by addition of increased concentration of sucrose influenced the *in vivo* rooting and establishment of cuttings (Wainwright and Scrace 1989). However, decreasing the sucrose concentration in the medium enhanced the photosynthetic ability, thereby improving survival of plantlets in rose plants. The preconditioning at various sucrose levels before acclimatization was found to have no effect on plant establishment, but it influenced plant quality. Sucrose is necessary in the medium for many species. In some cases, independent growth was not possible on a medium without sucrose during rooting. The cultivars of hybrid tea roses

Sonia and Super-star differed in their requirement for sucrose during rhizogenesis. The microshoots of Sonia showed maximum rooting (92.6%) with 40 g/L sucrose, whereas the best rooting (84.5%) for Super-star was recorded at the sucrose level of 25 g/L. Increased sucrose levels improved rooting and root quality; however, a slight declining trend was noticed with the highest level (40 g/L) studied.

Rooting is an energy-consuming process, as any other morphogenetic process, and requires a high level of carbon source. The variation in their specific requirement for media might be because of genotypic effect. With a decreasing sucrose level, there was a proportional decrease in root initiation in apple, and shoots without sucrose failed to survive after being transferred to greenhouse environment (Zimmerman 1983). An increase in the size and number of starch granules was reported with an increase in the concentration of sucrose in the culture medium. It was reported by Hazarika, Nagaraju, and Parthasarathy (1999) that preconditioning of citrus microshoots *in vitro* with sucrose at a concentration of 3% provided optimum conditions for *ex vitro* survival and growth. Addition of sucrose to the medium resulted in a linear increase in biochemical constituents (reducing sugar, starch, and total chlorophyll). In a medium containing 3% sucrose, it was possible to obtain high-frequency *in vitro* shoot multiplication of *Plumbago indica*. By *in vitro* regeneration from the epicotyl segment of a plant grown *in vitro* using 3% sucrose in the medium, Kumar, Dhatt, and Gill (2001) achieved 68.8% *ex vitro* survival in kinnow mandarin. Multiple shoots were induced using shoot tips of gerbera in MS medium supplemented with 3% sucrose and other phytohormones and almost 100% survival rate was observed after the transfer. Mehta, Krishnamurthy, and Hazra (2001) found that increase in sucrose concentration from 2% to 4% in the medium enhances caulogenic response in tamarind plantlets from 34% to 48% in explants. The medium became brown on further increase in sucrose to 6%, which had a detrimental effect on the growth of shoots. It was observed that the liquid medium containing 9% sucrose and other phytohormones to be suitable for the growth of bulblets in isolated unrooted shoots. Within two months of inoculation, the high concentration of sucrose resulted in increase in the size of bulblets from less than 0.5 cm in diameter to about 1–1.5 cm. Higher plant growth, dry matter accumulation, and total leaf area were reported by Ticha, Cap, Pacovska, *et al.* (1998) under photomixotrophic conditions than under photoautotrophic conditions. Exogenous sucrose affected both biomass formation and photosynthesis positively. Photoinhibitory symptoms were prevented by the photomixotrophic growth of plantlets. Ticha, Cap, Pacovska, *et al.* (1998) deduced that the plant response was dependent on species because studies on gardenia showed that addition of sucrose enhanced the susceptibility to photoinhibition. It was suggested by Takayama and Misawa (1979) that changing osmotic potential could mediate the effect of sucrose on the bulb growth of *Lilium auratum*. It became apparent that high sucrose levels were more stressful for shoots, which showed reduced green leaves and poor development. The role of sugars was found to be osmotic. Sugars were also found to be the source of energy and carbon for shoot regeneration in tobacco callus from leaf parts of *Solanum melongena*. Unlike heterotrophic cells, photoautotrophic cells possess physiologically active and well-developed chloroplasts.

1.13.2   Photoautotrophic Micropropagation

It has been recently proposed that photoautotrophic micropropagation can reduce production cost and automatize the micropropagation process. Approximately 60% of the total production cost in the conventional micropropagation techniques is because of labour costs for multiplication, rooting, and acclimatization of plantlets. Hence, automation, robotization, or both at the culture stage can drastically reduce the production cost in micropropagation. In the photoautotrophic micropropagation, it is assumed that autotrophic cultures will have long-lived leaves and more photosynthetic productivity *ex vitro*. The aim is to modify culture-induced phenotype towards autotrophy in culture. To achieve this purpose, the oxygen concentration is decreased in the culture environment, which suppresses the photorespiration rate. Otherwise, sugar is reduced or eliminated from the medium, and the photosynthetic photon flux and the carbon dioxide concentrations can be raised. Increase in light intensity cannot alone increase the net photosynthetic rate for cultures at their CO_2 compensation point. The added advantage of the photoautotrophic tissue culture system is that the risk of microbial contamination is reduced when sugar is removed from the medium. Plantlets grow *in vitro* in larger numbers under photoautotrophic conditions than under heterotrophic conditions when *in vitro* environment is rightly controlled for stimulating photosynthesis. Hence, it is possible to reduce plant growth regulating substances, vitamins, and other organic substances because some of these will be endogenously produced in abundant quantities by the photoautotrophically growing plantlets.

Short, Warburton, and Roberts (1987) found that by culturing chrysanthemum meristems on a sucrose-free medium, the growth by photoautotrophy could be enhanced. Under this regime, the rates of photosynthesis in plantlet culture were comparable with those observed in seedlings. Hence, by employing this method, it is possible to successfully transfer tissue culture derived plants to soil condition, thereby preventing the necessity for any hardening regime. The sucrose promoted the growth of plantlets in the medium, reduced photosynthesis, and decreased *in vitro* hardening. Photoautotrophic plantlets had more abundant and longer root hairs than mixotrophic plantlets. Owing to photorespiration in a normal atmospheric concentration of O_2 (21%) and CO_2 (345 mM/mol), photoautotrophically generated C-3 plants lose up to 50% of carbon fixed photosynthetically. However, decreasing O_2 concentration represses photorespiration, and almost complete repression is observed at 2% O_2. It is possible to use a larger culture vessel in the photoautotrophic tissue culture method, without the risk of losing plantlets because of contamination. A large vessel aids in automation or robotization and automatic environmental control.

1.13.3 Growth Retardants

Plant growth retardants generally shorten the internode of higher plants *in vivo*. They also show some additional effects such as decrease in leaf size, intensification of the green colouration of leaves, and thickening of roots. Smith, Roberts, Mottley, *et al.* (1991) suggested several growth retardants that can be utilized in micropropagation to minimize damage by wilting without any harmful side effects. Paclobutrazol has been shown to reduce leaf water potential in *Malus domestica*. It impedes kaurene oxidase and thus prevents the oxidative metabolism from ent-kaurene to ent-kaurenoic acid in the pathway leading to the production of gibberellic acid. Paclobutrazol acts as an active growth retardant in a large number of species. In the rooting medium, at concentrations of 0.5–4 mg/L, paclobutrazol can result in reduced stomatal apertures, increased epicuticular wax, thickened roots, shortened stems, reduced wilting after transfer to compost, and also enhanced chlorophyll concentration per unit area of the leaf. In apple trees, it was found to effectively inhibit shoot growth and regulate various metabolic processes. On treatment with paclobutrazol, the following results were reported: shift in the partitioning of assimilates from leaves to roots; increased carbohydrate in all parts of apple seedlings; enhanced amounts of chlorophyll, soluble protein, and mineral elements in leaf tissues; increased root respiration; decreased cell wall polysaccharide and water loss; and build-up of water stress-induced ABA.

The clonal propagation of Nash Court variety of *Lapageria rosea* could be successfully achieved by a rhizome bud proliferation stage, followed by the stage of adventitious root development, all in the presence of paclobutrazol. The formation of rhizome buds can be induced from aerial shoot auxillary buds in *L. rosea* by adding the GA3 biosynthesis inhibitor paclobutrazol in the culture medium. There was dose-related decrease in wilting up to a concentration of 3 mg/L when plantlets were treated with paclobutrazol, flurprimidol, triapenthenol, chlorophonium chloride, uniconazol, and ancymidol. In *Phaseolus vulgaris*, triadimephon enhances both stomatal resistance and shoot water potential, and it increases stomatal resistance in *Lycopersicon esculentum*. It was reported by Hazarika, Nagaraju, and Parthasarthy (1999) that preconditioning citrus microshoots with paclobutrazol leads to higher *ex vitro* survival. This is caused by regulation of stomatal function and increase in epicuticular wax per unit area of leaf, which result in intensifying internode length, thickening root, and reducing leaf dehydration, in addition to increased chlorophyll synthesis. Addition of paclobutrazol in the growth medium produced stomata smaller than the normal size and with minimum apertures, probably because of a general reduction in cell expansion caused by anti-GA3-like activity.

However, micropropagated plants are less responsive to paclobutrazol than the seedlings. This is possibly because of an increase in the secondary roots of seedlings at early stages than micropropagated plantlets, which helps in the absorption of paclobutrazol by plants.

1.13.4 Reduced Humidity

Plants growing under lower relative humidity suffer from fewer transpiration and translocation problems *ex vitro* and have persistent normal leaves. The typical anatomical features of herbaceous plants growing under conditions of abundant moisture include low deposition of surface wax, stomatal abnormalities, and a non-continuous cuticle. By raising the vapour pressure gradient between leaf and atmosphere, this typical *in vitro* anatomy can be prevented. Relative humidity has been experimentally lowered *in vitro* with varying results. The experiments used a range of methods, such as using desiccants, opening culture containers, coating the medium with oily materials, using special closures that facilitate water loss, adjusting culture closures, or cooling container bottoms. It was found that a relative humidity of 85% reduced the multiplication rate of carnation. On increasing the sugar or agar concentration or adding osmotic agents such as PEG to the medium, the relative humidity is also lowered, and, in some cases, the increase served the same purpose as desiccant. High mortality and fewer roots were reported when silica gel and lanolin oil were used to reduce humidity in chrysanthemum. Reduction in relative humidity improved the survival rates of plants after transplantation. Reducing the humidity inside the culture vessel improves the internal structure of plantlets and gives a more successful establishment in the glasshouse. In cauliflower and chrysanthemum, optimum growth and *in vitro* hardening occurred when plantlets were cultured at 80% relative humidity. Chrysanthemum and sugar beet leaves, which were induced and de-developed at relative humidity of less than 100%, exhibited increased epicuticular wax, enhanced stomatal functioning, and reduced leaf dehydration.

1.13.5 Antitranspirants

Mixed results were observed when antitranspirants were used to reduce water loss during acclimatization. *Ex vitro* survival or performance was not found to be promoted by the use of antitranspirants. Two possible reasons cited were phytotoxicity and interference with photosynthesis. Agents that cover leaf surface, such as glycerol, paraffin, and grease, have been found to promote *ex vitro* survival of several herbaceous species. However, these have not been evaluated over long periods or examined on woody species. The stomatal transpiration of micropropagated cauliflower plantlets reduced substantially on spraying 10 mM ABA on persistent leaves. However, ABA significantly enhanced the stomatal resistance on leaves of seedling plants and the new leaves of acclimatized plants removed from culture. About 56% of the stomata on leaves of plants grown *in vitro* were closed following 4 h of mannitol-induced water stress and after 1 h of Folicote application to micropropagated apples, resulting in equal or better plant survival in a potting medium compared to acclimatization under mist. In tomato plants, phenylmercuric acetate (PMA) and CCC caused stomatal closure and delayed wilting. Stomatal closure was observed on the application of alachlor, and a nearly complete closure was noticed with PMA application. In plants treated with B-9 (1500 ppm) and PMA (20 ppm), the stomatal resistance was found to be high. On the contrary, alachlor (20 ppm), sunguard (0.02%), China clay (6% w/v), and silica powder (6% w/v) showed moderate stomatal resistance compared to control. In these treatments, relative water content exhibited the reverse trend.

Hazarika, Nagaraju, and Parthasarthy (1999) reported that 2 mL/L of 8 HQ was effective in controlling loss of water from citrus plantlets grown *in vitro*, consequently promoting in *ex vitro* survival. Tissue cultured walnut plantlets were acclimatized using latex polymer, and it was found that the survival rate of plants immersed in latex was higher than that of the control. On treating with latex, plantlets gained significantly more dry matter, apparently because their newly formed leaves could photosynthesize under favourable conditions of the open environment for a longer period than the control. The latex dipping technique is a simple method for the acclimatization of micropropagated walnut plantlets. Exogenous application of ABA to cell cultures can result in rapid hardening of cells to a considerable level. Less dramatic hardening responses were observed on the application of ABA to whole plants in comparison to cells. Insufficient uptake, rapid metabolism, and microbial degradation are attributed as possible causes for minor hardiness in whole plants as a result of ABA application. However, the extent to which low-temperature acclimation conditions induced hardening in plantlets could not be achieved by the treatment of ABA alone. This study suggests that, in addition to ABA, other factors are also responsible for hardening. In addition, ABA failed to maintain the hardiness levels of cold acclimation treatments, and plantlets de-acclimated to –9°C in media containing BAP and ABA. However, whether a pre-exposure to ABA before cold acclimation could maintain hardiness level is not yet known.

1.13.6   Simultaneous Rooting and Acclimatization

A number of commercial laboratories do not root microcuttings *in vitro* because it is labour-intensive and expensive. It has been estimated that the process of rooting *in vitro* incur approximately 35%–75% of the total cost of micropropagation. When cauliflower shoots were rooted *in vitro*, abnormality was found in the transition zone between root and shoots. At the time of removal of plantlets from the culture, vascular connections were found to be thinner and poorly formed. Thus, water uptake from the root into the shoot was reduced. The vascular connections were more significant after acclimatization, but uptake of water remained less than the seedling. An approach that combined the advantages of *in vitro* and *ex vitro* rooting was found to be successful for apples. It was possible to accomplish acclimatization and hardening in tea micropropagation as a one-step process within a short span of time before transplanting. Factors such as soil pH (4.0–6.4), shoot size, plant growth regulator treatment (IBA 500 mg/L for 30 min), CO_2 enrichment, optimization of time of harvesting of microshoots, and light (15 mM/m^2/s) conditions in specially constructed hardening chambers significantly contributed to hardening in tea micropropagation. Tea shoots were directly rooted when the cut ends were dipped in IBA (50 mg/L for 30 min) and subsequently planted in a soil–peat moss mixture (at 1:1 ratio). The percentage of survival in the field was higher for shoots directly rooted in soil than those rooted under *in vitro* conditions. Labour costs can be substantially reduced by *ex vitro* rooting, which can be carried out on a stonewool substrate moistened with auxin solution during the induction phase. Nonetheless, *in vivo* conditions during root formation are not so significant if an optimal plant material is obtained at the end of the *in vitro* cycle.

Sahay and Varma (2000) utilized *Piriformospora indica* as a potential agent to acclimatize micropropagated tobacco and brinjal. Under *in vitro* conditions, *P. indica* was grown in the dark in a culture bottle containing minimal medium for seven days. Then the regenerated shoots and plantlets (having roots and shoots) were co-cultured for 10–15 days. Unlike untreated controls, the morphology of the inoculated plants exhibited better revival and regeneration. During the acclimatization process, a better survival rate (more than 90%) was shown by *P. indica* inoculated plants than the untreated control plantlets (about 62%). This is because root colonization empowered them with extra-molecular weapons to tackle the situation. As *P. indica* promotes growth like arbuscular mycorrhizal fungi (AMF), it possibly induces the plant–microbe interaction similar to AMF during the root colonization, and the synthesis and expression of defence-related proteins and enzymes get stimulated in a controlled manner. However, when a pathogen comes in contact with these plants, the production of defence mechanism related secondary metabolites is enhanced. Chemicals such as phytoalexins, isoflavonoids, glyceollin, coumestrol, coumestin, and isosojagol provide protection to plants. Thus, they enhance the survival rate of plants.

In photoautotrophic micropropagation, rooting and acclimatization *in vitro, ex vitro,* or both can be easily achieved. It was possible to achieve plant regeneration from chlorophyllous root segments developed from *in vitro* rooted plants of *Holostemma annulare*. The survival rate was found to be 80% following a hardening period of four weeks on adjustment of the humidity conditions inside the mist chamber by removing the polythene covering for 1 h during the first week and increasing the exposure time in subsequent weeks. In the micropropagation of citrus and curry leaf, *ex vitro* rooting and acclimatization can be simultaneously achieved using soilrite alone or soilrite with farmyard manure as a carrier. Using these carriers, Parthasarathy, Nagaraju, Hazarika, *et al.* (1999) obtained better rooting (80%–91%) and very high *ex vitro* survival (90%–97%) in citrus, while Singh, Ray, Bhattacharjee, *et al.* (1994) recorded 60% survival in citrus. Hazarika (2003) carried out *ex vitro* acclimatization of *Aegle marmelos* using 2 cm long microshoots from proliferating cultures rooted in soilrite after giving them pulse treatment with IBA or NAA at 10 ppm for 2 min. Up to 79.46% rooting was found, which was higher than the conventional rooting carried out in an agar-based medium.

1.14   SHOOT TIP CULTURE

In the meristem tip culture, the organized apex of the shoot is excised from a selected donor plant for *in vitro* culture. The culture conditions are regulated to allow only the organized outgrowth of the apex directly into a shoot, without the interference of any adventitious organs. The excised meristem tip is generally small (often 1 mm in length), which is removed by sterile dissection under a microscope. The explant includes the apical dome and a few youngest leaf primordia and it does not contain any differentiated provascular or vascular tissues. There is a major advantage of working with a small explant.

It has the potential of excluding pathogenic organisms that the donor plants from the *in vitro* culture might contain. Another advantage is the inherent genetic stability of the technique, as plantlet generation is from a differentiated apical meristem. The plantlet production from adventitious meristems can be avoided. Callus tissue formation and adventitious organogenesis can be prevented by development of shoots directly from the meristem, which minimizes genetic instability and somaclonal variation. If there is no necessity of virus elimination, the less demanding related technique of shoot tip culture may be more convenient for plant propagation. In this method, even though the explant is still a dissected shoot apex, it is much larger in size, which can be easily removed, and contains a relatively large number of developing leaf primordia. The length of the explant is mainly 3–20 mm, and *in vitro* development can be regulated to allow direct outgrowth of the organized apex. The axillary buds of *in vitro* plantlets developed from meristem tip culture may also be used as a secondary propagule. The *in vitro* plantlet may be divided into segments when it has developed expanded internodes. Each segment contains a small leaf and an even smaller axillary bud. On placing these nodal explants on a fresh culture medium, the axillary bud grows directly into a new plantlet, and the process can be repeated further. This procedure adds a high propagation rate to the original meristem tip culture technique, and these techniques together form the basis of micropropagation, which is very useful for the horticulture industry. However, there is a possibility that callus tissue may be produced on certain parts of the growing explant, particularly at the surface damaged by excision. Under such circumstances, it is acceptable if the callus is slow growing and localized, and the callus mass and any organized development on it can be removed at the first available opportunity. Laboratory research on cauliflower plants regenerated from wound callus and floral meristems *in vitro* reported normal phenotypes for all plants produced by meristem culture, whereas those of callus origin had rosette and stunted forms with both glossy and serrated leaves. The abnormal plants derived from callus tissue and normal plants generated from meristem culture had the same DNA levels when measured by scanning microdensitometry, but there was considerable difference when the isozymes of acid phosphatase were investigated by gel electrophoresis. This suggests that the adventitious origin of the shoots introduces a range of unacceptable variations at the gene level.

Meristem culture technique preserves the exact arrangement of cell layers, which is necessary to maintain a chimeral genetic structure. The surface layers of the developing meristem have different genetic background in a typical chimera, and their contribution, in a particular arrangement, to the plant organs results in the desired characteristics, for example, the colour of flower in some African violets. The chimeral pattern will be maintained as long as the integrity of the meristem remains intact and development is normal *in vitro*. However, if callus tissues are allowed to develop and consequently shoots proliferate from adventitious origins, all the chimeral layers of the original explant might not be expressed in the required form in the adventitious shoots. It is possible to employ the meristem culture technique in situations in which the donor plant is infected with viral, bacterial, or fungal pathogens, although the symptoms of the infection may or may not be expressed. The basis of eradication is that pathogenic particles are unlikely to be present at the terminal region of the shoot meristem above the zone of vascular differentiation. So when a sufficiently small explant is collected from an infected donor and grown *in vitro*, the derived culture will likely be free of pathogens. Once screened and certified, such cultures can guarantee a disease-free stock for further propagation. The meristem tip method, together with heat therapy, can improve the effectiveness of disease elimination, or antiviral chemotherapeutic agents may be examined. Irrespective of the techniques employed for virus eradication, the size of the explant is crucial to achieve success. Although the smallest explants are less likely to be successful during *in vitro* culture, they will produce the largest percentage of virus-free materials when entire plants are grown in a glasshouse or field. This is clearly demonstrated by the efforts to eliminate white clover mosaic virus (WCMV) from *Trifolium pratense*.

If the use of meristem culture alone fails to produce any virus-free plants, one has to consider temperature stress treatment of donor plants and the use of antiviral agents. There are no prescriptive procedures for these treatments. In heat therapy, donor plants are exposed to temperatures of 30–37°C, and the treatment might continue for several weeks. If *in vitro* cultures provide the donor material, heat treatment of cultures is possible. The absence of virus particles from the meristems of treated plants can be attributed to various factors, including restricted movement of particles to apical regions, a thermally induced block on viral RNA synthesis, and inactivation of virus particles. Extended low-temperature treatment of the donor material might also produce effective results. While subjecting donor plants to temperature stresses, it is essential to study the physiology and developmental behaviour of the plants concerned. Accordingly, the maximum stress can be applied and an acceptable pattern of growth can be maintained. It is necessary to have the taxonomic knowledge of the investigated plant for selecting possible treatments from the literature because close or distant relatives can be identified.

Ribavirin, also known as virazole, is a widely used antiviral chemical added to the culture medium as an additive. It is a guanosine analogue with broad-spectrum activity against animal viruses and also apparently active against plant virus replication in whole plants. Increased ribavirin concentration and increased length of culture incubation generally improve the effectiveness of virus elimination, but they slow down the growth. Phytotoxicity might also be evident at high concentrations. The advantages of meristem culture are as follows:

- *In vitro* clonal propagation with maximal genetic stability
- Ability to remove viral, bacterial, and fungal pathogens from donor plants
- Meristem tip acting as a practical propagule for cryopreservation and other techniques of culture storage
- A technique for proper micropropagation of the chimeric material
- Cultures often acceptable for international transport with respect to quarantine regulations
- The steps described earlier can be applied to a broad range of plant subjects, although formulations of media and reagents may require a change.

1.15   RAPID CLONAL PROPAGATION AND PRODUCTION OF VIRUS-FREE PLANTS

There are various methods of vegetative propagation such as using naturally detachable structures (bulbs and corns), planting many kinds of cuttings, and budding and grafting. All these methods lead to the production of clones, which are genetically uniform plants.

In 1946, Ball developed complete plants of *Lupinus* and *Tropaeolum* (nasturtiums) using shoot apex culture. In this technique, the minute growing point (apical meristem), along with the first one or two leaf primordia, is carefully cut from the tip of a shoot. The larger developing leaf primordia enclose and protect the growing point, thus preventing any external infection from reaching the growing point. Hence, the shoot apex can be carefully transferred to a sterilized nutrient medium, where it can grow without any risk of fungal or bacterial infection. The apical meristem, along with the two leaf primordia, may be only about 0.1 mm in length; however, some researchers have used a shoot tip of longer length (about 5–10 mm). A major advancement was made by Morel and Martin (1952), who reported that virus-free dahlias can be obtained through shoot apex culture. The tip tends to have less amount of virus than the lower part of the stem because the conducting tissue in a plant stem has not yet differentiated in the growing point at the tip of the shoot. Hence, it may be possible to prevent virus infection present lower down the plant by taking a tiny terminal piece from the tip. The smaller the size of the explant, the higher the chances of virus removal.

Walkey, Webb, Bolland, *et al.* (1987) reported that viruses present in a shoot apex can be eliminated by culturing the shoot apices. To produce plants devoid of a particular pathogen, such as a virus, actively growing shoot apices should be used and their growth can be enhanced by the application of GA3 and through etiolation, fertilization, and irrigation. The parent plant can be heat treated (i.e., grown at a temperature ranging from 35–40°C for a period from a few hours to a few years) to have a greater likelihood of obtaining uninfected plant tissues. However, such experiments are usually carried out at a research institute. Virus-free plants produce more crops than virus-infected plants of the same clone. From a few plants, millions of strawberry plantlets can be produced in a year by tissue culture techniques. More than 50 viral and 8 mycoplasmal diseases have been reported in strawberry plants. In a few experiments, it was possible to produce strawberry plants free of some of these viruses. Virus-free plantlets derived from meristems can be successfully preserved in cold storage. Recently, it has been reported that cryopreservation of strawberry plantlets in liquid nitrogen (–196°C) is possible. Such a procedure should further improve the chances of eliminating viruses.

1.15.1 Shoot Apex Grafting

Another method of generating virus-free citrus clones is micrografting virus-free shoot apices on pathogen-free seedling rootstock plants growing in sterile culture. This procedure leads to the elimination of the issue of juvenility associated with the use of nucellar seedlings. This technique is also useful for woody plants that are not rootable.

1.15.2 Rapid Propagation

Morel demonstrated in 1960 that orchids could be propagated rapidly by the shoot apex culture technique. It is possible to initiate protocorm development from a single shoot tip. If the protocorm is excised into sections before it produces a root and shoot or kept agitated in a liquid culture solution, a large number of protocorms can be obtained. The subculturing or shaking should be stopped to develop complete plants. In various ornamental plant species, such as carnation, chrysanthemum, dahlia, potato, and asparagus, this shoot apex culture technique, referred to as mericloning, has been successfully applied.

Although earlier experiments with woody plant species were not successful, some success has been achieved with several species in recent times. For instance, Jones, Pontikis, and Hopgood (1979) reported that around 3000 apple rootstock plantlets can be reproduced in eight months from a single 5–8 mm shoot apex explant. They claimed that the use of phloroglucinol induced the growth of lateral buds in the axils of developing leaf primordia, and the cultures were consistently multiplied fourfold to fivefold every three weeks. A number of studies have reported that a wide range of plants have the ability to be propagated through plant tissue culture. Currently, there are two methods of plant propagation *in vitro*, which are as follows:

1. Induction of adventitious shoots from different types of explants, including stem tips, followed by rooting once the new shoots are formed.

2. Induction of axillary branching from buds in the leaf primordia axils of stem tip explants by overcoming apical dominance. Formation of roots can be induced once new shoots are developed.

A third approach is somatic cell embryogenesis. In this process, embryos are produced from cells other than the zygote (fertilized egg). Somatic cell embryogenesis has been successfully carried out in monoembryonic citrus and grapes, where adventitious embryos were induced to form tissue cultured from nucellus.

1.16   EMBRYO CULTURE AND EMBRYO RESCUE

Conventional breeding techniques are mostly used for improvement of crops. However, it sometimes becomes essential to supplement these methods with plant tissue culture techniques to enhance their efficiency or achieve an objective that is not possible by employing only conventional methods. Embryo culture is routinely used these days for the recovery of hybrid plants from their distant crosses, for example, recovery of hybrids from *Hordeum vulgare* × *Agropyron repens* and *H. vulgare* × *Triticum* sp. Viable seeds are developed in the case of *Triticale*, a rare hybrid between *Triticum* and *Secale*. However, most of the tetraploid and hexaploid wheat varieties possess two dominant genes, *Kr1* and *Kr2*, which prevent the development of seeds in crosses with *Secale*. The hybrid seeds are poorly developed and minute and show poor germination. By embryo culture, 50%–70% hybrid seedlings can be obtained. It was not possible to obtain hybrid seedlings from *Triticum aestivum* × *H. vulgare*. However, in embryo culture, on crossing *H. vulgare* or *T. aestivum* (used as male) with *H. bulbosum* (used as female), the chromosome complement of *H. bulbosum* gets eliminated from the developing embryo. Most of the seedlings obtained from such crosses are haploid, which possesses only one set of chromosomes, either from *H. vulgare* or from *T. aestivum* parent. Embryo culture is also beneficial for reducing breeding cycle, propagating orchids, and overcoming seed dormancy.

1.16.1   Embryo-rescue Techniques

Plants are bred through hybridization and selection. A breeder crosses plants to combine the best characteristics from two different plants in the offspring. The breeder selects the offspring that possesses a combination of the best features of both parent plants. This process relies heavily on the production of offspring, that is, on the formation of viable seeds. If the number of offspring produced is less, it will limit breeders to select new varieties. In other words, no fertilization will take place after pollination. Fertilization occurs many times, but the embryo produced dies at an early stage of development owing to various reasons. Methods to rescue unripe seeds or embryos from adult plants were developed in the early 1900s. This was particularly carried out with seeds having very long dormancy period or which were particularly heterogeneous. With growing advancements in tissue culture techniques, this method was also utilized to rescue those embryos in ovules which were fertilized but did not develop into viable seeds. At first, whole ovaries were used for tissue culture and seedlings were obtained from embryos that would have died at a later stage of development. Later, with the development of advanced ovule and embryo culture techniques, it became possible to recover embryos that died at an early stage of development. A combination of these techniques is generally used. Ovary pieces are put in tissue culture and then ovule or embryo culture is used. Embryo rescue techniques are employed for obtaining interspecific and intergeneric hybrids. Using this method, valuable varieties have been developed in potatoes, vegetables, and many ornamental plants. Many new lily hybrids have been produced using this procedure. Moreover, a large number of the new *Alstroemeria* varieties have been rescued with the aid of embryo culture methods. For a number of years, SBW International has been carrying out such projects and has produced many interesting interspecific and intergeneric hybrids. Modern plant breeding cannot be imagined without this technique.

1.17   PROTOPLAST CULTURE AND FUSION

Totipotency is the fundamental difference between plant and animal cells. Plant cells are totipotent, which forms the foundation of plant tissue culture. The presence of cell wall in plants is another important difference. This led to the most significant development in the field of plant tissue culture, that is, the isolation, culture, and fusion of protoplasts. Protoplasts are plant cells from which the cell wall has been removed. Besides being used for genetic recombination through somatic cell fusion, cultured protoplasts can also be employed for introducing foreign DNA, cell organelles, bacterial, and viral particles. Therefore, protoplast culture has attained an important position in plant biotechnology.

1.17.1   Isolation of Protoplast

Plant protoplasts were first isolated by von Klercker in 1892. He plasmolysed onion bulb scales in hypertonic solution to isolate plant protoplasts. The yield of protoplasts by this mechanical procedure was low, and the procedure could be utilized only for highly vacuolated and non-meristematic cells. In 1960, Cocking isolated intact protoplasts using the cell wall degrading enzyme cellulase were obtained from the fungus *Myrothecium verrucaria*. Purified cell wall degrading enzymes such as macerozyme, cellulase, and hemicellulase became commercially available by 1968, which led to further progress in the enzymatic isolation of protoplasts. Isolating protoplast through the enzymatic method can be classified into two categories:

1. **Sequential enzymatic** There are two steps in the sequential enzymatic method. First, macerated plant tissues are incubated with pectinase to obtain single cells. The single cells are then treated with cellulase to isolate protoplast.

2. **Mixed enzymatic** In the mixed enzymatic method, the processes of separation of cells and degradation of their walls to isolate protoplasts are carried out simultaneously. Protoplasts are isolated by treating plant tissues with a mixture of pectinase and cellulase.

Protoplasts have been extracted from all plant parts such as cotyledons, shoot tips, and flower petals. However, leaf mesophyll tissue is mostly preferred because it has high regeneration potential. Typically, the youngest fully expanded leaves from young plants or seedlings are used as a source of protoplast. The protoplast yield improves on preconditioning the plants in darkness or cold (4–10°C) for 24–72 h before isolation. The leaves are sterilized, chopped into small pieces, and floated on a solution that is osmotically adjusted at 20–25°C for 1–24 h. In this step, the plasmodesmata are broken down and water comes out of cells, resulting in shrinking away of cell contents from cell walls. As such, cells are able to retain their integrity even after the cell wall is removed. After this step, the leaf pieces are incubated in darkness with digestive enzymes in a shaker (30–50 rpm). Incubation temperature and time vary across species. For osmotic stabilization of protoplasts, the incubation solution should contain osmotic substances such as mannitol, sorbitol, glucose, sucrose, fructose, and other typical salts of plant nutrient media. For this purpose, osmotic concentrations of the enzyme mixture and the subsequent media are raised by adding mannitol or sorbitol. Adding 50–100 mM/L of $CaCl_2$ enhances the stability of plasma membrane, and pH should be maintained in the 5–6 range.

To minimize fluctuation in the pH value during the digestion, the buffer often contains 3 mM phosphate. Cell suspension cultures are used in the case of monocots, where mesophyll cannot be the source, for isolating protoplasts. By slow speed centrifugation for 10 min, the released protoplasts are separated. They are then washed two to three times to remove enzymes and debris before being transferred to the culture medium.

1.17.2   Protoplast Viability and Plating Density

The viability of protoplasts is estimated before culturing by staining with FDA. Under UV light, viable protoplasts show green fluorescence. These protoplasts should be observed within 5–15 min after the FDA treatment because the FDA dissociates from the membrane after this time. Metabolites leach out of the leaky membrane of some protoplasts. However, the metabolites can be absorbed back if there is a sufficient density of protoplasts. Therefore, a minimum plating density is necessary for the growth to begin and it can be measured by a haemocytometer. Single or less protoplast culture is required for genetic engineering, and either conditional medium or feeder layer is used for this purpose. In the conditional medium, plant cells are already grown and so metabolites have leached into it. After filtering, this medium is used for growing isolated protoplasts. The feeder layer is prepared by plating solid media with protoplasts, followed by irradiation. This inactivates the nucleus, but protoplasts are viable. Now protoplasts can be plated at a lower density on this feeder layer.

1.17.3   Protoplast Culture

The shape of freshly isolated protoplasts is spherical because they are not bound by cell wall. Under suitable conditions, viable protoplasts regenerate a new cell wall within 48–96 h after isolation. The new cell wall can be observed by staining with calcofluor white. Under UV light, protoplasts with the new cell wall fluoresce bluish white. The protoplasts that do not regenerate a wall remain undivided and eventually die. Moreover, all the healthy protoplasts may not divide and, hence, plating efficiency is determined to estimate the cell vigour. Plating efficiency is defined as the number of dividing protoplasts divided by the total number of protoplasts plated. Protoplasts having the potential to divide undergo first division within 2–7 days of isolation. As protoplasts are delicate in nature, modifications are necessary in

MS or B5 media or any other culture medium used for regeneration of organs from explants. In addition to higher osmotica, the inorganic salt concentration is adjusted (ammonium nitrate concentration is decreased and calcium level is increased). The amounts of organic components, vitamins, and plant growth regulators are increased to expedite cell wall synthesis. As protoplasts are sensitive to light, they are cultured in diffuse light for the initial 4–7 days. After protoplasts have generated new cell wall and undergone division, they are put back in the normal medium in which plants regenerate by somatic embryogenesis or shoot formation. Plants have been successfully regenerated from protoplasts in the case of alfalfa, tobacco, carrot, and tomato, but cereals still pose problems.

1.17.4   Protoplast Fusion and Somatic Hybridization

Purified protoplasts, once isolated from any two different sources (can be different tissues, plants or species, or genera), can be fused to form somatic hybrids. This non-conventional technique of genetic recombination involving fusion of protoplasts under *in vitro* conditions and the consequent development of their product into a hybrid plant is termed somatic hybridization. Carlson, Smith, and Dearing (1972) produced the first somatic hybrid plant by fusing the protoplasts of *Nicotiana glauca* and *N. langsdorfii*. Protoplasts from different species are induced to fuse by the use of a variety of fusogens or electrical manipulation that promotes membrane instability. Some widely used fusion inducing agents are sodium nitrate (used by Carlson), high pH/Ca^{2+} concentration, and PEG treatment. As sodium nitrate treatment results in low heterokaryon formation, high pH, and high Ca^{2+} concentration, it is suitable for a few plant species. Polyethylene glycol treatment is the most favoured method because of its reproducible high frequency of heterokaryon formation and low toxicity to most cell types. Treatment with PEG is reported to be most effective when it is used in the presence of high pH/Ca^{2+}. It increases the heterokaryon formation and their survival. Electrofusion is considered to be a more selective, simpler, quick, and non-toxic method. This method uses electric shock or short pulses of high voltage to induce membrane fusion between two cells. Electrofusion is reportedly used in producing many useful somatic hybrid plants, such as *Nicotiana plumbaginifolia* (+) *N. tobacum* and *Solanum tuberosum* (+) *S. charcoense* (resistant to Colorado potato beetle).

1.17.5   Somatic Hybridization

After fusion treatment, the protoplast population becomes a heterogeneous mixture of unfused parents, homokaryons, heterokaryons, and fused protoplasts with two independent nuclei. This necessitates the selection of stable hybrids. Hence, it becomes important to incorporate some identification and selection systems into each parental cell line before fusion. Generally, it is most convenient to use two parental cell lines with different requirements as selective screens because only those cells that possess the complementary traits of both parents, that is, the fused cell lines, will thrive. A few selection criteria currently in use include resistance to antibiotics and herbicides and the ability to grow on specific amino acid analogues. In addition, two stains fluorescein isothiocyanate (FITC) and rhodamine isothiocyanate (RITC) have been used as they do not adversely affect the viability of cells. On observing under a fluorescent microscope, FITC stained cells appear green and RITC stained protoplasts appear red, whereas fused cells containing both RITC and FITC appear fluorescent yellow.

1.17.5.1   Application of somatic hybridization

Somatic hybridization is applied in the following cases:

(i) **Genetic recombination in asexual or sterile plants:** The impediment of reproduction in haploid, triploid, and aneuploid plants has been overcome by protoplast fusion. Moreover, this approach can be used to recombine genomes of asexually reproducing plants. For instance, when protoplasts isolated from dihaploid potato clones are fused with the protoplasts of *S. brevidens*, hybrids of practical breeding value are produced.

(ii) **Genetic recombination between sexually incompatible species:** The incompatibility barriers in sexual recombination at interspecific or intergeneric levels can also be overcome by somatic hybridization. Somatic hybrids are used for transferring beneficial genes (e.g., disease resistant or abiotic stress resistant varieties). Also, Datura hybrids (*D. innoxia + D. discolor, D. innoxia + D. stramonium*) are obtained by fusing their diploid mesophyll protoplast. These hybrids show heterosis and higher scopolamine (alkaloid) content, which is 20%–25% higher than that in parent species, and, therefore, are industrially important.

(iii) **Cytoplasm transfer:** By somatic hybridization, the time required for transfer of cytoplasm can be shortened from 6–7 years in backcross method to 1 year at present. This method allows transfer of cytoplasm between sexually incompatible species. Cytoplasmic hybrids (or cybrids) contain nucleus from one species but cytoplasm from both the species are involved in fusion. The nucleus of the other parent is irradiated. This technique has been employed to transfer two desirable traits, that is, cytoplasmic male sterility (CMS) and resistance to atrazine herbicide. Cytoplasmic genes coding for both the traits in *Brassica* have been transferred to different crops such as tobacco and rice.

1.18   *IN VITRO* MUTATION METHODS

To improve yield, quality, and resistance to diseases and pests, as well as to enhance the attractiveness of flowers and ornamental plants, conventional mutation techniques have often been employed. More than 1700 mutant varieties involving 154 species have been officially released. In many countries, mutant varieties of a few economically important crops (e.g., barley, durum wheat, and cotton) are grown in majority of cultivated areas. Mutation techniques have become a major means for breeding ornamental plants, such as *Alstroemeria*, *Begonia*, *Chrysanthemum*, carnation, dahlia, and *Streptocarpus*. In both seeds and vegetatively propagated plants, *in vitro* techniques, such as shoot organogenesis, somatic embryogenesis, anther culture, and protoplast fusion, are used to overcome some of the drawbacks inherent to the mutation techniques. Together with induced mutations, *in vitro* culture can accelerate breeding programmes, ranging from the generation of variability, through selection, to the multiplication of desired genotypes. The expression of induced mutations in the pure homozygote, which is achieved through microspore, anther, or ovary culture, can enable quick recovery of desired traits. In combination with *in vitro* culture techniques, mutations may be the only method available for improving an existing cultivar in some vegetatively propagated species. Many molecular studies are dependent on the introduction and identification of mutants in 'model species' for the development and subsequent saturation of genetic maps, understanding of developmental genetics, and explanation of biochemical pathways. After being identified and isolated, the genes encoding agronomically valuable characters can be either introduced directly into crop plants or used as probes to find similar genes in crop species. Advances made in these technologies are likely to provide improved methods for the selection of desired mutants.

Many disruptive techniques are used by conventional breeding to produce mutations in crops. The currently available methods for creating mutations in plants, such as radiation or harsh chemical treatment, are generally used in the conventional breeding and they are reported to introduce extensive changes to plant chromosomes. In comparison to genetic engineering techniques, these processes are less refined, and, in some cases, they cause extensive chromosome alteration in plants.

Chemical mutagenesis is an example of a non-GM technique used by conventional breeders. To make Linola™, a special kind of food (linseed oil) developed in Australia, the chemical ethyl methane sulphonate was used. The toxic chemical sodium azide was used to produce malting barley mutants used in beer production. These mutant barleys produce reduced amounts of beer haze and include the barley variety Caminant, which is mostly used in Denmark. The success of Danish lager can be probably attributed to this mutation in barley varieties used in its production.

Radiation mutagenesis is another conventional breeding technique. This technique has been used to produce the durum wheat variety Creso (used to make pasta in Italy), the rice variety Amaroo (sown in more than 60% of the total rice producing areas in Australia), and the Rio Star® grapefruit (popular in the United States) (Ahloowalia, Maluszynski, and Nichterlein 2004; IAEA 2008; NAS 2004). Using ionizing radiation, hundreds of different mutant varieties of rice have been developed. Calrose 76 is one of these varieties and it was produced in California. It is a short stature mutant developed by exposing rice to intense gamma rays from radioactive cobalt. Radiation is also used to sterilize medical instruments (van Harten 1998).

The fact that radiation can cause extensive changes to chromosomes has been thoroughly documented in the peer-reviewed scientific literature. These changes are known to be similar to those caused by insertion of transgenic DNA. According to a recent detailed study of rice, the frequency of altered gene expression is higher after conventional radiation treatment (to induce a mutation) than after genetic engineering (to introduce a transgene). Many of these facts about practical breeding of crops using mutation techniques are known to plant breeders, geneticists, and biologists.

More than 3000 new plant varieties have been developed using radiation treatment. These varieties are cultivated for food and fodder. It is commonly known that radiation has more disruptive effects on chromosomal structure than the manipulations used to make transgenic plants. Radiation treatment of crop plants is not reported to have caused any health effects among consumers, despite having been used commercially for several decades.

1.18.1   Virus Elimination and Pathogen Indexing

It is illogical to use the term pathogen free, which is often used for plants grown in tissue culture. Pathogen free means the plant material is free from all known and unknown pathogens. However, the material is probably free from only those pathogens for which it has been indexed or tested. Moreover, any test is as good as the sensitivity and dependability of the available testing or indexing systems. It should be noted that non-recognized pathogens might still be present. Murashige (1979) prepared a list of specific pathogen-free (SPF) plants developed *in vitro* from infected plants (Table 1.1). This list includes only the diseases caused by viruses. Instances of plants freed from bacteria, fungi, nematodes, and other pathogens through tissue culture methods are not included in the list.

1.18.1.1   Time

It may be a misleading concept that tissue culture systems will produce a saleable product within a short period of time, especially while working with shoot apex cultures obtained from virus-infected plants. If any commercial firm is willing to produce quality plants from virus-infected stock by shoot apex culture, it has to invest at least 2–5 years.

1.18.1.2   Virus infection

Indexing prior to heat therapy and meristem tip culture showed that more than 95% of commercial geranium cultures were infected by viruses. Although some viruses that caused dramatic foliar symptoms were rogued out of commercial propagation, other viruses that did not show foliar symptoms severely affected flowering. Therefore, it is commercially worthwhile to remove these viruses from geraniums.

Elimination of viruses is a complex and variable process, which involves heat treatment of plants from which vascular wilt pathogens have been removed. It is followed by shoot apex culture and indexing. While indexing young plantlets for known viruses, it should be ensured that plantlets are individually isolated to prevent reinfection. Those plants that index negatively should undergo exhaustive selection for plant quality and only plants of the highest quality should be retained for immediate use. It is essential to thoroughly check the plants for horticultural and commercial qualities.

The chosen 'elite mother plants' constitute the source of a 'nucleus block', which must be maintained in an isolated greenhouse with regular virus indexing. Cuttings are taken from the nucleus block to form the 'increase block'. Cuttings are then taken from the 'increase block' for 'stock plants', which eventually provide cuttings for commercial purposes. Elite mother plants should be produced every year. Many scientists have produced shoot apex-cultured and virus-indexed negative plants. But unless these plants are of superior quality and can be commercially multiplied, their generation will remain an academic exercise.

1.18.2 Other Organisms

Most bacterial and fungal organisms can be easily detected during the cultural process, and the infected cultures are discarded without verifying the pathogenicity of the contamination. *Erwinia*, *Pseudomonas*, and *Bacillus* spp. are examples of some more persistent contaminating pathogens. However, fungal and bacterial contamination cannot be always detected at an early stage. Moreover, apparently clean cultures can possibly develop symptoms in subsequent subcultures, for example, a cloudy liquid medium, a milky white haze in the agar surrounding the explant, or a rapid fall in vigour of the plant in culture. Once it is found that a culture of an ornamental tropical foliage plant is free from known pathogens, it is important to maintain the mother block stock *in vitro*, though the variability problem must be carefully monitored. The market value of many plants has significantly increased because of the availability of their more commercially desirable form (enhanced vigour, prolific branching, more shoots ending in flowers, early flowering, and denser foliage), following tissue culture propagation. These improved qualities have resulted from the removal of bacterial and fungal pathogens, as well as viruses.

Tissue culture is extensively used for mass multiplication in commercial nurseries. However, no efforts are made to obtain clones free from specific pathogens having detrimental effects on plant growth and quality. Commercial producers should attempt to develop better quality plant materials. Poorer quality plants are produced because of the lack of knowledge, training, and research. The potentials of producing plants free from specific pathogens are not utilized to their optimum extent. Tissue culture has developed to its full potential in certain cases, such as carnations, chrysanthemums, and strawberries, with consequent economic benefits on production and quality. Local flower growers import indexed elite mother plants, generated from shoot apices, and then multiply them using normal cutting techniques.

1.19   SELECTION OF HYBRID CELLS AND REGENERATION OF HYBRID PLANTS

After treatment with a fusion inducing agent (fusogen), the protoplast suspension recovered consists of the following cell types: (i) unfused protoplasts of the two strains/species, (ii) fusion products derived from two or more protoplasts of the same species (homokaryons),

Table 1.1 List of excluded plants and pathogens

Crop	Pathogen excluded
Allium sativum L.	Garlic mosaic virus
Armoracia lapathifolia	Turnip mosaic virus
Asparagus officinalis	Viruses A, B, and C
Brassica oleracea	Cauliflower and turnip mosaic viruses
Caladium hortulanum Birdsey	Dasheen mosaic virus
Cattleya hybrids	Cymbidium mosaic virus
Chrysanthemum morifolium Ramat	Chrysanthemum B. vein mottle, green flower viruses
Citrus sp.	Citrus exocortis viroid; stubborn *Spiroplasma*; cachexia, concave gun, Dweet mottle virus, psoriasis
Colocasia esculenta L.	Dasheen mosaic virus
Cymbidium sp.	Cymbidium mosaic virus
Dahlia sp.	Dahlia mosaic virus
Dianthus barbatus L. and *D. Caryophyllus* L.	Carnation latent, mottle, ringspot, vein mottle

and (iii) hybrid protoplasts produced by fusion between one or more protoplasts of each of the two species (heterokaryotic). In experiments involving somatic hybridization, only those hybrids or heterokaryotic protoplasts are of interest that are particularly generated from fusion between one protoplast of each of the two species. However, they constitute only a small part of the population (usually 0.5%–10%). Hence, they should be identified and isolated using an effective strategy. This step is known as the selection of hybrid cells. It is the most important step and still an active area of research. Some strategies that have been employed for selecting hybrid protoplasts are discussed next.

Visual markers, such as pigmentation of the parental protoplasts, may be used for identifying hybrid cells under a microscope, which are then mechanically isolated and cultured. For example, protoplasts of one species might appear green and vacuolated (from mesophyll cells), whereas those of the other might be non-vacuolated and non-green (from cell cultures). In the case of unavailability of such characteristics, protoplasts of two parental species might be separately labelled using different fluorescent agents. This approach consumes more time and requires considerable skills and efforts. Many researchers have tried to devise systems to specifically select hybrid cells. These systems make use of some properties (typically deficiencies) of the parental species that are not manifested in the hybrid cells owing to their genetic systems being complementary. These attributes, including sensitivity to constituents of the culture medium, temperature, antimetabolites, and inability to yield an essential biochemical (auxotrophic mutants), might be naturally present in the parental species or artificially introduced by means of mutagenesis or genetic engineering. For instance, calli were produced when protoplasts of *Petunia hybrida* were cultured in MS medium. On the contrary, the protoplasts of *P. parodii* produce only small cell colonies. In addition, actinomycin D (1 µg mL/L) interferes with the cell division of *P. hybrida* protoplasts, but it does not affect those of *P. parodii*. Hence, protoplasts of both these *Petunia* species do not produce macroscopic colonies (calli) in MS medium containing actinomycin D (µg mL/L). However, in this medium, their hybrid cells (*P. hybrida* + *P. parodii*; somatic hybrids are denoted by '+' sign) undergo normal division and produce macroscopic colonies. The natural properties of the two parental species are exploited in this selection approach, which is simple, highly effective, and least demanding. The hybrid cells show complementation between the two parental species.

Using genetic engineering, resistance has been recently transferred to an antibiotic/herbicide from both the fusion parents. The hybrid cells are selected using a medium containing both concerned antibiotics and herbicides. Another widely applicable strategy is culturing the entire protoplast population without making any selection of hybrid cells. All types of protoplasts form calli. Later, the hybrid calli are identified based on chromosome constitution, callus morphology, and the banding patterns of proteins and enzymes. The identification may be delayed in some cases till plants are regenerated.

When the hybrid calli are obtained, plants are regenerated from them. This is a precondition for their exploitation in improvement of crops. Apart from possessing some beneficial properties, the hybrid plants should be fertile (at least partially) in order to be useful in breeding schemes. The culture methods have been modified to successfully regenerate plants from a number of somatic hybrids.

1.20   SYMMETRIC AND ASYMMETRIC HYBRIDS

1.20.1   Symmetric Hybrids

Many symmetric hybrids have been generated from sexually incompatible species. A number of these symmetric hybrids are fertile and express useful genes, which have been employed in breeding programmes. Symmetric hybrids offer the following advantages:

- It is possible to produce somatic hybrids between a non-flowering and a male sterile line of crop species.
- Some allopolyploid crops might have narrow genetic base, for example, in the case of *B. napus*. The genetic base of a species is defined as the total amount of genetic variation present in the crop. In the case of symmetric hybrids, the genetic base can be widened markedly by synthesizing the species.
- Desirable characteristics can be transferred from a related species to a crop species by the use of symmetric hybrids. A commercially useful gene transfer was from *N. rustica* to *N. tabacum*. Somatic hybrids obtained from these two species were backcrossed thrice to *N. tabacum*. Resistances to high alkaloid content and black rot

were transferred from *N. rustica* fusion parent. Selection in the backcross generations resulted in the production of the varieties Delgold and Chang. Delgold became a popular variety in Ontario, Canada.

- Some somatic hybrids possess commercially useful features. A few somatic hybrids produced between dihaploid clones of *S. tuberosum* ($2n = 2x = 24$) and its wild relative *S. phureja* ($2n = 2x = 24$) gave significantly higher tuber yields compared to the two fusion parents. A hybrid clone produced three times as much tuber as the commercial potato variety Bintje. In the Netherlands, farmers are employing somatic hybridization for potato improvement.

- Somatic hybridization can be employed to produce certain materials that might be useful in genetic, biochemical, physiological, and other studies.

1.20.2 Asymmetric Hybrids

Many somatic hybrids contain the full somatic complement of one parent species and the chromosomes of the other parent species are almost completely lost during the preceding mitotic divisions. These hybrids are known as asymmetric hybrids. Studies have suggested that such hybrids will probably exhibit introgression of chromosome segments from the eliminated genomes. Asymmetric hybrids can be generated even from those species that generally produce symmetric hybrids. In this technique, protoplasts of one of the parent species are irradiated with an appropriate dose of X-rays or gamma rays to cause extensive breakage of chromosomes. This approach can significantly increase the introgression of chromosome segments.

1.21 CYBRID PRODUCTION AND ITS APPLICATION IN CROP IMPROVEMENT

The hybrids which contain nucleus of one parent and cytoplasm from both parents are called cybrids. The process resulting in the development of cybrids is termed cybridization. During cybridization and heterokaryon formation, the segregation of nuclei is stimulated so that one protoplast contributes to the cytoplasm while the other to the nucleus alone. Protoplasts are rendered inactive and non-dividing when they are irradiated with X-rays and gamma rays. Use of metabolic inhibitors also makes the protoplasts inactive and non-dividing. Some genetic characters in certain plants are cytoplasmically controlled, such as some kinds of male sterility and resistance to specific antibiotics and herbicides. Hence, cybrids are essential for transferring CMS and antibiotic and herbicide resistance in agriculturally useful plants. Cybrids of *B. raphanus* have been developed that contain the nucleus of *B. napus*, chloroplasts of atrazine-resistant *B. campestris*, and male sterility from *Raphanus sativus*.

Applications of somatic hybridization and cybridization are given below:

- Somatic cell fusion seems to be the only method for recombining two different parental genomes in plants that cannot sexually reproduce.

- Protoplasts fusion of sexually sterile (haploid, triploid, and aneuploid) plants can result in the production of fertile diploids and polyploids.

- Barriers of sexual incompatibility can be overcome by somatic cell fusion. Somatic hybrids between two incompatible species have also been applied in agriculture or industry.

- Somatic cell fusion is used to study cytoplasmic genes and their activities. This study can be applied in plant breeding experiments.

1.22   CRYOBIOLOGY OF PLANT CELL CULTURE AND ESTABLISHMENT OF GENE BANK

The genetic material, especially its molecular and chemical composition, inherited and transmitted from one generation to the next is called germplasm. Alternatively, germplasm refers to the sum total of all the genes present in a crop and its related species. It is usually characterized by a collection of different strains and species. Germplasm contains a variety of genotypes which are required for the development of new and advanced genetic stocks, varieties, and hybrids. The basic indispensable ingredient of all breeding programmes is the germplasm; therefore, the collection, evaluation, and conservation of germplasm are considered essential. To develop high-yielding crop varieties resistant to biotic and abiotic stresses, it is necessary to conserve germplasms of different crops and their wild and weedy relatives.

1.22.1   *In situ* Conservation

In situ (on-site) conservation refers to the conservation of wild plants in their natural habitats where they grow and evolve naturally without the interference of human beings. The wild populations are regenerated and dispersed naturally by wild animals, wind, and water. An intricate relationship, which is often interdependent, is established between different species and other environmental components (such as pests and diseases). The evolution is influenced by only environmental pressures, and any changes in one component affect the other component. This dynamic co-evolution results in greater diversity and better adapted germplasm, provided the changes are not drastic. Forests and other wild plant species are often conserved by establishing protected areas of different categories, such as national parks, gene sanctuaries, and nature reserves. However, this approach of conservation suffers from limitations as there is a risk of material loss in the event of environmental hazards.

1.22.2   *Ex situ* Conservation

In *ex situ* (off-site) conservation, there is maintenance of germplasm outside the natural habitat in specially designed facilities. For both cultivated and wild plants, this is the principal method of conserving plant genetic resources. Seeds are the predominant form of germplasm conservation *ex situ*. Therefore, many field crops and vegetables that produce orthodox (desiccation-tolerant) seeds are preserved in gene banks by reducing their moisture content (3%–7%) and storing them at low temperature and humidity.

1.22.3   *In vitro* Conservation

In vitro techniques are most effective for germplasm conservation in the case of both vegetatively propagated crops and those crops that produce desiccation-sensitive or recalcitrant seeds (which lose their viability after being dried below a critical limit). This method, which is based on tissue culture, has been employed for conserving somaclonal and gametoclonal variations in cultures, materials from endangered plants, plants with medicinal values, storage of pollen, and storage of meristem cultures for the generation of disease-free plants and genetically engineered materials.

Germplasm can be stored *in vitro* in a number of forms at various stages of development. The forms may include isolated protoplasts, cells from suspension or callus cultures, meristem tips, somatic embryos, shoot tips, or propagules at various stages of development. On the basis of culture growth, *in vitro* methods of germplasm conservation can be classified into the following two categories:

1. **Slow-growth cultures** A limited growth of the culture is allowed in a slow-growth culture. This is an effective, simple, and economic technique that can be used for all species if shoot tips/nodal explants are available. In this method, growth is suspended by cold storage or reducing oxygen levels available to the culture. This method requires serial subculturing so that cultures are renewed at periodic intervals. There are disadvantages of storing germplasm by repeated culturing. In the storage of germplasm by serial subculture, there is a risk of pathogen contamination and introduction of genetic changes.

2. **Cryopreservation** In cryopreservation, the growth in plant cell and tissue culture is suspended by storing it at ultra-low temperature (–196°C) in liquid nitrogen, although the viability of the culture is retained. This method is also called freeze preservation and is the most common and effective technique for indefinite storage. For conserving germplasm, only shoot tips and buds are used for cryopreservation. However, cells, tissues, protoplasts, and somatic embryos can also be cryopreserved for other tissue culture processes.

1.22.4 Factors Affecting Viability of Cells Frozen for Cryopreservation

The following factors affect the viability of cryopreserved cells:

- **Physiological state of material** Cryopreservation of plant cell cultures is best performed with cells that are at the late lag or early exponential growth phase. After thawing, these cytoplasm-rich cells retain their viability and again start growing from the actively dividing meristematic cells. However, in shoot tips and embryos, tissues are large and contain highly vacuolated cells. These cells are damaged by freezing and fail to survive and recover.

- **Prefreezing treatment** Cells become hardened and their survival rates increase on conditioning treatment before freezing. Hardening is achieved by growing culture in the presence of a cryoprotectant, at low temperature (4°C) (for cold dormant species), or in the presence of osmotic agents such as sucrose. These treatments alter the cell water content, metabolite content, or membrane permeabilities.

- **Cryoprotectants** Cryoprotectants are chemicals that aid in withstanding low temperature. The most frequently used cryoprotectant for plants is dimethyl sulphoxide (DMSO). Plasmolysis of cells is prevented by gradually adding about 5%–10% of DMSO. Other common cryoprotectants include glycerol, polyvinylpyrollidone, and PEG.

- **Thawing rate and reculture** The preserved samples can survive better upon rapid thawing from −196°C to about 22°C. Thawing rapidly minimizes the harmful effects of ice crystal formation (crystallization of cell water during freezing). For reculturing, the thawed samples need specific conditions, such as high osmoticum, dim light, gibberellic acid, and AC in the medium, for increased recovery rates.

1.22.5   Methods of Cryopreservation

Cells of different species vary in their sensitivity to low temperature. Generally, the sample to be preserved is first treated with an appropriate cryoprotectant and then frozen by rapid freezing method or controlled freezing method. These methods are described as follows:

- **Rapid freezing** In this method, the vials containing plant materials are dipped in liquid nitrogen, which results in a rapid decrease in the temperature at a rate of 200°C/min. The survival rate is low in this hard treatment. However, this technique has been successfully employed for conserving germplasm of a variety of species in the cases where plant materials with small size and low water content have been selected.

- **Controlled freezing** In this method, the plant material is cooled to an intermediate temperature (−20°C) from room temperature, maintained at that temperature for 30 min, and then dipped into liquid nitrogen for rapid freezing. This method can be applied to a wide range of plant materials, such as buds, shoot apices, and suspension cultures.

1.22.6   Advantages and Disadvantages of Cryopreservation

The advantages of cryopreservation are as follows:

- Indefinite preservation as metabolism is stopped.
- Low maintenance cost as only liquid nitrogen has to be replaced.
- No contamination.
- Can be applied to all species amenable to tissue culture.

The disadvantages of cryopreservation include the following:

- Sophisticated equipment and facilities needed.
- Expertise required.
- Cells and tissues get spoiled because of ice crystal formation or high solute concentration during desiccation.

1.23   PRODUCTION OF VIRUS-FREE PLANTS USING MERISTEM CULTURE

Plants are susceptible to viral diseases, which lower their yield and quality. Virus-infected plants can be difficult to treat and cure. Hence, plant breeders always prefer to develop virus-free plants. In some ornamental plants, it has been possible to produce virus-free plants through tissue culture for commercial purposes. For this, plants are regenerated from cultured tissues isolated from virus-free plants, virus-free calli, meristems that are not infected,[1] and meristems treated with heat shock (34–36°C). Efforts have been made to remove viruses from infected plants by using chemicals for treating the culture medium. For example, adding CKs inhibited the multiplication of some viruses. Of all the culture techniques, the meristem tip culture method is the most effective method for the elimination of viruses and other pathogens. Viruses have been removed from a variety of economically important plant species, which resulted in a significant increase in yield (e.g., potato virus X from potato and mosaic virus from cassava). However, these virus-free plants are not disease resistant. Hence, stock plants should be maintained so that virus-free plants can be generated whenever required.

---

[1] For eliminating virus, the size of the meristem is very critical because most of the viruses exist by establishing a gradient in plant tissues. Regeneration of virus-free plants through cultures is inversely proportional to the size of the meristem used.

1.24 ANTHER, POLLEN, AND OVARY CULTURE FOR PRODUCTION OF HAPLOID PLANTS AND HOMOZYGOUS LINES

Anther culture refers to *in vitro* culturing of anthers on an appropriate nutrient medium to regenerate haploid plantlets. Bergner first discovered haploid plants in *D. stramonium* in 1921. Guha and Maheshwari (1964) obtained embryos from the anthers of *D. innoxia* grown *in vitro*. Since then haploid plants have been produced through anther culture in more than 170 species.

1.24.1 Haploid Production

A haploid plant is a sporophyte that has a gametophytic chromosome number. Haploid plants can be produced *in vitro* by the following techniques:

- **Delayed pollination**   This may not lead to fertilization. Only female genome develops to produce a haploid plant.
- **Temperature shock**   Temperature extremes (both high and low) are employed to prevent syngamy or inactivate pollen. This results in the initiation of haploidy.
- **Irradiation effect**   In this technique, chromosomal breakage is induced in pollen cells by X-rays and UV rays, which makes them sterile and, in turn, results in haploid production.
- **Chemical treatment**   Treatment with chemicals such as colchicines, maleic hydrazide, and toluene blue facilitates chromosomal elimination.
- **Genome elimination by distant hybridization**   In distant crosses such as intergeneric and interspecific crosses, one of the parental genomes is selectively removed during the developmental process, consequently producing haploid plants.

Androgenesis is the method of producing a haploid plant in which the egg cell is inactivated, resulting in the presence of only the male genome. Gynogenesis is the procedure which results in the development of embryo that contains only maternal chromosomes owing to the inactivation of pollen. Among all the above methods, anther culture is most popular and successful for haploid production.

1.24.1.1   Anther culture procedure

The following steps are involved in the anther culture technique:

- **Experimental material**   Flower buds of the suitable stage (varies with species) are excised from young healthy plants developed under controlled conditions and used as experimental material.

- **Disinfestation, excision, and culture of anther**   In a laminar air flow chamber, the flower buds are surface sterilized. This is followed by excision of anther from the buds. An anther is squashed in acetocarmine and observed under a microscope to determine the stage of pollen development. Anthers should be carefully excised from flower buds because any injury to anther during excision may result in callusing, leading to a mixture of diploids, haploids, and aneuploids.

- **Culture medium conditions**   Anthers are usually cultured on a solid agar medium. In this medium, they develop into embryoids under alternate light and dark periods. The medium should contain sucrose for stimulating embryogenesis.

- **Haploid plants**   In species undergoing direct androgenesis (i.e., in species that develop through embryoid formation), small plants develop in three to eight weeks after culturing. These plants are then transferred to a rooting medium containing a small amount of auxin and low salt. In species that undergoes indirect androgenesis involving callus formation, callus is isolated from the anther and put in a regeneration medium containing an appropriate ratio of CK to auxin. In both the cases, the haploid plants produced are transplanted into separate small pots and kept under controlled conditions in a greenhouse.

- **Diploidization of haploid plants**   Haploid plants produced through anther culture can grow till the flowering stage and it cannot be perpetuated. These plants are haploid and contain only one set of homologous chromosomes of the diploid species, and so they cannot produce viable gametes. Hence, seeds are not set for further perpetuation. Therefore, the chromosome number of haploids should be doubled to get homozygous diploids or dihaploid plants, which are then transferred to a culture medium for further propagation.

1.24.1.2   *Application of haploid production*

Diploidization of haploid plants leads to rapid attainment of homozygous traits in doubled haploids. Hence, these anther-derived haploid plants are employed in breeding and improvement of crop species.

- **Production of homozygous lines**   Haploids are generally employed in the production of homozygous lines, which might be used as cultivars or utilized in breeding programmes. For instance, doubled-haploids are used for the rapid creation of inbred lines in a hybrid maize programme. The anthers of F1 hybrids of a desirable cross constitute ideal breeding materials for generating anther-derived homozygous plants or doubled-haploids in which good complementary characters of the parents are combined in a single generation. Superior plants are selected from the doubled-haploid (DH) plants (Figure 1.3). This approach is termed the hybrid sorting in which recombinant superior gametes get selected since the heterozygous gene combination in the F1 hybrid is converted into homozygous combinations. Conventional breeding by pedigree/bulk method requires 10 years to develop a variety. In contrast, it takes only 4–5 years to develop a variety by haploid breeding. Moreover, selection among DH lines decreases the size of the breeding population.

- **Gametoclonal variation**   The variation observed in haploid plants having gametic chromosome number developing from culture derived from anthers is termed the gametoclonal variation. Such variations lead to obtaining desirable traits, which are selected at the haploid level and then diploidized to develop homozygous plants to be released as new varieties.

- **Selection of desirable mutants**   In haploids, even recessive mutations are manifested. This is contrary to diploids in which they are expressed only in segregating single plant progeny in M_2 generation. Hence, in several crops, desirable mutants possessing traits such as resistance to diseases, antibiotics, and salts have been derived from haploids produced through anther culture. For instance, tobacco mutants resistant to black shank disease and wheat lines resistant to scab (*Fusarium graminearum*) have been developed as improved cultivars.

However, haploid plants also suffer from the following drawbacks:

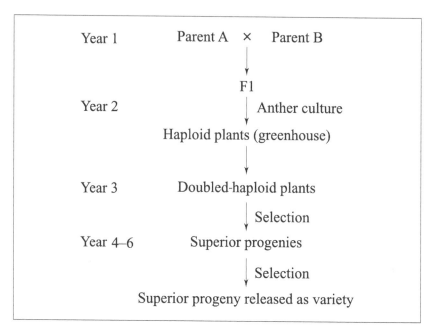

Fig. 1.3   *Anther culture-derived haploid plants for hybrid sorting*

- Many species are not yet responsive to haploid production.
- Deleterious mutations might be induced during the *in vitro* phase.
- Plants are also produced which possess higher or lower gametic chromosome number, and this necessitates cytological analysis.
- The occurrence of gametoclonal variation limits the utilization of anther derived embryos for genetic transformation or gene transfer.

The major advantages of cell culture systems over the conventional cultivation of whole plants are as follows:

- Greater and rapid yields can be obtained from a small amount of plant material required for culture initiation. This is contrary to the large amounts of mature plant tissues processed to get low yields of final product. For instance, shikonin produced from cell culture has 20% more dry weight than that produced from plants.
- If a plant species is facing the risk of extinction or is limited in supply (for example, *L. erythrorhizon*), *in vitro* production of secondary metabolites can rescue the species from extinction.
- Continuous supply of metabolites can be ensured through controlled environmental conditions in cell culture. In conventional systems,

source plants may be location specific, seasonal, and also prone to environmental degradation. For plants requiring a longer period to mature, *in vitro* culture is more economical.

- Low-cost precursors are added as substrates to cell cultures to obtain high-cost products, thus reducing labour, cost, and time. In addition, specific substrates can be biotransformed to more useful products by single-step reactions *in vitro*.

- Novel compounds that are not found naturally in plants can be produced from mutant cell lines. For instance, cell suspension cultures of *Rauwolfia serpentina* produce novel glucosides of ajmaline (alkaloid).

REFERENCES

Ahloowalia, B. S, M. Maluszynski, and K. Nichterlein. 2004. Global impact of mutation-derived varieties. *Euphytica* 135: 187–204

Belling, J., M. E. Farnham, and A. D. Bergner. 1921. A haploid mutant in the Jimsom Weed. *Datura stramonium. Science* 55:646–47

Carlson, P. S, H. H. Smith, and R. D. Dearing. 1972. Parasexual interspecific plant hybridization. *Proceedings of the National Academy of Sciences of the USA.* 69: 2292–94

Charles, C. 1851–1908. Chronological reference marks - Pasteur Institute. Archived from the original on 19 December 2006. Retrieved on 19 January 2007

Fowler, M. W. 1983. *Plant Biotechnology,* S.H. Mantell, and H. Smith (eds), pp.3–37, Cambridge: Cambridge University Press

Gamborg, O. L., R. A. Miller, and K. Ojima 1968. Nutrient requirements of suspension culture of soybean root cells. *Ex. Cell. Res.* 50: 15–158

Goldberg, R. B. 1988. Plants: novel developmental process. *Science* 240: 1460–67

Gorut, B. W. W. and S. Millam. 1885. Photosynthetic development of micropropagated strawberry plantlets following transplanting. *Ann. Bot.* 55: 129–131

Guha, S. and S. C. Maheshwari. 1964. In vitro production of embryos from anters of datura. *Nature,* p. 204, p. 497

Hazarika, B. N. 2003. Acclimatization of tissue-cultured plants. *Current Science* 85: 1704–12

Hazarika, B. N., V. Nagaraju, and V. A. Parthasarathy. 1999. Acclimatization technique of citrus plantlets from in vitro. *Adv. Plant Sci.* 12: 97–102

IAEA. 2008. Mutant plants can boost yields, resistance: IAEA conference (Vienna, Austria)

Jones, O. P., C. A. Pontikis, and M. E. Hopgood. 1979. Propagation in vitro of five apple scion cultivars. *J. Hort. Sci.* 54: 155–58

Kumar, K., A. S. Dhatt, and M. I. S. Gill. 2001. In vitro plant regeneration in kinnow mandarin (*Citrus nobilis Lour* × *C. deliciosa Tenora*). *Indian Journal of Horticulture* 58: 299–302

Lloyd, G. and B. McCown. 1980. Commercially feasible micropropagation of mountain laural (*Kalmla latlfolia*) by use of shoot tip cultures. *Cmb. Proc. Intl Soc.* 30:421–27

Mehta, U. J., K. V. Krishnamurthy, and S. Hazra. 2001. Regeneration of plants via adventitious bud formation from zygotic embryo axis of tamarind (*Tamarindus indica* L.). *Current Science* 81: 1530–33

Morel, G. and C. Martin. 1952. Virus-free Dahlia through meristem culture. C. R. Hebd. *Seances Acad. Sci.* Paris 235: 1324–25

Murashige, T. 1979. Principles of rapid propagation. In *Proc. Symp. on Propagation of Higher Plants through Tissue Culture: A Bridge between Research and Application*, University of Tennessee, Knoxville, Tennessee

Murashige, T. and F. Skoog. 1962. A revised medium for rapid growth and bioassays with tobacco tissue cultures. *Physiol. Plant* 15: 473–79

NAS. 2004. NAS Report Safety of Genetically Engineered Foods: Approaches to Assessing Unintended Health Effects (2004) by Food and Nutrition Board (FNB), Institute of Medicine (IOM), Board on Agriculture and Natural Resources (BANR), Board on Life Sciences (BLS)

Nitsch, J. P. and C. Nitsch. 1969. Haploid plants from pollen grains. *Science* 163: 85–87

Parthasarathy, V. A. and V. Nagaraj. 1992. Bud establishment and shoot proliferation *J. Tree Science* 11(2): 146–48

Parthasarathy, V. A., V. Nagaraju, B. N. Hazarika, and A. Baruah. 1999. An efficient method of acclimatization of micropropagated plantlets of citrus. *Tropical Agriculture* 76: 147–49

Sahay, N. S. and A. Varma. 2000. A biological approach towards increasing the rates of survival of micropropagated plants. *Current Science* 78: 126–9

Saiprasad, G. V. S. 2001. Artificial seeds and their applications. *Resonance* May: 39–47

Schenk, R. U. and A. C. Hildebrandt. 1972. Medium and techniques for induction and growth of monocotyledonous and dicotyledonous plant cell cultures. *Can. J. Bot.* 50: 199–204

Short, K. C., J. Warburton, and A. V. Roberts. 1987. In vitro hardening of cultured cauliflower and chrysanthemum to humidity. *Acta. Hortic.* 212: 329–40

Singh, S., B. K. Ray, S. Bhattacharjee, and P. C. Deka. 1994. In vitro propagation of *Citrus reticulata* Blanco and *Citrus limon* Bunn. *Hort. Sci.* 29: 214–16

Skoog, F. and R. A. Miller. 1957. Chemical regulations of growth and organ formation in plant tissue culture in vitro. *Sym. Soc. Exp. Biol.* 11: 118–31

Smith, E. F., A. V. Roberts, J. Mottley, and S. Denness. 1991. The preparation in vitro of chrysanthemum for transplantation to soil. 4. The effect of eleven growth retardants on wilting. *Plant Cell Tiss. Org. Cult.* 27: 309–13

Takayama, S. and M. Misawa. 1979. Differentiation of Lillium bulb scales grown in vitro: effect of various cultural conditions. *Physiologia Plantarum* 46: 184–190

Ticha, I., F. Cap, D. Pacovska, P. Hofman, D. Haisel, V. Capkova, and C. Schafer. 1998. Culture on sugar medium enhances photosynthetic capacity and high light resistance of plantlets grown in vitro. *Physiologia Plantarum* 102: 155–162

van Harten, A. M. 1998. *Mutation Breeding: theory and practical applications.* Cambridge: Cambridge University Press

von Klercker, J. 1892. Oefrers, Ventenskapsakad Foerh., *Stockholm* 49: 463–75

Wainwright, H. and J. Scrace. 1989. Influence of in vitro preconditioning with carbohydrates during the rooting of microcuttings on in vivo establishment. *Scientia Horticulturae* 38: 261–67

Walkey, D. G. A., J. W. Webb, C. J. Bolland, and A. Miller. 1987. Production of virus-free garlic (*Allium sativum* L.) and shallot (*A. ascalonicum* L.) by meristem-tip culture. *J. Hort. Sci.* 62: 211–20

White, P. R. 1963. *A Handbook of Plant Tissue Culture*, p. 345. Lancaster: Jacques Cotteil Press

Zimmerman, R. H. 1983. Factors affecting in vitro propagation of apple cuttings. *Acta Horticulturae* 131: 171–78

Tissue-culture Applications

2.1 REGENERATION

The ability of plants to regenerate was discovered long before the discovery of regeneration in animals. The cutting of twigs and leaves of many plants to become independent plants is a common phenomenon. In 1902, while reporting his pioneering experiments on plant tissue culture to the German Academy of Science, G. Haberlandt predicted that by using vegetative cells, embryos can be successfully cultivated. It is possible that if Haberlandt had easy access to IAA and kinetin, he might have realized his prediction and plant tissue culture might have progressed even further today. Two phenomena led to the successful culture of plant cells: first is the discovery of plant hormones auxin and cytokinin. Second, in 1957, F. Skoog and C. O. Miller revealed that by varying the proportion of hormones in the nutrient medium, the regeneration of roots and shoots could be manipulated in the cultured cells. When genetically engineering crops, the single cells or protoplasts undergo molecular manipulations initially. The effort must be completed with reconstituted plants. At present, there are two methods by which plants can be regenerated from isolated cells. In the first method, plants are obtained through a sequence of shoot formation, and consequently by rooting the shoot. In the second method, initiation of embryos takes place, which are structures with differentiated shoots and roots.

In both the cases, a series of nutrient formulations are necessary. When obtaining a separate shoot and root, a critical cell mass, or callus, is essential for any organ formation. In order to develop a callus, the supply of auxin and in some cases cytokinin to the culture medium becomes essential. The callus differentiates into a shoot when it is transferred to a medium that contains a relatively high level of cytokinin and a low level of auxin. To generate roots, shoots of suitable sizes are separated and sub-cultured into another medium that does not contain cytokinin but in which a little amount of auxin is present. The presence of adenine and tyrosine as additional supplements helps in enhancing the shoot-initiation step. Another method of improving rooting is by including cofactors such as phloroglucinol and caffeic acid in the medium as well as reducing the concentration of salts in the medium. In the second method, that is the embryo method, the cells are first stimulated for division and then induced for embryo formation. For the cell division and induction phases, the medium often contains an auxin such as 2,4-D; a high level of nitrogen, preferably NH_4^+; and sufficient potassium. The embryo method has been used in lettuce, where only those cells are selected that can survive exposure to oxalic acid and ethylene gas. Oxalic acid is a toxin produced by lettuce drop organisms, and ethylene gas is responsible for injury in them, resulting in russet spotting. To select characteristics other than disease resistance, selection pressures can be applied. For example, select pressures can be applied to select traits such as cold and heat tolerance, and tolerance to chemicals such as salts and herbicides.

In order to improve crops, scientists have crossed plants with desirable characteristics. Plants being crossed should be sexually compatible. This restriction can be removed by protoplast fusion. Protoplasts of any two plants can be fused irrespective of species they belong to. Fused protoplasts can also be regenerated into whole plants. Somatic hybrids or asexual crosses have been possible because fused protoplasts have been regenerated into whole plants. Theoretically, it means that protoplast fusion can lead to potential gene pool of a plant to every other biological organism. Protoplast fusion helps in improving crop yield. Perhaps every plant cell is totipotential or has the capacity to regenerate into the whole plant. The difficulty of this characteristic varies from species to species and even within cells of the same plant. Certain cells respond more to manipulations than others, resulting in embryogenesis or organogenesis. J. G. Torrey called these manipulatable cells meristemoid. The meristemoids look very similar to the apical meristem cells or embryos as present in seeds. Non-meristemoid cells sometimes differentiate *in vitro* into meristemoids, for example tobacco stem pith, carrot root phloem, potato leaf mesophyll, and citrus nucellus. However, the basis for this differentiation still remains ambiguous. Meanwhile, by observing a few key relationships between plant growth phase and totipotency, the chances of obtaining plants in cell cultures can be increased. The regenerative capacity of cells in adult plants is less than that of the plants in their juvenile phase of development, that is, not yet competent to flower. The younger the plant or the organ in the juvenile sources, the higher their tendency to contain regenerative cells. Before the cell culture is initiated, it must be seen that the seasonal climatic requirements of the donor plant or organ are satisfied. In a poorly regenerating tissue, highly regenerative cell lines can be derived by repeatedly selecting the trait during the course of sub-cultures. In the future, molecular biology may be able to discover methods by which selective control of gene repression and de-repression can be done. Then, the expression of totipotency can perhaps be achieved in any plant cell, whether meristemoid or non-meristemoid, regardless of the plant species or variety.

2.2 PRODUCTION OF HAPLOIDS

In flowering plants, androgenesis is a distinct biological process that explains the biological basis for the production of a dihaploid plant by single-cell microspore embryogenesis. Apart from reducing the breeding cycle, the process helps in fixing agronomic traits in the homozygous state, such as recessive genes for disease resistance. In modern crop breeding, the most desirable dihaploid variation has already been developed and utilized, including in rice, wheat, barley, maize, rapeseed, cotton, sun flower, and coffee. Such developments, however, comprise both known and unknown factors. Donor plants, media composition, genotypic variation, and handling of cultures are some factors that may impact the response of androgenesis. To further improve crops, pollen-specific genes, promoter and transgenic dihaploids (homozygous), gene expression, proteomics, and translational regulation and post-translational modification of genes may be used. The homozygous (isogenic) lines will provide unique genetic material for mapping populations to be used in functional genomics and molecular breeding.

In the stamens of flowering plants, the male reproductive processes take place. Haploid male spores or microspores are produced when diploid cells undergo meiosis. Normally, when microspores undergo mitotic division, they differentiate into multicellular male gametophytes or pollen grains. Androgenesis is the process by which pollen grains are not allowed to develop and instead forced towards a somatic pathway. *In vitro* microspores can undergo androgenesis via callus formation or direct embryogenesis, resulting in the formation of haploids. Callus-derived plants show genetic variation and polysomy, and so they are generally undesirable. In crop improvement, the main technique for haploid induction is anther culture (AC). This technique has been simplified by culturing the whole or parts of inflorescences. An alternative to this technique is the culturing of isolated or shed microspores. However, limited reports are available on the cultures of isolated microspores in relation to AC, which has received better *in vitro* responses. One of the aims of modern plant breeding practices has been to accelerate the production of homozygous lines of plants as it normally requires at least six inbreeding generations. The accidental, but valuable, discovery of androgenic haploidy by Guha and Maheshwari (1964) and production of rice haploids by Niizeki and Oono in 1968 showed impressive advancements. Haploid gametes can produce homozygous lines of offspring in just one generation, which allows an easy selection of phenotypes for quantitative characters.

2.2.1 Microspore Stage

In most cases, microspores in their early to mid uninucleate stages are best suited for androgenic response. The response is maximum in maize anthers that contain microspores in late uninucleate to early binucleate stages. In the case of dicot species, microspore embryogenesis is most suited in unicellular to early bicellular stages of pollen (for example, Brassica microspore).

Pre-culture treatment: Stress treatment of anthers before the culture strongly influences the induction of microspores to sporophytic pathway instead of gametophytic pathway. Response to cold or heat treatment has been reported to be genotype dependent. Apart from providing nutrition to the microspores, the nutrient medium also directs the pathway of embryo development. With the onset of the development of embryo, the pH of the media (especially liquid media) dramatically changes. To maintain the pH and balance of micronutrients, it is critical to change the composition of the media or replenish them. The chemically defined N6 medium and the MS medium are the two familiar basal media that are generally used with modifications. Anthers are suspended on 10 mL aliquots of media in 50 mm sterile plastic dishes, or 10–15 anthers are cultured in 24 wells, each containing 1.5 mL of the media. The nitrogen composition of a culture medium plays a major role in androgenesis. In many cereal species embryo development gets enhanced when the concentration of glutamine is increased and ammonium nitrate is decreased. In wheat microspore-embryogenesis response was shown to be better in the presence of high concentration of sucrose. Plant regeneration improved in barley when Ficoll was added to the liquid medium. The rate of haploid induction improved further in barley and wheat by modifying the media and using Ficoll 400. Again embryo induction and plant regeneration have been found to be dramatically enhanced when sucrose was replaced with maltose; however, the concentration needs to be suitably adjusted for each crop. Several researchers are using maltose for improving cereal androgenesis, protoplast culture, and plant regeneration after it was successfully used in the liquid medium of barley. In rice greater incidence of androgenesis can be induced by the use of abscisic acid (ABA) or potato media as compared to standard media. In the maturation of microspore-derived embryos, the osmotic pressure of the medium may play an important role.

Microspore treatment procedures for inducing androgenesis in the medium can broadly be classified into four categories.

1. The simplest way is to culture the inflorescences with mid- to uninucleate microspores in the liquid or solid medium. In crops such as barley direct embryos can be obtained from microspores derived from the anthers of the spikelets.

2. The shed pollen culture is the second method where anthers are cultured on the liquid/solid (agar) medium. After a few days, microspores shed freely from the anthers and divide and develop further in major cereals. Such microspores often start dividing within the anthers before their release. The embryos thus formed

float on the top of the liquid medium containing Ficoll and show androgenesis (direct plant regeneration).

3. In the mechanical isolation of microspores, protoplasts can be cultured. However, the response of plant regeneration is limited in this procedure.

4. It involves culturing the anther on the surface of the agar medium. Orientation of anthers is not important in regulating embryo development. They can be plated on the agar medium after the removal of anthers individually by hand or by using a suction pump.

In microspore-derived plants, the haploid set of chromosomes spontaneously double under culture conditions. However, depending on the different genotypes of a cultivar, the percentage of doubling varies in different crops. Barley shows the highest incidence (up to 87%) of spontaneous doubling of haploid chromosomes. It is followed by rice (up to 72%), wheat (up to 50%), and maize (6.3%). Generally, to double the chromosomes, colchicine is added to the medium at the whole-plant level for barley, rice, maize, and also in wheat in some cases. Colchicine is added to the medium before there is a significant increase in the gametophytic chromosome number as a result of the first microspore mitosis. Di-haploid plants are formed when microspore undergoes mitotic division and then endomitosis or nuclear fusion. Spontaneous doubling of the chromosomes along with specific pathways influence direct microspore embryogenesis, which leads to regeneration of fertile plants in barley and wheat. Aneuploidy, dihaploidy, and polyploidy are often exhibited by microspore-derived calli and embryoids.

Many factors influence the regeneration of fertile green plants from cultured microspores. These include type of nutrient media, genotype, culture vessels, donor plants condition, sources of carbohydrate, phytohormones, reduced nitrogen (glutamine), and handling of cultures. Individual factors such as co-culture of the ovary or ovary-conditioned media become crucial in the induction of microspore embryogenesis for wheat. However, these factors may not have any influence on genotypes with inherent potential for high-frequency plant regeneration, such as *in vitro* friendly barley genotype 'igri', which gives 50 green plants per cultured anther. Indra Vasil popularized embryogenic culture developed from immature embryos, which is very useful in cell-culture development. From day 1 to subsequent sub-culture, many changes take place in the culture. So a critical requirement is thorough observation of the cultures from induction to embryo development. Conditioning of cultures and their further development is furthered by replenishing the media to avoid depletion of some necessary micronutrients and balancing the pH. Another critical stage is embryo maturation. This is owing to transferring the developed cereal embryos to a regeneration medium at correct time, lowering the concentration of carbohydrates, and increasing the relative cytokinin to auxin levels. A simple regeneration medium works as is evident from the germination of embryos derived from encapsulated microspores in barley and wheat.

A common feature of plantlets derived from microspore is albinism. The genetic background of the donor plants is a crucial factor that affects the extent of albinism. Green plants are produced by delaying or arresting nuclear synchronization. This can be achieved by using Ficoll in the liquid medium and by cold pre-treatment in general. Albino barley do not contain mature chloroplasts while in albino rice 23S and 16S rRNA are absent. In general, deleted forms of the plastid genome are present in albino plants, such as wheat, barley, and rice. Different plants have different size and location of the deletions. Some albino plants do not have the region that codes the *rbcL* gene in the plastid genome.

Synthetic or artificial seeds possess somatic embryos in a protective coating, such as calcium alginate, that need to be used efficiently for the conversion of plants. Artificial seeds can be developed only when perfect somatic embryos are produced. The gel encapsulation system is composed of calcium alginate made from brown algae. The selected embryos and sodium alginate are mixed, and a single embryo is dropped into a bath of calcium salts, which results in a single somatic embryo encased in a clear, hydrated bead. The hardness of the gel beads gives protection to the fragile embryo. As in the case of an artificial endosperm, the capsule gel can also act as a potential reservoir of nutrients. After storing them in a cold room, artificial seeds of barley and wheat derived from microspores were germinated into normal plants. Somatic embryogenesis has been reported in all early major monocot and dicot species and a few gymnosperms.

Breeding programmes are complemented by biotechnological tools in several ways. One such way is identifying target genes (or mapped genes of agronomic importance) with the help of DNA markers. This process is called marker-assisted selection (MAS). Nuclear-encoded genes control anther culturability, which is a quantitative trait. However, previous genetic studies on haploid merely investigated the differences in responses among varieties and the heritability of traits such as callus induction and plant regeneration. After the introduction of the MAS system, these characteristics can now be identified at the molecular level too. A molecular map made from a doubled haploid (DH) line population of AC was used for the survey and analysis of the quantitative trait loci (QTL) responsible for the culturability of anthers. Callus induction, green plant differentiation frequency, albino plant differentiation frequency, and green plantlet yield frequency are the parameters for four traits which show a continuous distribution among the DH lines. Five QTLs were identified on chromosomes for the callus induction frequency. Chromosomes 1 and 9 were found to possess two QTLs for green plantlet differentiation frequency, whereas chromosome has a major QTL for albino plantlet differentiation. Green plantlet yield frequency had no independent QTL. In the case of rice haploid breeding and establishing permanent DH populations for molecular mapping, these results help in selecting parents with high response to AC. The anther culturability of Japonica/Indica cross derived DH lines with distorted segregation on chromosomes was investigated to ascertain the link between chromosomal regions showing distorted segregation and anther culturability. It was found that one region on chromosome controlled callus formation from microspores and another region controlled the proportion of green to albino regenerated plants. In both the regions, the Nipponbare (Japonica parent) allele had a positive influence. However, there was no significant effect of the three regions on chromosomes on anther culturability.

Genes controlling the production of certain enzymes essential for the QTLs, associated with green plant regeneration, located on chromosomes 3 and 10, were mapped using recombinant inbred lines from a cross between Milyang and Gihobleo. The QTL on chromosome 10, tightly associated to the three markers, was repeatedly detected using three AC methods. On the basis of the cultivars and two F2 populations, RZ400, one of the three markers, could effectively identify genotypes with good (>10%) and poor (<3%) regenerability. The screening of rice germplasm for anther culturability is enabled by this marker as well as its introgression into elite lines in breeding programmes. Genes that control the production of certain enzymes essential for the metabolism of differentiating cells, tissues, and organs influence the growth and development of plants from callus cultures. Peroxidases, associated with many physiological processes including morphogenesis, are metalloprotein enzymes having porphyrin-bound iron. The amount of peroxidase present in four indica cultivars and all possible F1 combinations were quantified. Regeneration and embryogenic potentials of calli can be identified by using isozyme markers such as peroxidase and utilized for selecting high regenerating calli.

2.2.2   Dihaploids in Genomics

Genomics implies sequencing of DNA, routine use of DNA microarray technology to analyse the gene expression profile at the mRNA level, and enhanced informatics tools to organize and analyse such data. DH lines are beneficial for genetic analysis, especially quantitative traits. In cultivated rice, some important agronomic traits are affected by QTLs. The development of molecular markers from segregating populations, F2, or backcross populations have facilitated studies on QTL. However, replication of these studies are difficult for obtaining accurate phenotypic values required for precise QTL mapping. Despite the many advantages of using recombinant inbred lines in QTL studies, development of such population will take a long time. Several studies have constructed genetic maps and located QTLs using DH populations.

As DH lines are homozygous, they can be propagated without further segregation. This property makes possible the exact measurement of quantitative traits by repeated trials and a decrease in the environmental component of the total phenotypic variance. However, there is a possibility of distorted segregation of RFLP-markers-derived DH populations on using AC-derived materials.

Molecular and biochemical analyses confirmed a high degree of genetic stability of clones. Therefore, although AC may modify the performance of microspore-derived plants to some extent, it will not dramatically influence their use in plant breeding and genetic engineering programmes.

Tiller angle is one trait that has a great significance in the high-yield breeding of rice. Resistance to disease is reduced by extremely small tiller angle, whereas for a high yield of crops, a large tiller angle is undesirable. On the basis of the linkage map constructed from a female parent of a DH population, which has a spreading plant type, and a male parent having a compact plant type, chromosomes 9 and 11 were found to have two major QTLs and chromosome 9 with one minor QTL. Similar studies using DHs have investigated QTLs for the length of top internodes, plant height and days to heading, ratooning ability and grain yield traits, and cold tolerance of seedlings.

Phenotype selection for rice root traits can be done through molecular and genetic markers. It is difficult to breed varieties having enhanced root penetration ability through hardpans and other root traits because it is laborious and time consuming to screen numerous genotypes under field conditions. The evaluation of root traits becomes further difficult owing to non-uniform and inconsistent soil compaction throughout rice fields. DH populations have been utilized for studies on QTLs for rice root characteristics such as root vitality, constitutive root morphology such as deep root morphology and root thickness, osmotic adjustment, root penetration index, basal root thickness, penetrated root thickness, root pulling force, total root dry weight, penetrated root dry weight, and penetrated root length. A DH population can be considered permanently stable as its genetic structure is fixed. Thus, for detecting QTLs and evaluating interactions between genotypes and the environment, it can be grown at different times and locations, that is, the phenotypic expression level of QTLs in different environments. This technique has helped in identifying 22 QTLs for 6 agronomic traits in rice at 3 different locations (environments). Traits such as heading date and plant height were more sensitive to the environment, while QTLs for spikelets and grains per panicle were common across the environment. Researchers have also used DH rice populations in the QTL studies on rice grain quality, grain shape, paste viscosity characteristics, aromatic traits, and brown plant-hopper resistance.

Genetic male sterile composite crosses were used for the accumulation and fixation of marker genes and AC technique was employed. Segregation ratio for male sterility and five or six marker genes were depicted in the dihaploid plants induced from the AC of the composite crossed plants.

In genetic research, aneuploids (plants having extra chromosomes besides the normal haploid chromosome complement) are used for investigating genetic imbalance resulting from additional chromosomes at the haploid and diploid levels. Aneuploids are also used for studying chromosomal behaviour in meiosis and genome construction in rice. Although, aneuploid production is difficult in rice, haploid plants with one extra chromosome ($n + 1$) can be developed with the help of AC. Similarly, aneuploids and tetrasomics have been derived from the AC of trisomic rice plants. These aneuploids can be utilized for assigning individual chromosomes with DNA markers. Genetic and cytological studies have also been done by researching AC-derived auto-pentaploid rice plants, meiotic behaviour, and morphological features. Restriction fragment length polymorphism (RFLP) and random amplified polymorphic DNA (RAPD) are the new tools of marker-assisted breeding. These have been effectively utilized together with DH lines to analyse molecular genomes. RFLP markers for different resistance genes, including the resistance gene *ym4*, have been discovered in barley. Moreover, concerning *ym4* an isozyme as well as different RAPD markers, including the very tightly linked marker OP-Z04H660, are known. However, this primer shows a quite complex banding pattern and exhibits an additional band of about 660 bp in susceptible lines. Although OP-Z04H660 is regarded as well suited for MAS in DH lines, experiments were carried out to convert it into a more specific marker that distinguishes between homozygous and heterozygous genotypes. This is because OP-Z04H660 is inherited in a dominant manner and is connected to the resistance allele in the repulsion phase, and it does not facilitate the identification of heterozygous susceptible plants in F2, which will segregate resistant plants in the offspring. Genetic and molecular maps as well as QTLs of major cereals (rice is in the advanced level) are now available. Exact location of such genes and their cloning could lead to further use in breeding cereals. *Zm-13* from maize is the most extensively studied pollen-specific gene. This gene is expressed in the final stages of microspore development. A similar gene *PS1* has been cloned from rice and is shown to be expressed in rice.

2.3 GERMPLASM CONSERVATION

More than one-third of all plant species across the globe (approximately 100,000 plants) are currently endangered or facing extinction in the wild. Thus, for classical and modern (genetic engineering) plant-breeding programmes, plant biodiversity has to be preserved. Moreover, pharmaceutical, food, and crop protection industries are also benefitted from this. Since the 1970s, many landraces and wild relatives of cultivated crops have been kept in the existing banks. In national, regional, international, and private gene banks around the globe, there are around 6 million samples of plant genetic resources. Storing desiccated seeds at low temperature is a favourable method of preserving plant germplasm. However, it is not suitable for crops that do not produce seeds (such as bananas), plants with recalcitrant seeds (that is, non-orthodox seeds that cannot be dried to the extent that moisture contents are low enough for storage, for instance, many tropical trees), and plants that propagate vegetatively for preserving the unique genomic constitution of cultivars (such as fruits and many timber and ornamental trees). It is also not viable to preserve valuable germplasms only in field collections because they can be lost in the form of genetic erosion, owing to pests, diseases, and adverse weather conditions. Other methods such as maintaining clonal orchards and *in vitro* collections are laborious and expensive. There is also a risk of contamination, human error, or somaclonal variation (that is, mutations occurring during tissue culture, which increase with repeated sub-culturing) which increases the risk of losing germplasm accession. Hence, long-term conservation of plant genetic resources can be done by freeze preservation or cryopreservation at an ultra-low temperature (−196°C, the temperature of liquid nitrogen). This is because cryopreservation helps in arresting biochemical and major physical processes and preserving for a long time specific characteristics of plant tissues, such as medicinal and alkaloid-producing cell lines, hairy root cultures, and genetically transformed and transformation-competent culture lines. It also helps to successfully eradicate viruses from plum, banana, and grapes. Regular application of cryopreservation is still limited for preserving plant biodiversity despite its development for a large number of recalcitrant seeds and *in vitro* tissues/organs (Hitmi, Sallanon, and Barthomeuf 1997).

2.3.1 Theoretical Basis of Plant Cryopreservation

Intra-cellular ice crystal formation causes irreversible damage to cell membranes and destroys their semipermeability. Hence, they should be avoided in cryopreservation. Some species have the ability to adapt against the formation of ice crystals at sub-zero temperatures. They do this by synthesizing specific substances (such as sugars, proline, and proteins) that can decrease the freezing point in living plant cells and result in 'supercooling'. However, in the case of ultra-low temperatures of cryopreservation (– 196°C), such 'avoidance' of crystallization is not possible, while still maintaining a minimal moisture level needed to maintain viability. Vitrification is the physical process by which an aqueous solution transits into an amorphous and glassy (that is, non-crystalline) state. It prevents the formation of crystals without an extreme reduction in cellular water. For a cell to vitrify, two requirements should be met: (i) rapid freezing rates and (ii) concentrated cellular solution. For obtaining rapid freezing rates (6°C/s), explants enclosed in a cryovial are normally plunged into liquid nitrogen. One can obtain higher cooling rates by enclosing the meristems in semen straws (attaining cooling rates of about 60°C/s), or utilizing a 'droplet freezing protocol', in which the material is placed on aluminium foil strips, which are directly plunged into liquid nitrogen (cooling rates of 130°C/s). Concentration of cell cytosol can be achieved by air drying, freeze dehydration, application of penetrating or non-penetrating substances (cryoprotectants), or adaptive metabolism (hardening). The water content must be reduced to at least 20%–30% for a solution to be vitrified at high cooling rates. The following techniques are applied for dehydration:

1. **Air drying:** Generally, samples are dried with the help of sterile airflow in a laminar airflow cabinet. Temperature and air humidity

cannot be controlled and both strongly influence the evaporation rate. Air-drying method that uses closed vials having a fixed quantity of silica gel is more reproducible.

2. **Freeze dehydration:** During slow cooling, crystallization is initiated in the extra-cellular spaces. This is because ice-nucleating agents are rarely present in plant cells. Water that contributed to extracellular solution undergoes transition into ice. The remaining solution becomes more and more concentrated to the cell (hypertonic). Osmotic equilibrium is restored when cellular water leaves the protoplast and cellular dehydration occurs. On the basis of the type and physiological state of the plant material, freezing rate ranging from 0.5°C/min to 2°C/min are applied. Computer-driven cooling devices, stirred methanol baths, and propanol containers held at –70°C help in obtaining these slow-cooling rates.

3. **Non-penetrating cryoprotective substances:** One can obtain osmotic dehydration through non-penetrating cryoprotective substances, which include sugars and sugar alcohols, and high-molecular-weight additives, such as polyethylene glycol (PEG).

4. **Penetrating cryoprotective substances:** Dimethyl sulphoxide (DMSO) and glycerol are two commonly used penetrating cryoprotective agents. DMSO is preferred in many cases owing to its ability of extreme rapid penetration into cells. Where DMSO toxicity becomes an issue, glycerol or amino acids (for example, proline) are utilized.

5. **Adaptive metabolism (hardening):** To survive unfavourable environment, a process called hardening has been developed. Environmental parameters such as shortening of day length and reduction in temperature cause unfavourable environment stress. Osmotic changes and ABA treatments have similar results. As the levels of proteins, sugars, glycerol, proline, and glycine betaine increase in the cell owing to hardening, the osmotic value of the cell solutes also rises. Solution and mechanical effects enable most hydrated tissues to withstand dehydration to moisture contents, which is needed for vitrification (20%–30%). However, pollen, seeds, and somatic embryos of most orthodox seed species are exceptions. Thus, dehydration tolerance has become more important than freeze tolerance for a successful cryopreservation. Chemical

cryoprotection with sugars, amino acids, DMSO, glycerol, etc., can induce this tolerance. However, mode of action of these substances is still far from being known. Alternatively, adaptive metabolism also induces dehydration tolerance. For instance, it has often been observed that cold acclimation in nature leads to the aggregation of specific proteins, sugars, polyamines, and other compounds that protect cell components from drying. Moreover, it has been reported that alterations in membrane composition influence both their flexibility and permeability (Gonzalés-Benito, Iriondo, Pita, *et al.* 1995).

2.3.2 Application of Cryopreservation to Herbaceous Species

1. **Cell suspensions and callus cultures:** Slow-cooling protocol often cryopreserves cell suspension and callus cultures. The cryopreservation of the non-organized tissues, apart from long-term storage of genetic diversity, preserve certain features of tissues that can disappear during normal *in vitro* maintenance. Storing embryogenic callus lines in liquid nitrogen does not affect their morphogenetic potential. In addition, it has been proven that cryopreservation did not influence the expression of a foreign gene named *sam* in transgenic *Papaver somniferum* cells. It is even useful for pyrethrin biosynthesis by *Chrysanthemum cinerariaefolium* cell cultures. Moreover, cryopreservation does not influence the production of regenerable protoplasts, as seen for *Festuca* and *Lolium* species of rice. Transgenic plants can be produce regenerable cell cultures. Cryopreserved rice and maize calli have been a constant source of the latter. It was shown that rice calli stored in liquid nitrogen had a higher competence for transformation, which was indicated by the levels of transient gene expression. Comparable levels of transient expression, as well as stable transformation, were observed for cryopreserved and non-cryopreserved suspension cells in banana and grape. For coffee, oil palm, and banana, large-scale cryopreservation of non-organized cultures has been reported.

2. **Cryopreservation of pollen:** Preservation of pollen is required for a number of activities such as facilitating crosses in breeding programmes, distributing and exchanging germplasm among

locations such as cultivars with different flowering periods, and preserving nuclear genes of germplasm, including research in basic physiology, fertility, biotechnology, and biochemistry, involving gene expression, transformation, and *in vitro* fertilization. The application of existing cryopreservation techniques of pollen from many crops is still confined to a few research centres.

3. **Cryopreservation of meristematic tissues:** For the cryopreservation of vegetatively propagated species, such as fruit trees and many root and tuber crops, the most common explants are meristematic tissues. Organized tissues such as meristems are often preferred over non-organized tissues, such as calli and cell suspensions, as the former have a lower chance of somaclonal variation. From 2000 onwards, most reports are on encapsulation/ dehydration or vitrification methods. In view of its rather severe freeze dehydration, the classical slow-cooling method is presumed to be good in preserving the integrity of individual cells, but less efficient in preserving the tissue integrity essential for the survival of meristem. Large-scale application of cryopreservation of meristems (shoot tips) is limited to fruit crop germplasm collections. The number of accessions, in the case of herbaceous species stored in liquid nitrogen, is significantly smaller, but is continuously increasing.

Using the droplet freezing method, meristems of 519 old varieties of potatoes were cryopreserved at the German Collection of Micro-organisms and Cell Cultures (DSMZ, Braunschweig, Germany). At the International Potato Centre (CIP, Lima, Peru), the vitrification protocol was used to preserve 345 potato accessions (Panta, personal communication). Utilizing the droplet vitrification method, 306 banana accessions are currently safely stored at U. Leuven, Belgium. This represents more than one-fourth of the total banana collection worldwide. Considerable efforts have also been made for preserving cassava, garlic, mint, and Australian endangered species.

4. **Cryopreservation of seeds:** Most common agricultural and horticultural species seeds are tolerant to desiccation and exposure to liquid nitrogen. At $-20°C$ cryogenic storage of orthodox seeds can be seen as an alternative to the traditional storage. The longevity

of seeds of some plant species at −20°C is only a few years, which can be significantly increased by storing in the vapour phase of liquid nitrogen. The ultra-low temperatures can be avoided by simply drying to decrease the moisture content to 5%–10%. An example of orthodox seed cryopreservation is celery.

2.3.3 Application of Cryopreservation to Woody Species

Many forest angiosperms (for example, *Acer* spp., *Quercus* spp., chestnut, horsechestnut, and many tropical species) lack orthodox seeds, and the period of their conservability is very limited. Till date, seed and field collections have been the only reliable option of long-term germplasm preservation of woody species. Fruit trees are mainly vegetatively propagated. The conservation of these trees require a huge number of accessions in clonal orchards, including old and newly selected cultivars, local varieties, and wild material. For example, preservation of the European *Prunus* germplasm requires maintaining more than 30,000 accessions in the field repositories of 21 countries. Although in assisting woody plant conservation programmes, clonal orchards have a pre-eminent role to play, maintenance of these species requires huge areas of land and involves high running costs (mainly for pruning operations, weed and pest management, and irrigation). Moreover, they are susceptible to environmental stresses (such as heavy frosts and flooding) and the perils of pests, diseases, and genetic alterations. Thus, preserved trees should essentially be monitored on a periodic basis. Nowadays, collection, molecular characterization, disease indexing, pathogen elimination, propagation, documentation, preservation, and exchange of disease-free plant genetic resources can be done through a wide range of biotechnological techniques. For long-term storage of woody plant by germplasm, cryopreservation is a better option to seed and in-field banks. Traditional *in situ* and *ex situ* approaches to tree germplasm preservation cannot be replaced by storing specimens at −196°C. Instead, it should be considered an alternative approach to develop a multi-option modus operandi for conserving and using gene banks and providing assurance against the accidental loss of plant genetic resources.

2.3.4   Organ and Tissues from *in vitro* Culture Used for Cryopreservation

Advancements in tissue-culture technology have enabled testing of a wide range of organs and tissues for storage in liquid nitrogen. Three categories of herbaceous species are commonly utilized for the cryopreservation of woody plants:

1. **Shoot tips:** These differentiated organs are used for preserving vegetatively propagated plants (such as various fruit and timber tree cultivars) for which genetic fidelity must be maintained. Shoot tips (1–2 mm, on an average) are collected from apical or axillary buds excised from shoot cultures grown *in vitro*. To obtain apical meristem, surrounded by some of the original leaf primordia and leaflets, the bud is handled under sterile conditions.

2. **Seeds or isolated embryo axes:**   Species that are mainly reproduced by seeds are preserved by these methods. Cryopreservation forms an important option for the long-term conservation of genetic resources for species that have non- or sub-orthodox seeds, and cannot be kept in traditional seed banks.

3. **Embryogenic callus:** This is the most powerful *in vitro* morphogenetic system. Cryopreservation of embryogenic cultures is an important long-term method of preserving valuable germplasm of woody plants. In seed-propagated species, plants with non-orthodox seeds can be preserved by cryopreservation of embryogenic cultures. In species that propagate vegetatively, embryogenic cultures are not helpful in the preservation of germplasms, because somaclonal variation cannot be prevented. However, using cryopreservation, valuable embryogenic lines used in bioengineering can be maintained. For example, preventing the loss of embryogenic potential resulting from repeated sub-culturing or allowing the storage of transgenic material during field trials.

2.3.5   Cryopreservation of Hardwood Trees

Recently, the continuous improvement of vitrification/one-step freezing technique has led to it being increasingly applied to a number of hardwood species. The introduction of PVS2 solution for cryopreservation of nucellar cells in *Citrus sinensis* prompted wide application of the cryogenic technique to woody species. The PVS2 solution has been successfully utilized for cryopreserving shoot tips from various economically important hardwood species (such as *Malus, Pyrus, Prunus,* and *Populus* spp., and *Vitis vinifera*). The solution has protected hardwood-tree shoot tips from damage caused by ultra-rapid freezing and increased the survival rates by 50%. In the case of *Prunus jamasakura* and *Populus alba*, 90% of shoot-tip survival, the highest percentage, have been reported. When PVS2 solution is loaded on explants, post-thaw survival is dependent on the treatment duration. The treatment should be long enough for sufficient cell dehydration to take place, without any cytotoxic effects. The exposure time ranges from 20 to 120 min when the solution is applied at 25°C. Otherwise, the risk of toxicity can be minimized by using a chilled solution (0°C). Rapid warming in a water bath is necessary after storage in liquid nitrogen to prevent recrystallization and ensure appropriate recovery of the vitrified material. For woody species, the proposed thawing temperatures are in the range of 20–40°C. For instance, after cryopreservation in *Populus* spp., the shoot tips were thawed at different temperatures (30–50°C). Survival differences were negligible in the case of *P. canescens*, while *P. alba* showed shoot tips best survival after being warmed at 40°C. In majority of cases of both the species, rewarming shoot tips at 40°C led to obtaining healthy, well-developed shoots. Shoot tips of nine different genera of hardwood species were cryopreserved, including *Malus, Pyrus,* and *Prunus*, using 'encapsulation/dehydration' technique. Cryopreserving 50% of the species after encapsulation–dehydration procedure led to 80% survival or more in shoot-tips. Encapsulation/vitrification is a technique that follows the combination of explant encapsulation with the utilization of a vitrification mixture. This method has already been used successfully for the cryopreservation of the shoot tip of apple and plum. There is not enough information on cryopreservation of embryogenic calli and somatic embryos of hardwood tree species. In recent times, 'vitrification/one-step freezing' procedure has helped to increase the number of temperate hardwood species, including *Castanea sativa, Fraxinus angustifolia, Quercus suber, Olea europaea,* and *Aesculus hippocastanum*. Successful cryopreservation of embryogenic calli of some forest species and fruit tropical species has been done. Encapsulation and dehydration followed by direct plunge into liquid nitrogen or treatment with the PVS2 have been done for the explants obtained from isolated somatic embryos or samples of embryogenic callus. In the case of hardwood species, apart from vitrification/one-step freezing method, application of slow-cooling technique has been sporadically reported. However, shoot-tip survival attains higher variability with slow cooling as compared to vitrification method, ranging from a minimum of 34% (*Juglans regia*) up to a maximum of 92% (*Malus* spp.). An encapsulation/dehydration protocol has been proposed for walnut somatic embryos, followed by slow cooling and immersion of the beads in liquid nitrogen.

2.3.6 Cryopreservation of Softwood Trees

The advanced technology of cryopreservation of embryogenic cultures of conifers has already been successfully applied to many species of *Picea*, *Pinus*, *Larix*, *Abies*, and *Pseudotsuga*. The cryopreservation protocol based on slow-cooling technology has been applied to more than 5000 genotypes of 14 conifer species in British Columbia alone in the United States. However, reports dealing with cryopreservation of conifer shoot tips do not exist till date.

2.3.7 Other Cryopreservation Procedures

A cryopreservation method based on the original procedure for dormant vegetative buds by Sakai, Hirai, and Niino (2009) was applied to 1915 accessions of apple. For this, scions collected in winter were desiccated in a cold room at −5°C to 30% moisture content and very slowly cooled (at the rate of 1°C/h) to −30°C; they were retained in that room for 24 h before being transferred to the vapour phase of liquid nitrogen (−160°C). More than 90% of the accessions exhibited a survival rate higher than 30% upon grafting onto rootstocks after retrieval from storage. For the cryopreservation of winter-dormant axillary buds of persimmon (*Diospyros kaki*), directly collected from the field, a more classic 'vitrification/one-step freezing' approach has been proposed. Finally, seed cryopreservation can be done by applying dehydration/one-step freezing procedures to intact seeds or embryonic axes of numerous woody species.

2.3.8 Cryostored Woody Plant Germplasm

Examples of woody plant repositories utilizing the cryopreservation method are now available. Besides the collection of conifer embryogenic lines in British Columbia (United States), the following cryobanks can be mentioned:

- The National Seed Storage Laboratory (NSSL) of Fort Collins (United States), with about 2100 accessions of apple (dormant buds)
- The National Clonal Germplasm Repository (NCGR) of Corvallis (United States), with more than 100 accessions of pear (shoot tips)
- The Association Forêt Cellulose (AFOCEL) of Nangis (France), with more than 100 accessions of elm (dormant buds)
- The National Institute of Agrobiological Resources (NIAR) of Ibaraki (Japan), with about 50 accessions of mulberry

Some tropical and sub-tropical woody species are also presently being cryopreserved, for example, at ORSTOM, now Institut de Recherche pour le Développement (IRD), France, which has 80 accessions of oil palm, and at the National Bureau of Plant Genetic Resources (NBPGR), India, which has numerous accessions of citrus, jackfruit, almond, Litchi, and tea.

2.3.9 Genetic Integrity of Plants from Cryopreservation

Introduction, manipulation, and regeneration of plants *in vitro* enabled numerous investigations on somaclonal variation. This is a problem that also concerns conservation of germplasm at ultra-low temperatures. The peculiarities of this (for example, blocked metabolism of cells and absence of sub-cultures) decreases the risks of genetic and epigenetic alterations to a minimum. In contrast, cryoinjury may result when tissues are exposed to physical, chemical, and physiological stresses during cryopreservation. Moreover, particular reactions such as free radical formation and molecular damage owing to ionizing radiation may result in threats to genetic stability that can still occur at $-196°C$, as well as from the common practice of utilizing up to 10% DMSO as a cryoprotectant. Although limited number of detailed studies on these aspects is available, however lack of substantial evidence of cytological, morphological, or genetic alterations owing to cryopreservation is promising.

Even after the introduction of slow-cooling approach in the 1970s, cryopreservation of plant tissues was not studied widely for a long time owing to highly complex procedures and expensive cryofreezers. In the early 1990s, cryostorage of genetic resources became a realistic target for most plant species. This was a result of the development of 'new' and simplified cryopreservation protocols based on preventing intra- and extracellular ice crystals through cell vitrification and direct immersion of explants in liquid nitrogen. Despite vitrification protocol being regarded as a standardized protocol, many studies are still being carried out only in the framework of academic studies and involve only one or a few accessions per plant species. The unavailability of efficient cryopreservation protocols remains the main drawback for a wider application of plant cryopreservation for many species. The preparation phase of tissues towards dehydration (specifically sugar and/or cold treatments) and the length of explant treatment with the vitrification solution are the two most significant parameters that should be optimized for each species and tissues. The procedures should be simplified and standardized so that technology becomes available to a wide range of institutions and companies of both private and public sectors.

Moreover, a firm physicochemical background of cryopreservation is required to facilitate the development of even more efficient cryopreservation protocols. This can be possible only through fundamental studies involving both thermal analysis and an extensive examination of the parameters that can affect the cryobehaviour, such as membrane composition, oxidative stress, oxidative stress, and cryoprotective proteins. From 2010 onwards, these parameters are being examined for different plant species in the framework of a European project (CRYMCEPT).

2.4 PRODUCTION OF SECONDARY METABOLITES FROM PLANT CELL CULTURE

In plant cell cultures, the yield of secondary metabolites can be improved by following several strategies. Shikonin and berberine showed success when approaches such as screening and selection of high-yielding cell lines and optimizing growth and production media were followed.

Many new approaches, for instance, culturing of differentiated cells (such as shoots, roots, and hairy roots), induction by elicitors, and metabolic engineering, have been developed. The culture of differentiated cells, in levels comparable to that of the plant, has been possible. However, large-scale production has proved to be a major constraint in bioreactors. For biosynthetic studies, such systems are very useful.

The application of elicitors has been successful in several cases. It, however, remains confined to a certain kind of compound, which most likely acts as phytoalexins, for each plant. Hence, the focus is more on metabolic engineering. Several possibilities can be envisaged to apply metabolic engineering to increase yields:

- Increase activity of enzymes that are limiting in a pathway.
- Induce expression of regulatory genes.
- Block competitive pathways.
- Block catabolism.

The first two approaches require the yield of active enzymes encoded by gene expression, while the latter two require the down-regulation of genes through their antisense. In all the cases, knowledge of respective biosynthetic pathway is required at the level of products, enzymes, and genes, as well as regulation on all these levels, including compartmentalization and transport. In the plant, the latter enzyme serves all cytochrome P-450 enzymes; among others, Geraniol-10-hydroxylase (GlOH) is an important enzyme in the terpenoid part of the alkaloid biosynthesis.

The expression of *fdc* gene results in an active enzyme in tobacco plants. These plants release up to 1% of dry weight (DW) in tryptamine, that is, similar levels as found for the indole alkaloids derived from tryptamine in plants generating such compounds. The activity of anthranilate synthase, which is the first committed enzyme of the tryptophan pathway from chorismate, did not increase in tryptamine-generating transgenic tobacco plants. With their normal primary metabolic machinery, plants can apparently produce about 1% of their DW in tryptophan-derived compounds. In *Catheranthus roseus*, the introduction of the *rdc* gene resulted in callus cultures exhibiting up to 10-fold increase in TDC activity and a concomitant rise in tryptamine levels, but no significant increase in strictosidine or other alkaloids. This supports another observation that the terpenoid part of the pathway is also a constraining factor. Both *tdc* and *sss* genes are single-copy genes, which are repressed by auxins and induced by elicitation. They might be controlled by one transacting factor because of this similar regulation pattern. Thus, it is of great interest to identify this factor and the encoding gene, as such a regulatory gene possibly also controls further steps in the biosynthetic pathway. By cloning such a regulatory gene, the need to clone a large number of genes coding for every individual protein can be avoided.

Upon elicitation in *C. roseus*, a 2550% increase is noticed in the activity of AS, TDC, and SSS, whereas the activity of phenylalanine ammonia-lyase (PAL) reduces and that of chalcone synthase remains unaffected. The major change, however, is seen in the content of phenolics and 2,3-dihydroxybenzoic acid (DHBA) is the major compound formed. A strong induction of isochorismate synthase is observed at the same time. The enzyme converts chorismate into isochorismate. It is likely to be a major defence compound in bacteria, as this pathway is known to lead to DHBA, which exhibits antimicrobial activity, specifically in combination with UV light and anti-feedant properties. Chorismate and mevalonic acid are used as a precursor for phytoalexin biosynthesis in *Cinchona* cell cultures (anthraquinones). Mevalonate is utilized for the mterpene biosynthesis after elicitation in *Tabernaemontana* species, which channels away this precursor from the alkaloid pathway. Blocking such pathways with the help of antisense genes might enhance the availability of these precursors in the pathway for alkaloid biosynthesis.

Another significant factor in the accumulation of alkaloids is catabolism. Studies feeding ^{14}C-labelled alkaloids to *Tabernaemontana divaricata* and *C. roseus* cell cultures revealed that the rate of catabolism equals the de novo biosynthesis at a certain point of the growth phase. Antisense gene technology can be used to eventually block catabolism and increase production after identifying enzymes involved and cloning of the encoding genes.

Biosynthesis of secondary metabolites can be studied through plant cell cultures. Generation of these compounds is minuscule for commercialization in most cases. To enhance yields, metabolic engineering provides promising perspectives but needs the knowledge of the regulation of the secondary metabolite pathways working on the levels of products, enzymes, and genes, including aspects such as transport and compartmentation. A joint colaboration of several disciplines is required, for example, phytochemistry, plant physiology, cell biology, and molecular biology, for understanding the regulation pathway.

In plants, secondary metabolites have far more restricted distribution than primary metabolites; most of the time, a compound is present only in a few species or in a few varieties within a species. They still may have an ecological role though their role in plant metabolism is not clear; for example, as sexual attractants for pollinating insects or in defence mechanisms against predators.

The extraction of secondary metabolites is often difficult as they mostly accumulate in plants in small quantities and sometimes in specialized cells. Most of these compounds are of commercial importance such as medicinal substances, fragrances, food additives (pigments, flavouring and aromatic compounds), and pesticides.

Extraction of secondary metabolites from plants still holds commercial importance despite gaining success in organic synthesis or semi-synthesis of a variety of compounds similar to those produced by plants. Production of a large number of these metabolites is still not economical. In several cases, it is seen that consumers prefer natural products in comparison to artificial products. Both objective and subjective reasons explain that natural extraction is the only option for a variety of aromas or fragrances, which result from a mixture of numerous different compounds (for example, jasmine and strawberry), or for biochemicals with complex molecular structures (for example, some alkaloids and glycosides).

However, great interest has been taken for the development of alternatives to the intact plant for the generation of plant secondary metabolites. Originally, it centred on the application of tissue and cell cultures. The current approach involves the application of molecular biology techniques to stimulate the metabolic pathways leading to the production and extraction of specific compounds. In the past, the use of plant cell and tissue cultures had been the focus of research, for commercially producing a wide range of secondary metabolites in a way similar to the production of antibiotics with the help of bacteria and fungi. Plant tissue cultures were first formulated in 1939–40. The first patent was filed by the American pharmaceutical company Pfizer Inc. for the production of metabolites by mass cell cultures. In the late 1960s, it was discovered that plant cell cultures have the potential to generate useful compounds, especially for developing drugs. A large number of cultures, however, could not synthesize products characteristic of the parent plant. For example, synthesis of morphinan, tropane, and quinoline alkaloids was possible only at extremely low levels in cell cultures (Berlin 1986). That is why, the trend of research declined after a surge in interest. In 1976, the outstanding metabolic capacities of plant cells were demonstrated that highlighted the spontaneous variability of plant cell biosynthetic capacity which could explain the contradictory results obtained earlier. To identify high-yielding cultures and for its use on an industrial scale, this natural variability can be utilized. Since the late 1970s, an increasing number of patent applications have been filed on research and development in this area, especially by scientific and corporate sectors of Germany and Japan. In 1983, Mitsui Petrochemical Industries Ltd, for the first time, produced an anti-inflammatory and antibacterial dye, shikonin, on an industrial scale with the help of plant cell cultures. Shikonin is still the only plant product to be produced on a commercial scale by cell cultures though this was thought to be a major breakthrough. The requirement of biochemical and molecular biology research on the production of secondary metabolites in plant cell cultures has increased with the need for increasing the production. Research findings in this field could successfully manipulate secondary metabolism and increase the quantity of the compounds significantly sought. It is believed that any substance of plant origin can be generated by cell cultures. Therefore, a wide range of compounds (may be several hundreds), such as alkaloids, flavonoids, terpenes, steroids, glycosides, can be synthesized, with complex chemical structures through plant cell-culture technology. Moreover cell lines can be possibly identified that can generate compounds in equal or higher amounts than the plant from which these are derived. New molecules have been created by cell cultures that were not previously found in plants or were synthesized chemically. Hence, this technology is a genuinely new means of producing novel metabolites. Finally, through a variety of reactions such as hydrogenation, dehydrogenation, isomerization, glycosylation, hydroxylation, opening of a ring, and addition of carbon atoms, it is possible for plant cells to transform natural or artificial compounds introduced into the cultures. Furthermore, large-scale production of plant cells would result in a stricter control of product quality as well as their continuous production without depending on the variations of natural production resulting from climate and sociopolitical changes in their countries of origin.

The following sequential stages or developments are involved in the techniques of plant cell cultures: selection among wild plants of a high-producing one; *in vitro* culture or callogenesis, which involves the selection and stabilization of producing calli with a view to identifying a high-producing line or strain; maximizing callus or cell suspension, culture conditions and isolation of the best-producing line; industrial scaling-up; mass cultivation in bioreactors; and downstream processing, that is, extraction and purification of the compounds sought.

Plant cell cultures or suspensions can increase the production of secondary metabolites. These involve application of biotic or abiotic elicitors that can induce the metabolic pathways as in the intact plant; addition of a precursor of the desired compound in the culture medium for enhancing its production or inducing changes in the flux of carbon to favour the activation of pathways leading to the compound sought, that is, alteration of controls of secondary metabolism pathways; creation of new genotypes through genetic engineering or protoplast fusion, but this presupposes the identification of the genes encoding the key enzymes of secondary metabolic pathways and their expression once introduced in the plant cells; utilization of mutagens to increase the variability already existing in living cells; and use of root cultures.

Mutagens can increase the naturally occurring diversity among the cell clones. It can also create a new higher-producing lines or strains. Also, improvement could be seen in the capacity of cell synthesis when cell fusion and gene transfer are applied. However, the knowledge of the metabolic pathways and their genetic control governs the results of these efforts.

2.4.1 Alteration of Controls of Secondary Metabolism Pathways

For most secondary metabolism pathways, the proposed biosynthesis, deduced from chemical considerations and feeding experiments, should await verification through the recognition of the corresponding enzyme reactions. Thus, the absence of enzymology background restricts direct manipulations of the pathways. The study of cell suspension systems has provided the knowledge of enzymology relating to secondary metabolism pathways. Such enzymes can be obtained from organ cultures and intact plants.

Identification of regulatory or rate-limiting enzymes is another possible issue related to the alteration of secondary metabolism pathways. This is most possibly for enzymes at the start of a sequence or at the branching points of metabolic pathways. This is especially so when their absence restricts the de novo formation of the compounds or when the activity of these enzymes increases manifold under conditions that promote a major increase in product formation. It is also essential to understand whether there is a co-induction of other biosynthetic enzymes with the proposed regulatory enzyme or these enzymes are permanently present even in non-producing cell cultures. Such information helps in the evaluation of the possibility of specifically manipulating these secondary metabolism pathways for increasing the desired product's concentration.

A general assumption is that the regulatory genes occur as gene families, and the expression of each gene within one family is regulated by a different signal. Thus, direct alterations of such controls would be difficult to achieve. However, desired gene can be introduced into an intact plant using suitable vectors and expressed in a specific organ such as chloroplasts. For studying its interference with a secondary metabolism pathway, the gene coding for chalcone synthase (CHS) could be the first plant gene to be utilized in such a transformation process. Veltkamp and Mol (1986) have used genetic transformation experiments to repair anthocyanin biosynthesis in mutants that are deficient in CHS. Although examples of alteration of regulatory controls of secondary metabolism pathways through genetic engineering do not exist (with respect to increasing the rate of production of a desired compound in cell cultures), this seems to be the only encouraging approach. To achieve this end, the identification of genes encoding biosynthetic enzymes involved in regulatory function in secondary metabolism pathways is required. Another biotechnologically relevant approach would be the transfer of plant genes into bacterial cells and their expression for stereospecifically difficult biotransformations (instead of organic synthesis). The best tool (transformed plant cells, transformed microorganisms, or immobilized enzymes) for increasing the production of a specific compound could be decided only after the genes for the relevant enzymes are isolated and cloned.

2.4.2 Effects of Elicitors

In cell culture systems, abiotic and biotic elicitors play an important role in improving the yields of products. The induction of biosynthesis of flavonoids occurs through light via phytochrome or/and UV photoreceptors and after infection with phytopathogenic organisms or compounds that trigger the synthesis of antimicrobial compounds in plants. In cell suspensions of *Gossypiunz arboreum*, the synthesis of gossypol was induced by an extract of *Verticillium*, a parasitic fungus of plants. In plant cell cultures, researchers have found very specific elicitors that will significantly increase the quantities of morphinan, tropane, or quinoline alkaloids. Increase in the generation of morphinic alkaloids through poppy cells could be similarly increased; however, this is still an empirical issue and a single general rule cannot be proposed. As the elicitor needs to be added, this complicates its use.

2.4.3 Bioconversion

A wide range of substrates can be transformed by plant cells. They also perform reactions such as oxidation, reduction, hydroxylation, methylation, acylation, glucosylation, and amino-acylation. For instance, steviobioside, a glycoside that is 300 times sweeter than sucrose, is transformed from steviol by *Stevin rebaudiana* cells. It is used as a sweetener in Japan. Salicylic acid is another example in which cell cultures of *Mallotus japonica* glycosylate to produce a compound with higher analgesic power and better tolerance in stomach than aspirin (acetylsalicylate). From the leaves of *Digitalis lanata*, two cardiotonic compounds, digitoxin and digoxin, are isolated. Digoxin, extracted in smaller quantities, pharmaceutical properties, but cannot be generated either chemically or through microbial bioconversion. Digitoxin is extracted in large quantities.

The 12-hydroxylation of the cardiotonic drug beta-methyl-digitoxin into a more desirable, less toxic drug beta-methyl-digoxin through cell cultures of *D. lanata* is a bioconversion of high pharmaceutical interest. Higher-yielding cell lines have been isolated using selection techniques. So far the most effective bioconversion obtained has been 1 g/L during a cultivation period of 28 days.

Roots of tobacco plants have been successfully grown after discovering that tomato root cultures can unlimitedly grow in media having macro- and micro-nutrients, sucrose, and yeast extract. It was found that the rise in nicotine concentrations in tobacco root cultures is almost similar to (positively correlated) the increase in root tissue, as measured through length of roots, number of branches, and accumulation of dry weight. Hyoscyamine was also shown to be synthesized by root cultures of henbane (*Hyoscyamus niger*). Production of alkaloids was inversely correlated to the growth rate of callus cultures in *Datura* cells. Undifferentiated calli of *Atropa belladonna* do not generate the tropane alkaloid hyoscyamine. Production occurs when roots develop on the callus. When calli undergo embryogenesis after treatment with plant growth regulators, the cardiotonic glycosides of *Digitalis* are produced. Owing to all these observations, researchers now utilize continuously growing, organized plant tissue cultures or even cultures of organs such as roots for producing secondary metabolites.

It was found that a 'hairy root' disease, caused by the microorganism *Agrobacterium rhizogenes*, affected a variety of dicotyledonous species and proliferated fast-growing adventitious roots at the host's wound site (because a portion of the Ki plasmid of *A. rhizogenes* stably integrated into the plant genome). For studying and producing secondary metabolites, researchers at the Plant Biotechnology Group, Agriculture and Food Research Council at the Institute of Food Research, Norwich Laboratory, UK, utilized transformed root cultures as a biochemically and genetically stable system. It might be possible to directly exploit the resulting stable high-producing lines as cultures in bioreactors or as plants in the field after plant regeneration. Following inoculation of plant material with *A. rhizogenes* strain LBA9402, researchers developed axenic transformed root cultures of *Nicotiana rustica* and *Datura stramonium* (which produce nicotine tropane and the alkaloids hyoscyamine and scopolamine, respectively). The exact nature of the factors that lead to root formation is not properly understood. The transformed roots grow faster compared to plant standards; the pattern of growth is logarithmic with doubling times, which can be under 48 h and characterized by a high degree of branching. This growth is linked to the generation of the characteristic secondary metabolites that are similar to the parent plant species as regards both absolute amounts and spectrum of products.

Tropane, a class of alkaloids, was produced by more than 20 hairy root clones of *Hyoscyamus* sp. at concentrations similar to the whole plant or in normal root cultures. For over 40 monthly passages, production of alkaloids was stable in two selected clones. It, however, markedly decreased upon the induction of roots to form calli and then reappeared when the calli exhibited root differentiation.

After examining 29 hairy root clones of *Scopolia japonica*, two highly productive clones were isolated: clone S1, which gathered scopolamine to 0.5% (w/w) dry weight, and clone S22, which synthesized hyoscyamine at 1.3% (w/w) dry weight. Accumulation of hyoscyamine and scopolamine, respectively, in the range of 0.04–1.1% and 0.06–0.3% (w/w) dry weight was noticed in normal root cultures of several *Hyoscyanzus* species. It is not exceptional to observe the production of alkaloids derived from the tropane ring by root cultures, as well as that of beta-xanthine. Secondary metabolites of the family Asteraceae (1000 genera over and 15,000 species), that is, sesquiterpene lactones and polyacetylenes can be produced by hairy root cultures. Sesquiterpene lactones and polyacetylenes show strong action against fungi, bacteria, and nematodes. They are present in the roots, and infection with fungal pathogens may elicit their synthesis. In 1986, researchers established more than 30 normal and hairy root clones from the genera *Anzbrosia*, *Bidens*, *Rudbeckia*, and *Tagetes*. The growth of these clones was faster than their normal counterparts. They generated thiophene-like compounds similar to those of normal roots and showed stability during several monthly passages. Therefore, researchers concluded that it was possible for these organ cultures to synthesize polyacetylenes and their cyclic derivatives.

For the production of onion, cultures of onion roots in bioreactors are preferable to aroma cell suspensions. These differentiated tissues can directly produce complex chemical aromatic compounds. These compounds can be subsequently isolated through high-pressure liquid chromatography after being released into the culture medium. Several of the 13 different species of onions from Europe, Japan, and North America that were tested seem to be promising for producing onion aroma industrially.

For enhancing the production of metabolites, transformed roots are also being utilized to elucidate the control mechanisms of secondary metabolism pathways. For example, an approach has been developed that manipulates the metabolic pathway at the level of individual genes and thus individual enzymes. This involves identification of gene coding for under-expressed enzymes in culture and then the isolation of the gene and its reintroduction into the plant under the control of de-regulated, high-expression promoters for increasing carbon flux through the pathway. Besides putrescine methyl-transferase (PMT), ornithine decarboxylase (ODC) restricts the part of the metabolic pathway that is common to nicotine and hyoscyamine biosynthesis. Therefore, the gene coding for ODC was transferred from *Saccharomyces cerevisiae* into the transformed roots of *N. rustica*, where it is under the regulatory control of the cauliflower mosaic virus 35s promoter. After integration into the plant genome, the gene was successfully transcribed into a fully active enzyme. In the transformed cells, there was no effect on the pattern of expression of other enzymes responsible for the biosynthesis of alkaloids. An alternative approach is studying the coordinated regulation of the expression of the entire pathway to comprehend the temporary and developmental regulation of the expression of the group of genes. Thus, the manipulation of this regulation would optimize the expression of the whole pathway. To summarize, commercial application of root cultures to generate secondary metabolites depends on the following factors: scaling-up of production and recovery techniques, further knowledge of the regulatory signals that influence the production of compounds at more than 10% dry weight, and identification of high-value biochemicals in roots.

2.5 INDUSTRIAL PRODUCTION OF USEFUL BIOCHEMICALS BY HIGHER-PLANT CELL CULTURES

The production of natural substances and biochemicals is governed by economic considerations. Since the late 1960s, considerable progress has been made to remove the major constraints of industrial production. A benzophenanthridine alkaloid extracted from the roots of *P. somniferum* is sanguinarine. These plants require 3 to 4 years of maturation before the substance can be extracted. To produce large quantities of this alkaloid, cell cultures have been used. This alkaloid is used in toothpastes and mouth lotions to combat dental plaque and tooth decay. The production process utilizes the eliciting power of extracts of a fungus Botrytis, which induces sanguinarine and dihydrosanguinarine synthesis in plant cells. In pilot experiments, it was found that the production rate of alkaloids reached 3% of dry biomass in semi-continuous cultures, which can be elicited twice. These examples show that with correct equipment and processes analogous to those used for microorganisms, industrial manufacturing of plant cell biomass and secondary metabolites is possible.

Cells immobilized in a gel, which can permeate to the molecules of the nutrient medium or polymers (which preserves their metabolic capacity and makes them usable multiple times), have the advantage of extending the production time of cells (more than 6 months) and of making the cells catalyse the same reaction almost indefinitely. Since the early 1980s, active research has been carried out in this area. A bioreactor has been developed at the Department of Engineering, Tokyo University, for producing codeine, which is mostly used as an anti-cough agent in pharmaceuticals. In calcium alginate beads, poppy cells have been immobilized to catalyse the transformation of codeinone into codeine. With this technique, the Japanese researchers overcame the weaknesses of plant-cell bioreactors that arise from the instability of cells and low yields of the desired compounds. They reduced the size of cell clusters (2.5 mm in diameter), which increased their lifespan, and obtained yields of codeine equal to those in non-immobilized cells. As the products now arise in the medium itself, the utilization of immobilized cells should bypass the direct extraction of the compounds from the biomass. The production of caffeine, capsaicin, and berberine is an example of this approach. However, many metabolites still seem to gather in the cell vacuoles, and it is, therefore, crucial to gain further information on the mechanism of diffusion of these metabolites into the culture medium. Even though immobilized cell technology is a promising technique and aimed at reducing production costs, precise examples that would demonstrate a gain of productivity on an industrial scale still do not exist.

2.5.1 Prospects

Japan and Germany started the industrial production of secondary metabolites in the early 1990s using plant cell cultures. Shikonin remained the only marketed product until 1990. In Japan, seven private corporations created a common subsidiary for developing plant cell cultures. On the contrary, most North American and European companies have not shown much enthusiasm in this regard.

There are several factors for this. First, 2–3 years are required for the selection and stabilization of cell lines because somaclonal variation is difficult to be controlled and directed. Consequently, a tedious and time-consuming screening is required for a large number of lines. Second, the knowledge of biosynthetic pathways of secondary metabolites is limited, which explains certain failures. For instance, cell line isolation has not been possible with a good level of dimeric alkaloid production from *C. roseus*, although the plant possess these alkaloids in minute quantities (1 tonne of dry leaves yields 0.5–2 g of these compounds). Third, there are considerable difficulties in the extraction of the desired compounds, especially when the compounds are found as combined substances. However, technical and scientific difficulties can be overcome in fairly short periods resulting in lower industrial production costs. The identification of worthwhile targets along with economic and legal considerations are essential for securing the necessary investments and are also the most difficult. The generation of secondary metabolites through plant cell or tissue cultures should be competitive with other traditional means of production. For instance extraction from field-grown plants, chemical synthesis or semi-synthesis, alternative enzymatic processes, genetic engineering, microbial fermentation or improvement of the plant itself through somaclonal variation.

2.6 INDUSTRIAL PRODUCTION OF FOOD ADDITIVES AND FRAGRANCES

To make food palatable and attractive, food additives are used. The additive enhance or improve flavour, colour, and texture. Food technologies use these criteria with respect to texture, taste, and aroma of the foodstuff. To retain or imitate the taste and aroma of specific foodstuff and satisfy consumers' tastes, the application of natural or artificial aromas is essential, especially in certain products where they may be indispensable. Till a few dacades ago, aromas were extracted from plant raw materials, which were converted by fermentation, grinding, and heating. Later, chemical compounds in pure or mixture form completed the range of plant aromas. Nowadays, the existing ones are combined with new compounds and they are usually produced from enzymatic reactions or biotechnological methods, that is by using microbial and plant cells. The preference of fermentation techniques or chemical synthesis depends on their cost. The only means to produce the thickening additive xanthan is microbial synthesis. Since the late 1950s, national and international regulatory authorities have been questioning the safety of food additives, regarding their long-term use and consumption. Consumer associations have been exerting pressure on governmental bodies to enforce replacement of chemical or artificial additives with 'natural' ones obtained from plant or animal tissues. With new production processes, food industries can respond to the favourable opinion towards natural aromas and overcome the present limitations related to the vagaries of supply in the producing countries. They can also take advantage of the differences in price between compounds produced naturally and chemically. For example, the price of vanillin, when extracted from the vanilla pod, is about 20,000 FF (French franc) per kg, whereas if produced from lignin using chemical synthesis, it costs only 50 FF per kg. It is impossible to distinguish between the two binds of vanillin molecules through chemical or physical analyses even though this difference is considerable. Hopefully, eventual fraud will be easily detectable for aromas, with a restricted market, and will presumably be rigorously monitored by industrialists who will benefit most by conserving their existing sales. In the 1970s, consumers associations strongly criticized the use of food additives because the origins of most of the products were chemical synthesis and were unrelated to any naturally occurring material. Presently, 24 pigments have been authorized by the European Economic Community, out of which 10 are not of natural origin. As the colour of foodstuffs is associated with their acceptance, the number of pigments used for this purpose is bound to increase; however, the trend is towards replacement of artificial colourings with natural ones. The increased price for 'natural' products will be determined by the market. The biotechnological methods involved in producing natural food colorants include growing microorganisms, microalgae, and higher plant cells. The technique of extracting pigments from plants, for example, anthocyanins from red berries, betanine from beet, and curcumine from saffron, is old. Some pigments cannot be extracted before plants become adult; one such example is rocou, a pigment extracted from the seeds of a shrub, or of shikonin, extracted from the roots of *L. erythrorhizon*. In the industrial production of shikonin, plant cell cultures have been successful.

2.6.1 Aromas and Fragrances

Natural aromas include a mixture of numerous compounds. Researchers have identified more than 500 in roasted coffee beans and 200 in apple. Natural aromas are influenced by food conservation processes such as sterilization, pasteurization, and freezing. Enzymatic or chemical reactions may alter some aromas, which usually disappear if stored for a long time. This is why since the end of the 19th century substitutes have been sought. Coal or oil derivatives were used to manufacture artificial aromas and were added in very low concentrations (10^{-6} or ppm, and 10^{-9} or ppb). Today, either synthetic molecules identical to natural molecules are produced or biotechnological methods are used.

The advantages of the first category of aromas are their constant composition, independent of the season, and manufactured according to the market. Their disadvantage is that they do not constitute mixtures of substances that are similar to natural ones. Hence, the biotechnological routes are now followed. For example, it was reported that from cell cultures of *Tlaeobroma cacao* and *Coffea arabica*, respectively, the characteristic aromas of cocoa and coffee have been produced.

According to consulting firm Précepta, France there is an uptrend in the market of aromas and fragrances, compared with that of foodstuffs, which is growing at a slower pace. About 5%–6%, and sometimes up to 12%, of the total turnover is attributed to research and development in the aroma and fragrance sector, that is, one order of magnitude higher than in the food sector.

2.6.2 Aromachology: New Prospects for Fragrance Production and Commercialization

Aromachology is a field that aims to modify the physical and mental conditions of human beings by using scents and fragrances. It is different from aromatherapy which uses essential oils for therapeutic purposes. Companies are focusing on producing relaxing or stimulating compounds and are not likely to enter the medical research area by investing in aromachology. However, it is eventually turning out to be a promising area not only for fragrance-manufacturing companies but also for those that would like to seize this opportunity to widen their hold in the fragrance market. In the future, it is possible that people might purchase a perfume for its relaxing, stimulating, or physiologically specific effect instead of only its scent. In this respect, it is worth mentioning the report jointly made by the fragrance foundation IFF and the University of Cincinnati on their discovery of fragrances that can increase vigilance. Elsewhere, Duke University conducted research on the effects of fragrances on violence. The University of Tsukuba studies the significance of certain fragrances on the rates of recovery of athletes after strenuous activities, in collaboration with Shiseido. Very little information is available on the reaction of people to these developments. Cell cultures have tremendous potential in the production of useful substances.

Many laboratories conduct research and development programmes for finding possibilities to increase their yields. Moreover, even if some compounds are not worthy of development to the commercial stage, some biosynthetic compounds could be utilized as significant precursors for organic synthesis or could themselves be entirely new products. Major factors that influence the choice of the appropriate manufacturing technique, especially when deciding the favour of plant cell or tissue cultures, are market size of the end product, cost–benefit ratio of the production technology, competition with substitutes, and existence of other sources of supply. Thus, genetically engineered microbial cells may be able to produce metabolites more efficiently than plant cells or tissue cultures, as these are synthesized via simple enzymatic reactions under the regulation of a single gene. Identification of economically profitable targets is difficult because industrial secrecy or confidentiality prevails in this area and there is a lack of crosstalk between researchers and companies working in the field of aromatic and medicinal plants.

2.7   BIOREACTOR SYSTEMS

In vitro propagation depends on the proliferation of enhanced axillary buds and on the capacity of differentiated, often mature, plant cells to re-differentiate and generate new meristematic centres capable of regenerating fully normal plants. Two morphogenic pathways potentiate regeneration: organogenesis, which is the formation of unipolar organs, and somatic embryogenesis, which is the production of bipolar structures, somatic embryos with a root and a shoot meristem. Once isolated and cultured *in vitro*, all plant somatic cells are capable of expressing totipotency. The injured cells of the isolated explant that are present in its outer layers evolve ethylene, which makes the inner layers of cells to undergo dedifferentiation. When the coordinated control is lost, ensued by cell division, the formation of new gradients of endogenous phytohormones in the dedifferentiated cells increases cell divisions in response to the growth regulators added to the medium. Cell division can take place in one of two ways: in an unorganized pattern with the formation of callus or in an organized pattern with the formation of meristematic centres directly in the explant tissues. Re-differentiation in the callus tissue occurs when new physiological gradients are produced in the non-organized parenchymatous tissue, along with the generation of meristemoids (*in vitro* meristems), which can differentiate into organized structures.

and automated dispensing. In some plants, these techniques were reported to reduce hand manipulation, thereby decreasing *in vitro* plant production costs. Various aspects of propagation in several plant species in bioreactors and problems emerging during the operation of bioreactors have been recently reviewed. For plant cells, somatic embryos, and organ cultures, liquid media have been used in both agitated flasks and various types of bioreactors. Research on improving bioreactors for somatic embryogenesis has been reported for various plant species even though the application of bioreactors has been directed mainly for cell suspension cultures and production of secondary metabolites. Plant quality and multiplication rates of banana, coffee, and rubber can be improved by cultivating in liquid media and using a temporary immersion system with different frequencies of immersion. Bioreactors have been used for cultivating hairy roots mainly as a system for secondary metabolite production. Biomass growth would be improved by applying a simplified acoustic window mist bioreactor for transformed roots and carnation shoots. Information on the application of bioreactors as a system for plant propagation through the organogenic pathway, though limited to a few plant species, has been applied to many ornamental, vegetable, and fruit crop plants. But liquid cultures create problems associated with the abnormal development of plants, a phenomenon known as hyperhydricity (previously called vitrification), which results in poor plant development *in vitro* and later in *ex vitro*. There have been efforts to overcome hyperhydricity in liquid cultures by using growth retardants, inhibitors of gibberellin biosynthesis (which reduced hyperhydricity), decreased shoot elongation, and induced bud or meristem cluster formation. It has been shown that the clusters are an alternative propagation system for bioreactor cultures, which provide a biomass with limited leaf elongation. Recently, using clusters in a temporary immersion system (ebb and flow) in bioreactors, micropropagation of pineapple has been scaled up. Bioreactors are currently being used in the United States, Japan, Taiwan, Korea, Cuba, Costa Rica, Holland, Spain, Belgium, and France for commercially micropropagating ornamental and bulbous plants, pineapple, potato, and forest trees. The estimated cost per propagule unit of foliage plants has reportedly reduced from 17 cents to 6–7 cents in United States.

2.7.1 Plant Developmental Pathways in Bioreactors

Somatic embryogenesis: In 1958, Steward and Reinert reported the application of liquid cultures for cultivating somatic cells in recapitulating embryogeny in carrot (*Daucus carrota L.*). The assumption that isolation and bathing of the cells in a specific medium will simulate the zygotic embryo conditions in the ovary was the basis of the underlying concept for cell totipotency and the ability to renew growth and express morphogenesis. The bathing liquid medium was assumed to be a source for stimuli similar to those present in the zygotic embryos' immediate environment, which contains several nutrients and growth regulators. Besides, segregation of embryogenic from non-embryogenic cells magnified the expression of cell totipotency. By the initial utilization of an induction medium containing an auxin, usually 2,4-dichlorophenoxyacetic acid (2,4-D), and coconut water, somatic embryogenesis was achieved; the latter was a source of cytokinins, inositol, and reduced nitrogen compounds. The removal or lowering of the auxin concentration after the formation of the pro-embryonic clusters initiated a sequence of events similar to zygotic embryogeny. In the beginning, somatic embryogenesis was mainly noticed in members of the Umbelliferae cultured in liquid media. Since then it has been observed in gymnosperms and angiosperms, including many ornamental, vegetable, and field crops, as well as perennial woody plants. For the large-scale clonal propagation, the embryogenic pathway, instead of the organogenic pathway can act as a more efficient and productive system. Somatic embryogenesis may lead to less variation through chimerism. However, in direct regeneration, there is less likelihood of somaclonal variation but more pronounced in somatic embryos developed from continuously cultured callus tissue. The rooting stage required in the traditional *in vitro* bud or shoot propagation technology is obviated because the embryo contains both a root and an apical shoot meristem. Somatic embryos can be handled easily in scaled-up procedures owing to their small size. They can be sorted and separated through image analysis and dispensed by automated systems. They can also be encapsulated and either stored or planted directly with the help of mechanized systems. Somatic embryogenesis in liquid shake or bioreactor cultures has been observed in carrot, caraway, poinsettia, rubber, and spruce. However, successful encapsulation of somatic embryos and thus production of synthetic seeds from liquid shake or bioreactor cultures have been reported only for carrot, celery, and white spruce. The application of this procedure in several other species is presently under investigation.

Organogenic pathway: The commercial propagation of most plants is currently carried out through organogenic pathway in agar-gelled cultures, despite the protocols being lengthy and expensive. Reduction in hand manipulation and labour expenses can be achieved by mechanizing the process. The scaling up of the process is required using liquid cultures in bioreactors to amend it to automation. The problems of hyperhydricity of leaves and shoots restrict information regarding the application of bioreactors for unipolar structures such as protocorms, buds, or shoots. One of the earliest attempts in utilizing liquid cultures for micropropagation of buds was reported for orchids that produced protocorms with minimal shoot elongation during the liquid-culture stage. The protocorms can be separated and induced to generate new plants after sub-culture in agar-gelled medium. Organogenic pathway has been used in liquid shake cultures and bioreactors to propagate many other ornamental plants. A high proliferation rate was achieved without the phenomenon of hyperhydricity by regulating shoot growth and providing culture conditions that decreased abnormal leaf growth and increased the formation of bud or meristematic clusters, as seen in potato, gladiolus, and *Ornithogalum dubium*.

Bud or meristem clusters: Highly proliferative and rapid growth can be achieved by spherical meristematic or bud clusters in liquid cultures. The development of condensed organized structures has been reported in many plant species in which the shoots are reduced to buds or meristematic tissue. The clusters are made of densely packed meristematic cells, which actively divide and form new meristematic centres on the outer surface. The meristemoids surround loosely packed cells in the centre and exhibit some vascularization, as was found in liquid-cultured poplar clusters. However, the clusters are composed of condensed buds surrounding a central core with a cavity as is the case in banana. In 1967, Halperin explained the formation of pro-embryogenic masses (PEMs) in carrot root explants.

It resembled orchid's protocorm-like bodies formed due to the addition of 2,4-D to the liquid medium in which isolated buds were cultured in shake cultures. By adding paclobutrazol or ancymidol (gibberellin biosynthesis inhibitors) to the medium, protocorm-like clusters have been induced in liquid-cultured gladiolus buds and in many species of the complex Brodiaea. In chicory, nodules have been induced on leaves by including IBA and benzylaminopurine (BA) in the medium. In potato and banana, bud clusters are induced when the liquid medium contained a balanced ratio of kinetin and ancymidol. While working with several ornamental species, a several-fold increase in an organogenic biomass of clusters was observed that proliferated in bioreactors and were separated mechanically for further growth before being dispensed to agar-gelled cultures. The presence of BA and an inductive treatment of ancymidol for 24–48 h are required in the liquid medium for the production of clusters in Philodendron. The inductive treatment restricted the carryover dwarfing effect of ancymidol on the development of leaves and shoots after the clusters were transplanted to agar-gelled medium for further plant growth. Gladiolus clusters attained 60% of the vessel volume after 4–5 weeks when cultured in a disposable 2 L bioreactor. As the biomass in the bioreactor increases, the aggregates attaining 0.5–1.5 cm in diameter tend to sink to the bottom of the bioreactor and require an increased rate of aeration to get resuspended. An increase in the rate of aeration from 0.5 to 1.5 vvm (volume air/volume medium per minute) is the required rate for the recirculation of the cluster biomass. The rate of proliferation of potato and banana bud clusters in bubble and disposable plastic bioreactors was faster than that on an agar-solidified medium.

Growth regulators, when provided in a balanced ratio of promoting and retarding substances, are responsible for morphogenic signals and control the growth of the spherical meristematic or bud clusters. Various terms such as pro-PEM in carrot, protocorms in orchids, nodules in poplar, and radiata in pine, nubbins in daylilly, and meristematic or bud clusters in gladiolus, *Brodiaea*, *Nerine sarniensis*, fern, *Philodendron*, potato, and banana have been used to describe the spherical meristematic or bud clusters. The term cluster is used to describe these proliferative and rounded structures, which develop from buds or other proliferative tissues in liquid culture via organogenic or somatic embryogenesis pathways.

Anomalous plant morphogenesis: A liquid medium is required instead of an agar-gelled medium for scaling-up proliferation and biomass production in bioreactors. Anomalous morphogenesis is known to be caused in the culture of plants in liquid medium, resulting in plant hyperhydricity. The plants that develop in liquid media appear glassy and are fragile. They have succulent leaves or shoots and an underdeveloped root system. In liquid cultures, leaves are most severely affected because they have an unorganized mesophyll tissue composed mostly of spongy parenchyma tissue with large intercellular spaces, a deformed vascular tissue, and an abnormal epidermis. Moreover, in hyperhydric leaves, the epidermal tissue lacks a well-developed cuticle and possesses malfunctioning guard cells that fail to respond to closure signals. After transplantation, hyperhydricity causes the loss of *in vitro* developed leaves resulting in wilting and the eventual death of the whole plant.

Photosynthesis and transpiration, the two major processes performed by leaves, are not fully functional in hyperhydric state. This is one of the major reasons behind the poor performance of transplanted plants *ex vitro*. In several plants propagated *in vitro*, the anomalous morphology and anatomical aberration often seen in plants occur owing to deviation from the defined course of morphogenic events that are manifested in plants *in vivo*. Apparently, some of the deformations such as hyperhydricity, abnormal embryos, malformed leaves and shoots, recurrent embryogenesis, and many other disorders are because of the interruption or faulty timing of the signals associated with the normal sequence of organizational events known to exist *in vivo*. These severely manifested problems in liquid medium require further research to understand, help, and control plant morphogenic events in bioreactor cultures.

Plant cell and tissue growth in bioreactors: Aerated liquid cultures in bioreactors provide a better contact between the plant biomass and the medium, eliminate restrictions of gas exchange, regulate the composition of both the medium and the gaseous atmosphere, and allow manipulation of the plant biomass with respect to the medium volume. Optimal growth and sediment prevention take place when there is efficient circulation and mixing of the plant biomass, especially for cluster and embryogenic tissue. Various kinds of bioreactors with mechanical or gas-sparged mixing, involving stirring, circulation, and aeration, have been utilized for plant cell cultures. Mechanically stirred bioreactors are dependent on impellers, including a helical ribbon impeller, magnetic stirrers, or vibrating perforated plates. In bubble column or airlift bioreactors, aeration, mixing, and circulation are done when air enters the vessel from an opening on its side or bottom through a sparger and rises up; as the air bubbles rise, they lift the plant biomass and provide the required oxygen. For effective mixing, aeration, and dispersion of air bubbles in cell suspension cultures, and to prevent the formation of large cell aggregates, mechanically stirred bioreactors are used. The large-scale cultivation of plant cells, embryos, or organs, airlift or bubble column bioreactors and, to a lesser extent, mechanically stirred tank bioreactors have been utilized because of the lower shearing force properties of the former. Plant cells have relatively large size, sensitive cell walls that are highly prone to shearing forces, and are easily damaged. Moreover, the application of bubble column or airlift bioreactors was found to be adequate and advantageous for plant tissue cultured in liquid media as plant cells, unlike microbial cultures, do not need high oxygen content. Besides, it has been shown that because of lower shearing stress, mixing in bubble column or airlift bioreactors through gas sparging and the lack of impellers or blades are far less damaging for clusters than mechanical stirring.

Airlift bioreactors have the following advantages: simple construction of airlift bioreactors, lack of regions of high shear, reasonably high mass and heat transfer, and relatively high yields at low input rates. For embryogenic cell suspensions and providing foam-free cultures, a bubble-free oxygen supply bioreactor with silicone tubing has been found to be suitable. It has been found that an acoustic mist bioreactor significantly increases root biomass in hairy root culture.

For the large-scale, efficient cultures of both somatic embryos and organogenic plant tissues, the configuration and volume of the bioreactor must be determined depending on the mixing and aeration requirements of the specific plant or tissue propagated, as well as for minimizing the intensity of the shear stress. Contamination is a major problem encountered in large-scale liquid cultures. Fungi, bacteria, yeast, and insects are serious contaminants that cause heavy losses of plant material in commercial laboratories. The losses are even greater in scaled-up liquid cultures, because the source of contamination caused by the manipulation of the bioreactor apparatus depends on the many stages of preparation and maintenance of the equipment. In various laboratories, contamination risks are reduced by keeping the operation area sterile through a positive pressure airflow. Additional safeguards include constant screening of the plant tissue for contaminants and continuous indexing.

Physical and chemical factors in liquid cultures: Various parameters such as culture conditions, including gaseous atmosphere, oxygen supply and CO_2 exchange, pH, minerals, carbohydrates, growth regulators, liquid medium rheology, and cell density, must be changed to control plant morphogenesis and biomass growth in bioreactors.

Gaseous atmosphere: The atmosphere of the culture vessel is composed of mainly nitrogen (78%), oxygen (21%), and carbon dioxide (0.036%), and other gases. The volume of the vessel and the extent of ventilation are the two factors that influence the gas composition of the culture vessel. Plants consume carbon dioxide and release oxygen during photosynthesis, whereas evolve carbon dioxide and consume oxygen during respiration. Elevated levels of carbon dioxide were found in cultures during the dark period, and the levels were found to decrease in the light period if photoautotrophic conditions prevail. Additional components of the gaseous atmosphere *in vitro* include ethylene, ethanol, acetaldehyde, and other hydrocarbons. Most of the effects of carbon dioxide, oxygen, and C_2H_4 on plant growth *in vitro* have been reported for agar-gelled or cell suspension cultures. In bio reactors, gas flow can be manipulated to provide the desired levels of oxygen, carbon dioxide, and C_2H_4 and control the gaseous phase. The supplied air in airlift or bubble column bioreactors is used for both mixing and aeration, the importance of which was shown in potato culture. The induction of tubers was inhibited under continuously submerged conditions, and the development of microtubers was seen only after the elongated shoots reached the gaseous phase of the bioreactor. Enrichment with oxygen and manipulation of the hormonal and osmotic conditions were found to have no effect and could not describe the phenomenon. A two-phase culture, which substituted the growth medium with a tuber induction with 9% sucrose in the medium, induced the formation of tubers from shoots that developed above the medium and were exposed to the gaseous phase. These results highlight the significance of the gaseous phase in bioreactors for specific phases of development (Elleuch, Gazeau, David, *et al.* 1998).

Oxygen level: The levels of oxygen in liquid cultures depend on the presence of oxygen in the gas phase above, in the air bubbles inside the medium, and as dissolved oxygen in the medium. Air is released through a sparger located at the base of the bioreactor. The oxygen available for plant cells in liquid cultures, as determined by the oxygen transfer coefficient (kLa values), is the part that dissolves in water. The culture yield is influenced by its depletion as a function of the metabolic activity of the growing cell biomass. As plant cells have a lower metabolic rate compared to microbial cells and a slow doubling time, they require a lower oxygen supply. Normally, high aeration rates decrease the biomass growth. There might be variations in the requirements for oxygen from one species to another and oxygen must be continuously supplied to provide adequate aeration. This is because it influences metabolic activity and energy supply as well as anaerobic conditions. By agitation or stirring and through aeration, gas flow, and air bubble size, the level of oxygen in liquid cultures in bioreactors can be regulated. The application of a porous irrigation tube as a sparger generates fine bubbles, high kLa values, low mechanical stress, and a high growth rate. When the level of oxygen dropped below 10%, the development of poinsettia cell suspension in bioreactors gets inhibited. On elevating the level of oxygen to 80%, the cell number increased to 4.9×105 per mL as compared to 3.1×105 per mL at 40% oxygen. At 78% and 60% oxygen levels, respectively, somatic embryo development increased in alfalfa and poinsettia suspension cultures. It was found that high aeration rates inhibited cell growth in cell suspensions cultured in airlift bioreactors because of the effect of 'stripping' of the volatiles developed by the plant cells, which are apparently essential for cell growth. Under 20% oxygen concentration, the best production of somatic embryos was achieved from embryogenic cultures of *Eschscholzia californica* and carrot in a helical ribbon impeller bioreactor. Low level of oxygen (5%–10%) limited the production of biomass and somatic embryo, whereas high oxygen level (60%) enhanced undifferentiated biomass production. Information on the optimal dissolved oxygen requirements in large-scale liquid cultures can be found only by additional studies.

Effects of CO_2: The effects of carbon dioxide have been observed mainly for plants in agar-gelled cultures or for cell suspension cultures utilized for the production of secondary metabolites. Beneficial promotion of plant growth during plant acclimatization and transplanting *ex vivo* has been suggested by reports on the effects of carbon dioxide enrichment in agar-gelled sugar-free cultures. The role of carbon dioxide supply during the proliferation and multiplication stages in the media supplied with sucrose in bioreactors is debatable. It is implicit that carbon dioxide enrichment beyond 0.36% in the air supply is unnecessary if photoautotrophic conditions do not prevail. Some reports suggest that high aeration rates instead of excessive oxygen levels restricted growth and that the reduced growth might be the result of depletion of carbon dioxide or the removal of several culture volatiles, including carbon dioxide. The requirement for carbon dioxide was not associated with photosynthesis but with some other metabolic pathways responsible for amino acid biosynthesis. In poinsettia, the development of embryos in Erlenmeyer flasks was considerably higher compared to that in bioreactors. It has been proposed that the growth difference was associated with the differences in the gaseous atmosphere in the headspace. Biomass growth was not affected by the carbon dioxide enrichment in an illuminated bioreactor culture of *Brodiaea* clusters. A rise from 0.3% to 1% resulted in a similar growth value under the two carbon dioxide levels and the supply of photosynthetic photon-flux at the rate of 135 μmol/m^2/s. In Cyclamen persicum Mill, a correlation was found between high carbon dioxide levels and an increased production of PEMs.

Ethylene: The level of ethylene in the headspace in liquid cultures in flasks differs from continuously aerated bioreactor cultures. Most reports on ethylene are related to agar-gelled or cell suspension cultures. The rate of aeration affects the level of ethylene in various plant species cultured in bioreactors. The high rates of aeration are often needed at high biomass densities and cause 'stripping' of volatiles that are apparently necessary for some plants grown in culture. Ethylene does not affect the growth in the clusters of *Brodiaea* cultured in a liquid medium. Its level reduces in the presence of silver thiosulphate which is an inhibitor of ethylene action. In highly aerated bioreactor cultures, the level of ethylene was decreased from 0.38 to 0.12 mL/L without any effect on biomass growth, which was identical under both levels.

2.7.2 Consumption of Mineral Nutrients

For most plant species in agar-gelled or liquid cultures *in vitro*, revised MS medium or media with several partial modifications in the inorganic and organic constituents of MS are utilized. The availability of mineral nutrients is dependent on the kind of culture (whether agar-gelled or liquid), type and size of the plant biomass, and physical characteristic of the culture. The rate of absorption of various nutritional constituents depends on factors such as temperature, pH, aeration, light, concentration of minerals, medium volume, and viscosity of the medium. Plant cells developing in liquid cultures have better exposure to the medium components, and the uptake and consumption are quicker.

Liquid shaken and agar-gelled cultures often dehydrate owing to water evaporation to the headspace of the bioreactor and out of the vessel. Such water loss concentrates the medium. In the medium, the content of a certain component is a product of its concentration and the medium volume. The influence of the medium volume and the initial strength on potato have been studied. The level of nutrients in the medium is affected mainly by absorption rate and cell lysis in bioreactors in which either humidified air or condensers are utilized to avoid dehydration. It has been observed that differentiation and proliferation of micropropagated fern, gladiolus, and *Nerine* nodular clusters in bioreactors show better result on half-strength than on full-strength MS minerals (Ziv, unpublished). This was also applicable to Lilium bulblets that differentiated on bulb scales cultured in bioreactors. A decrease in pH to 4.5 or even lower and the subsequent increase to pH 5.5 have been attributed to the initial use of ammonium and the uptake of nitrate at a later stage. In many species, the first limiting factor in the development of biomass and somatic embryos is the depletion of NH^+. In carrot cultures in a helical ribbon impeller bioreactor, increasing the concentration of NH^+ to 15 mM caused maximum production of somatic embryos. A detailed study proved the achievement of nutrient uptake in *E. californica* embryonic cultures in a helical ribbon impeller bioreactor.

2.7.3 Carbohydrate Supply and Utilization

Carbohydrates act as a constant source of energy for plant cultures. The most commonly used carbohydrates *in vitro* are sucrose and, to a lesser extent, glucose, fructose, or sorbitol. In both agar-gelled and liquid cultures, sucrose can be removed rapidly from the medium. For example, from an initial level of 30 g/L, it can be completely depleted or reduced to a concentration of 5–10 g/L after 10–15 days. Sucrose hydrolysis causes the appearance of glucose and fructose in the medium due to sucrose hydrolysis can increase and reach the levels of 5–10 g/L. Cell suspensions of *C. roseus* cultured in a column airlift bioreactor exhibited a lag phase of 5 days, during which complete hydrolysis of sucrose to glucose and fructose resulted. Sucrose was also hydrolysed during the first 5 days in suspension cultures of alfalfa. After Day 5, most of the sugars are taken up and glucose is taken up preferentially over fructose. Researchers obtained a higher yield of alfalfa embryos when 30 g/L maltose combined with NH^+ was utilized instead of 30 g/L sucrose, which was applied in combination with various nitrogen sources. Somatic embryogenesis was enhanced and cell lysis reduced with the addition of mannitol in embryonic suspension cultures of celery. An increased frequency of singulated normal embryos was found with a higher number of embryos when 40 g/L mannitol was added. Embryogenic cultures of *E. californica*, grown in a helical ribbon impeller bioreactor, showed sucrose uptake after 100 h and the concentration was found to deplete after 600–800 h. The level of carbohydrate was found to deplete after 10–14 days in *Picea* species cultured in shaken flasks with a concentration of 30 mM glucose in the medium. A level of 60 mM sucrose resulted in the highest cell biomass and somatic embryo number in mechanically stirred bioreactors. The response of the growth of spruce somatic embryos to various concentrations of carbohydrates was found to be species specific, for example, the biomass of fern meristematic clusters in a bubble column bioreactor was found to increase with an increase in sucrose concentrations from 7.5 to 30 g/L, while still higher levels decreased cluster growth. A decrease in the chlorophyll content of the clusters and leaves was seen with elevated concentrations of sucrose. There was more than 50% decrease in biomass FW in bioreactor-cultured gladiolus clusters as a result of increase in sucrose concentration from 30 to 60 g/L. When 30 or 60 g/L sucrose was mixed with paclobutrazol, a further reduction in FW was noticed. On the contrary, 60 g/L sucrose induced a higher DW increment, which can be the result of an osmotic effect and the subsequent water status of the clusters. When cultured in the presence of growth retardants, gladiolus clusters exhibited a higher level of starch: 845 mg/g as compared to 585 mg/g DW in the control. A greater number of bulblets was produced in bioreactor-cultured bulb scales of Lilium at 30 g/L than at higher sucrose levels. But larger-size microbulbs were produced at 90 g/L than at 30 g/L sucrose in the medium.

2.7.4   Effects of pH

The initial range of pH in most plant cell cultures lies between 5.5 and 5.9. Changes in pH occur mostly during autoclaving and biomass growth in culture as most media are not buffered. A rapid drop in pH to 4.0–4.5 was found to take place within 24–48 h in cell suspension, organ, and embryogenic cultures because of initial ammonium uptake and acidification due to cell lysis. However, pH of the culture increased and reached a stable level of around pH 5.0–5.5 after a few days after the uptake of nitrates. In a culture of spruce species in a liquid medium, the pH levels were shown to increase to 6.5–6.8 after 14 days in culture. In embryogenic cultures of poinsettia, the initial pH levels stabilized between 5.3 and 5.7 after a period of 20 days in batch culture bioreactor. Oxygen levels influenced pH changes by 0.2 pH units. The pH dropped to 4.2 after 24 h and subsequently increased to 5.4 in fern cultured in an airlift bubble column bioreactor. The growth response was not affected by keeping the pH constant by titration with KOH. In alfalfa, the growth of somatic embryos was affected by the pH. A higher rate of embryo production was noticed at a constant pH of 5.5 than at a non-buffered medium or at lower pH levels.

The pH influenced carrot cell differentiation in a controlled bioreactor. At a pH of 4.3, the highest rate of embryo production was noticed. However, the development of embryos ceased before the embryos reached the torpedo stage and continued only at pH 5.8. Changes in the development of carrot embryo have been attributed to sugar uptake and ammonium depletion and can be associated with enzyme and metabolic activity at an optimal pH. Requirements of pH level maintenance appear to be dependent on species and developmental stage.

2.7.5   Effects of Growth Regulator

The proliferation and regeneration potential can be controlled effectively by using growth regulators in liquid cultures than in agar-gelled medium, owing to the direct contact of plant cells and aggregates with the medium. Lack of information on regulators seems to indicate the use of similar levels of growth regulators in both agar-gelled and liquid cultures even though availability in liquid medium appears to be better. In many species, rapid cell division was promoted in an auxin-containing medium where initially somatic embryogenesis was induced. Expression of the embryogenic potential was achieved in auxin-free media, or in a medium with low levels of auxin. Further, addition of ABA promoted the conversion and maturation of embryo into plants, which was found to control abnormal embryo and plant growth. In Lilium cultures, development of bulblets from scales was higher in the presence of BA than of kinetin in bioreactors, but high BA had an inhibiting effect on further growth of bulblets. High level of kinetin was also found to inhibit further growth of bulblets but stimulated bulblets differentiation. However, the addition of 0.1 mg/L naphthaleneacetic acid (NAA) was found to have an enhancing effect on kinetin activity, which was, however, restricted again in the presence of high sucrose levels (90 g/L). Auxins and cytokinins were used to induce pro-embryogenic clusters in embryogenic cultures of *Nerine*. A short exposure to 2-iso-pentenyladenine (2iP) promoted embryogenic expression and further sub-cultured in a growth-regulator-free medium. A major problem in liquid-cultured plants is the malformation of shoots (hyperhydricity); one of the solutions to hyperhydricity was induction of meristematic or bud clusters with arrested leaf growth. Relatively high levels of cytokinin and growth retardants that inhibit gibberellin biosynthesis was found to be the most effective method for reducing the shoot and leaf growth and promoting the formation of meristematic clusters. Information on ABA is scarce and was found to be effective mainly in the later stages of somatic embryo development, which promoted normal embryo growth and maturation of embryogenic cultures in carrot and alfalfa. The growth of callus and leaves in bulblets regenerated on bulb scales of *Lilium* in bioreactor cultures was restricted by ABA, resulting in singulated propagules for easier handling and storage. Effects of ethylene seem to depend on species. Information is not available and further research is required on the effects of growth regulators in bioreactor cultures to optimize culture conditions.

2.7.6   Effects of Temperature

The temperature of the liquid medium inside the bioreactor can be easily controlled with the help of a heating element in the vessel or by water circulation in a jacket enveloped outside the vessel. However, limited information is available regarding the effects of temperature on the cultures present in bioreactor, which is usually kept constant at 25°C, with small day-and-night fluctuations. Studies on the effects of temperature on potato tuber formation in an airlift bioreactor suggested that lower temperature caused a decrease in tuber size. A higher number of larger-sized tubers were observed at 25°C than at 17°C. In potato internode explants in liquid cultures, it was found that tuber formation was best during a 16 h photoperiod and at 18–15°C day-and-night temperature. FW increase was higher at 25°C than at 17°C when bulblets of *Nerine* were cultured in a liquid medium. However, rooting was better at 17°C when sub-cultured from liquid to an agar-gelled bulb induction medium.

2.8   GREENHOUSE TECHNOLOGY

India's food grain production has increased from 51 million tonnes in the 1950s to 206 million tonnes towards the end of the 21st century making it self-sufficient. However, as the agricultural land rapidly shrinks, it makes it difficult to meet the pressure of growing demand

and sustainable food grain production. So soil and crop management practices should be carefully monitored keeping local needs in mind and maintaining the pace of production. These constraints can be overcome with the help of precision farming (also called site-specific farming or prescription farming). This kind of farming uses tools such as global positioning system, geographical information system, remote sensing, and sensors to accurately assess the variability of crop production practices in space–time continuum.

Man has always known crop productivity can be largely improved by wisely modifying the environment. For example, the Romans knew that light-transmitting shelters provided a suitable environment. Moreover, controlled-environment agriculture (CEA) has been in practice for a long time. In precision farming, greenhouse technology has been considered the main part. In Indian horticulture greenhouse technology can increase the production of fresh fruits, vegetables, and flowers, thereby ensuring better nutrition and improved standard of living.

At present, greenhouse technology is considered a part of CEA, as it requires precise control of air, temperature, water, humidity, plant nutrition, carbon dioxide, and light. Greenhouse technology is 'framed or an inflated structure with a transparent or translucent material in which crops could be grown under at least partially controlled environment and which is large enough to permit persons to work within it to carry out cultural operations'.

Growth factors such as light, humidity, temperature, and carbon dioxide concentration control the microclimate of a greenhouse. During the cultivation period, these factors are controlled to the optimum level, which enhances productivity manifold.

2.8.1 Greenhouse Production

State-of-the-art protected cultivation technology ushered in India following the liberalized seed policy of the late 1980s, globalization of the Indian economy, and the economic reforms initiated in the early 1990s. However, greenhouse technology was used in commercial agriculture only after the technology improved in the late 1990s which enabled the products to be sold at competitive prices. Since the application of plastics in agriculture, greenhouse technology became a breakthrough in agricultural technology. Now market-driven quality parameters have been integrated with production system profits. Moreover, technology has advanced to the extent that a climate-control computer can produce the desired microclimate for a particular crop, keeping the benefit–cost ratio more than 1 (Table 2.1 lists areas under greenhouse in different countries).

In spite of its delayed entry, the Indian greenhouse industry has expanded rapidly with about 300–500 ha of area currently under protected cultivation. Copious sunshine during the year, especially in autumn and winter, helps the year-round production, without the need for artificial lighting and the increased cost of additional energy input. However, in the greenhouse, high summer temperatures have to be reduced with appropriate cost-effective cooling methods to boost the year-round production.

2.8.2 Design of Greenhouses

Depending on the highly varying climatic conditions in different parts of India, greenhouses and their supporting facilities have

Table 2.1 Area under greenhouse in different countries

Country	Area (ha)
Algeria	500
Australia	600
Belgium	2,400
Bulgaria	1,350
Canada	400
Chile	1,600
China	48,000
Columbia	2,600
Egypt	1,000
England	3,500
France	5,800
Greece	4,240
Hungary	5,500
India	500

to be built. The greenhouse designs must conform to local agro-climatic conditions as well as economic constraints. For example, naturally ventilated polyhouses are suitable for the southern plateau and the coastal regions, whereas both cooling and heating facilities are required in the northern plains due to composite climate. The costs of the structure, cladding, and temperature-control mechanism govern the initial cost of setting up a greenhouse. The operational cost depends primarily on the cost of maintaining temperature and humidity in the greenhouse. See Figure 2.1.

2.8.3 Low-cost Greenhouses for Nursery and Off-season Cultivation of Vegetables

For producing healthy seedlings and fresh vegetables at lower costs, nurseries and off-season cultivation of vegetables in cost-effective greenhouses have to be increased. Farmers have to fine tune greenhouse technology, structure, and design for developing high-quality planting materials that suit local climatic patterns. They also need to utilize suitable technology for producing healthy seedlings in plastic-perforated trays. For optimum plant growth, they can standardize new growing media such as soil-less culture.

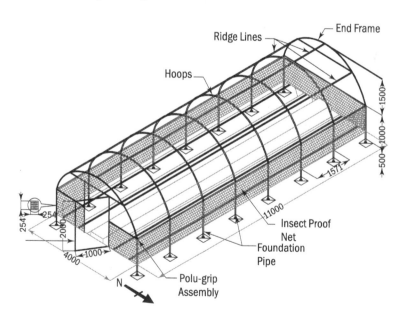

Fig. 2.1 *Greenhouse designs suitable for subtropical India*

2.8.4 Greenhouse Floriculture Production

The international demand for floriculture products has grown, especially for flowers produced through cut greenhouse technology, which has registered an annual growth of more than 11%. In view of this increasing demand, India has given floriculture a priority status in the export-oriented market of precision farming. Protected cultivation and special attention are required for producing cut flowers for export from India. But to meet sustainability requirements, international standards for cut flowers and automatic environmental gadgets are required. However, for following the 'hi-tech necessary steps to maintain the glory of floriculture' as a specialized industry in India, hand-to-hand support from the greenhouse industry is required. The imported modern greenhouse technology requires considerable amendment in the design to ensure cost-effective production and to remain competitive in the global market.

2.8.5 Greenhouse Cultivation in Precision Farming Development Centre, IIT Kharagpur

The Precision Farming Development Centre (PFDC), IIT Kharagpur, includes the design and development of low-cost plastic greenhouses for seedlings and the off-season cultivation of vegetables and their yield responses. See Figure 2.2. The centre has also estimated the water requirement for different vegetable crops. Moreover, it has initiated a detailed research to study the design aspects of greenhouse floriculture suitable for the climatic conditions in India's eastern parts. The centre has also undertaken studies on the variation of greenhouse microclimate as well as on the benefits of cost-effective cooling methods in reducing high summer temperatures. The comparison of yield in open filed, greenhouse and hydroponic systems are given in Table 2.2.

In maintaining the pace of precision farming in India, greenhouse technology has a tremendous role to play. Protected cultivation of vegetables and flowers has two-fold benefits: increased sustainability of agricultural production and improvement in the standards of living. The following factors can sustain the advantages of the greenhouse industry:

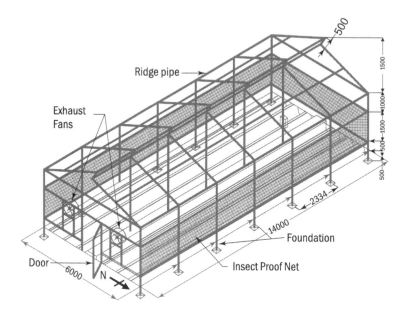

Fig. 2.2 *Greenhouse production of Gerbera in greenhouse at PFDC, IIT Kharagpur*

1. Improvement in domestic market facilities so that high prices can be offered for greenhouse products.
2. Exploration of high-value product alternatives such as propagating materials of export-oriented crops.
3. Development of vertical integration and joint ventures that Indian growers can adopt.
4. Establishment of support systems, including the government, universities, and the private sector, to help growers and the industry.

Table 2.2 Comparison of yield in open field, greenhouse, and hydroponic system

| | Yield (tonnes/ha) | | |
Crop	Greenhouse	Open field	Hydroponic
Tomato	150	50	187.5
Cucumber 180 8 250	180	8	250
Capsicum 110 100	110	100	—
Broccoli	15	7	—

In many plant species, meristematic centres are formed directly on the explant, with little or no callus formation, and can grow into either shoots or somatic embryos. After the initial stages of controlled re-differentiation, plant regeneration can follow through either organogenic or embryogenic pathways. The manipulation of the inorganic and organic constituents in the medium and the type of explant and the species regulate plant regeneration *in vitro*. In many species, successful regeneration from the callus or directly from the explants happens after a series of sub-cultures in different media, in a sequence often specific to the species, variety, or newly introduced genotype. The determining factors include the combination of concentration with respect to medium volume, composition of growth-stimulating and growth-inhibiting regulators in the medium, physiological status and competence of the cells, and ability of cells for morphogenetic expression. In agriculture and forestry, identical pathogen-free plants are produced by micropropagation. The *in vitro* propagation of axillary and/or adventitious buds as well as somatic embryos is known as micropropagation. But the technique is still expensive because of intensive hand manipulation of various culture phases and is not commercially used for all plant species. In addition, the initial stage of establishment and response is slow in some plants, and plant survival in the final stage *ex vitro* is often poor. This further decreases the micropropagation production potential (Dumet, Engelmann, Chabrillange, *et al.* 1994).

For efficient commercial micropropagation, there should be rapid and extensive proliferation along with the application of large-scale cultures for the multiplication phase. Moreover, normal development of plants during the acclimatization and hardening stage is necessary to ensure survival after transplanting to the greenhouse. The limitation imposed by existing conventional labour-intensive methods can be overcome by mechanization and automation of the micropropagation process. Special attention has been given to automation of the repeated cutting, separation, subculture, and transfer of buds, shoots, or plantlets during the multiplication and transplanting phases. Progress in the automation of tissue culture will be dependent on the application of liquid cultures in bioreactors. This will make possible fast proliferation, mechanized cutting, separation,

REFERENCES

Dumet, D., F. Engelmann, N. Chabrillange, and Y. Duvall. 1994. Effect of desiccation and storage temperature on the conservation of cultures of oil palm somatic embryos. *CryoLetters* 15: 85–90

Elleuch, H., C. Gazeau, H. David, and A. David. 1998. Cryopreservation does not affect the expression of a foreign same gene in transgenic *Papaver somniferum* cells. *Plant Cell Rep.* 18: 94–98

Engelmann, F. 2004. Plant cryopreservation: Progress and prospects. *In vitro Cell and Dev. Biol.-Plant* 40: 427–33

Florin, B., E. Brulard, and B. Lepage. 1999. Establishment of a cryopreserved coffee germplasm bank. Abstract Book 'Cryo'99, *World Congress of Cryobiology*'. Marseille, France, 12–15 July, p. 167

Gonzalés-Benito, M. E., J.M. Iriondo, J. M. Pita, and F. Perez-Garcia. 1995. Effects of seed cryopreservation and priming on germination in several cultivars of *Apium graveolens. Ann. Bot.* 75: 1–4

Hitmi, A., H. Sallanon and C. Barthomeuf. 1997. Cryopreservation of *Chrysanthemum cinerariaefolium* Vis. cells and its impact on their pyrethrin biosynthesis ability. *Plant Cell Rep.* 17: 60–64

Hunter, B. G. , L. A. Heaton, C. E. Bracker, and A. O. Jackson. 1986. Structural comparison of Poa semilatent virus and barley stripe mosaic virus. *Phytopathology* 76: 322

Kilby, N. J. and C. S. Hunter. 1986. *Proc. 6th Int'l Cong. of Plant Tissue and Cell Culture*, p. 352 , Univ. of Minnesota

Meijer, E. G. M., F. Vaniren, E. Schrijnemakers, L. A. M. Hensgens, M. Vanzijderveld, and R. A. Schilperoort. 1991. Retention of the capacity to produce plants from protoplasts in cryopreserved cell lines of rice (*Oryza sativa* L.). *Plant Cell Rep.* 10: 171–74

Sakai, A., D. Hirai, and T. Niino. 2009. Development of PVS-based vitrification and encapsulation- vitrification protocols. In *Plant Cryopreservation: a practical guide*, B. Reed (ed.), pp. 33-57. New York: Springer

Veltkamp, E. and N. M. Mol. 1986. The role of chalcone synthase in the regulation of flavonoid biosynthesis in developing oat primary leaves. *Arch. Biochem. Biophys.* 250: 364–72

Zenk, M.H., M. Rueffer, M. Mann, and B. Deusneumann. 1985. Benzyl-isoquinoline biosynthesis by cultivated plant cells and isolated enzymes. *Journal of Natural Products* 48: 725–38

Ziv, M. and D. Shemesh. 1996. Propagation and tuberization of potato bud clusters from bioreactor culture. *In vitro Cell. Dev. Biol. Plant.* 32: 31–36

Plant Transformation Technology

3.1 INTRODUCTION

The most common method for inserting alien genes into plant cells and subsequently regenerating transgenic plants is plant transformation mediated by *Agrobacterium tumefaciens*. *A. tumefaciens* is a soilborne plant pathogenic bacterium that infects wound sites in dicotyledonous plants and causes crown gall tumours. More than 90 years ago, the first evidence emerged indicating that this bacterium causes crown gall tumours. Since then many researchers have studied this neoplastic disease and its causative pathogen for various reasons. Initially, for a long period of time, scientists attempted to discover the mechanisms of crown gall tumour induction. Their aim was to understand the mechanisms of oncogenesis in general and ultimately use this knowledge for developing drugs for the treatment of cancer in humans and animals. However, this hypothesis was discarded, and it became evident that this tumour resulted from gene transfer to infected plant cells from *A. tumefaciens*.

A particular DNA segment (T-DNA) of the tumour-inducing (Ti) plasmid is transferred into the nucleus of infected cells by *A. tumefaciens* and it is then stably integrated into the host genome and transcribed, causing the crown gall disease. Two kinds of genes are found in T-DNA: (i) oncogenic genes, which encode for enzymes involved in the synthesis of auxins and cytokinins and are responsible for tumour formation, and (ii) genes that encode for the synthesis of opines. The compounds opines are produced by the condensation of amino acids and sugars and excreted by the crown gall cells, which are then consumed by *A. tumefaciens* as carbon and nitrogen sources. The genes for opine catabolism are located outside the T-DNA. These genes are involved in the transfer of T-DNA from the bacterium to the plant cell and are also involved in the bacterium–bacterium conjugative plasmid transfer.

Diseases known as crown gall and hairy roots are caused when virulent strains of *A. tumefaciens* and *A. rhizogenes*, respectively, interact with susceptible dicotyledonous plant cells. These strains contain a large megaplasmid (more than 200 kb) that is responsible for tumour induction. It is called Ti plasmid for *A. tumefaciens* and Ri in the case of *A. rhizogenes*. Ti plasmids are categorized based on opines, which the induced tumours produce and excrete. During infection, T-DNA (a mobile segment of Ti or Ri plasmid) gets integrated into the plant chromosome after it is transferred to the plant cell nucleus. The T-DNA fragment is flanked by 25 bp direct repeats that act as a *cis* element signal for the transfer apparatus. The cooperative action of proteins encoded by genes determined in the Ti plasmid virulence region (*vir* genes) and in the bacterial chromosome mediates the T-DNA transfer process. Also contained in the Ti plasmid are genes for the catabolism of opines produced by crown gall cells. Besides, Ti plasmid contains regions for conjugative transfer and for its own integrity and stability. The 30 kb virulence (*vir*) region is a regulon organized in six operons: *virA*, *virB*, *virD*, and *virG* are essential for the T-DNA transfer and *virC* and *virE* are essential for increasing the transfer efficiency (Hookyaas 1998). Various genetic elements determined by chromosomes have demonstrated their functional role in the attachment of *A. tumefaciens* to plant cells and in the colonization of bacteria: Loci *chvA* and *chvB* are involved in the excretion and synthesis of β-1,2 glucan (Cangelosi, Martinetti, Leigh Lee, *et al.* 1989); *chvE* is needed for the sugar enhancement of *vir* genes induction and bacterial chemotaxis; cell locus synthesizes cellulose fibrils; pscA (exoC) locus has a role in the synthesis of cyclic glucan and acid succinoglycan (Cangelosi, Hung, Puvanesarajah, *et al.* 1987; Cangelosi, Best, Martinetti, *et al.* 1991); and *att* locus is associated with cell surface proteins.

The initial results of the research on the process of T-DNA transfer to plant cells highlight three important facts:

1. The formation of tumour is a process in which plant cells are transformed because of the transfer and integration of T-DNA and the eventual expression of T-DNA genes.

2. T-DNA genes are transcribed only in plant cells and do not influence the transfer process.

3. It is possible to transfer any foreign DNA placed between T-DNA borders to plant cells, regardless of where it originates from.

These facts helped in the construction of the first bacterial and vector strain systems for plant transformation.

At the start of the last decade, the first information on transgenic tobacco plant expressing foreign genes appeared, although many molecular characteristics of this process were not known at that time (Herrera-Estrella, De Block, Messens, *et. al.* 1983). Since that breakthrough in the development of plant science, a significant progress has been achieved in understanding the process of transfer of genes to plant cells mediated by *Agrobacterium*. However, only dicotyledonous plants are naturally infected by *A. tumefaciens,* and a number of plants that are economically important, such as cereals, remained accessible for genetic manipulation for a long period. Alternative direct methods for transformation have been developed for these cases, for example, gene gun technology, polyethylene glycol-mediated transfer, microinjection, and protoplast and intact cell electroporation. But the transformation mediated by *Agrobacterium* has considerable advantages over direct transformation methods. *Agrobacterium*-mediated transformation reduces the copy number of the transgene, consequently minimizing problems with transgene co-suppression and instability. Besides, as this is a single-cell transformation system, it does not form mosaic plants, which are common with direct transformation.

Until recently when efficient, reproducible methodologies were developed for rice, corn, wheat, banana, and sugarcane, *Agrobacterium*-mediated gene transfer into monocotyledonous plants was not possible. During the last years, studies have been carried out on plant transformation using *A. tumefaciens* and the molecular mechanisms involved. Recently, detailed analyses of the strategies for practically applying this methodology have been published. This chapter presents the latest information on the mechanisms of *A. tumefaciens*-mediated gene transfer. The application of this method to the transformation of monocotyledonous plants has been analysed, taking sugarcane as a model.

3.1.1   *A. tumefaciens* T-DNA Transfer

The transfer of genes from *A. tumefaciens* to plant cells involves several necessary steps: (i) colonization of bacteria, (ii) induction of bacterial virulence system, (iii) formation of T-DNA transfer complex, (iv) transfer of T-DNA, and (v) integration of T-DNA into plant genome. This section presents a hypothetical model that depicts the most important stages of this process, supported by the recent experimental data and accepted hypothesis on T-DNA transfer bacterial colonization.

The earliest and essential step in the induction of tumour is bacterial colonization, which occurs when *A. tumefaciens* attaches itself to the plant cell surface. Research on mutagenesis has revealed that mutants that do not attach lose the tumour-inducing capacity. In the colonizing process, polysaccharides of *A. tumefaciens* cell surface play an important role. The attachment of bacteria can be prevented by applying lipopolysaccharides (LPS) solution from virulent strains to the plant tissue before interaction with virulent bacteria. An integral part of the outer membrane of bacteria, LPS contain the lipid A membrane anchor and the O-antigen polysaccharide. Similar to other plant associative rhizobiaceae bacteria, *A. tumefaciens* also produces capsular polysaccharides (K antigens), which lack lipid anchor and have a strong anionic nature and tight association with the cell. Some evidences indicate that capsular polysaccharides might have a specific role while interacting with the host plant. It was noticed in the case of *A. tumefaciens* that the attachment of the wild-type bacterium to plant cells and the production of an acidic polysaccharide were directly correlated.

The genes required for the successful attachment of bacteria to the plant cell are contained in the chromosomal 20 kb *att* locus. Using transposon insertion mutants, this locus has been extensively researched. If the culture medium was conditioned beforehand by incubating wild-type virulent bacterium with plant cells, insertions in the left 10 kb side of this region generated avirulent mutants that could restore its attachment capacity. This interaction produced and accumulated a complement that was absent in the mutant strain; this conditioned the medium and led to the attachment of the strain to the plant cell.

Contrary to this, when there was the mutational insertion in the 10 kb right side of the *att* locus, attachment capacity was irreversibly lost, which the conditioned medium could not restore. These results indicate that genes located at the left side of *att* locus influence molecular signalling events, while the right-side genes probably synthesize the fundamental components. The left side of *att* locus suggests that an operon consists of nine open reading frames (ORFs). Four of these ORFs show homology to the genes involved in the so-called periplasmic binding protein-dependent (or ABC) transporter system. The analysis of mutants proved the failure in producing and accumulating specific compounds critical for bacterial attachment. There is a possibility of the involvement of the ABC transporter encoding genes in the secretion of these substances or in the introduction into bacteria of some plant-originated activators for synthesis of the compound specific for attachment.

3.1.2 Induction of Bacterial Virulence System

Products encoded by the 30–40 kb *vir* region of the Ti plasmid mediates the T-DNA transfer. This region comprises six essential (*virA, virB, virC, virD, virE,* and *virG*) and two non-essential (*virF* and *virH*) operons. The number of genes in every operon is different: *virA, virG,* and *virF* have only 1 gene each; *virE, virC,* and *virH* have 2 genes each; and *virD* and *virB* have 4 and 11 genes, respectively. The only constitutive operons *virA* and *virG* code for a two-component (VirA–VirG) system and activate the transcription of other *vir* genes. The VirA–VirG system shares functional and structural similarities with other systems already discussed for other cellular mechanisms. VirA is a transmembrane dimeric sensor protein that detects signal molecules, mostly small phenolic compounds, released by wounded plants. Among the signals for VirA activation are acidic pH, phenolic compounds (such as acetosyringone), and certain monosaccharides that act with phenolic compounds in a synergistic manner. Structurally, VirA protein has three domains: one periplasmic or input domain and two transmembrane domains (TM1 and TM2). The periplasmic domain is essential for detecting monosaccharides. The domains TM1 and TM2 act as a transmitter (signalling) and a receiver (sensor), respectively.

An amphipathic helix with strong hydrophilic and hydrophobic regions is within the periplasmic domain adjacent to the TM2 domain. This structure folds the protein to be simultaneously aligned with the inner membrane and anchored in the membrane. This structure is characteristic for other transmembrane sensor proteins. TM2 is the kinase domain that has a significant role in VirA activation, phosphorylating itself on a conserved His-474 residue in response to signalling molecules released by wounded plant sites. Detection of monosaccharides by VirA is a key amplification system, which responds when the levels of phenolic compounds are low. Only

ChvE, the periplasmic sugar (glucose/galactose) binding protein that interacts with VirA, makes the induction of this system possible. The amphipathic helix is exposed to small phenolic compounds because of this interaction, which suggests a putative model for the VirA–ChvE interaction.

Activated VirA transfers its phosphate to a conserved aspartate residue of VirG, a cytoplasmic DNA-binding protein. Functioning as a transcriptional factor, VirG controls the expression of *vir* genes when it is phosphorylated by VirA. The C-terminal region influences DNA binding and shows virulence homology with the VirA receiver (sensor) domain. External factors, such as temperature and pH, also influence the activation of the *vir* system. At temperatures above 32°C, expression of *vir* genes stops because a conformational change in the folding of VirA inactivates its properties. A mutant form of VirG (VirGc) suppresses the effect of temperature on VirA. VirG (VirGc) activates the constitutive expression of *vir* genes. However, at that temperature, this mutant cannot confer to *Agrobacterium* the virulence capacity. This is likely because high temperature also affects the folding of other proteins that actively participate in the T-DNA transfer process.

3.2 HAIRY ROOT

Chemicals derived from plants are valuable sources for flavours, oils, dyes, pharmaceuticals, and resins. A large number of these economically valuable phytochemicals are secondary metabolites not necessary for plant growth. These are produced in small quantities and often accumulate in specialized tissues, such as trichomes. These compounds usually have complicated structures and exhibit chirality. Therefore, their organic synthesis is not cost-effective in many cases. Extraction from field grown plants has been the major method of producing these important secondary metabolites economically. To obtain a crop, traditional agricultural methods often require months to years depending on the species of plant. Moreover, many factors, including pathogens and climate changes, influence the levels of secondary metabolites.

For producing valuable secondary metabolites, plant cell suspension culture offers an alternative to agricultural processes. Some industrial-scale commercialized plant cell cultures are available.

However, producing secondary metabolites from cultures of plant cell suspension is the biggest challenge because secondary metabolites are mostly generated by specialized cells and at distinct stages of development. If the cells remain undifferentiated, some compounds are not synthesized (Berlin, Beier, Fecker, *et al.* 1985). Hence, undifferentiated plant cell cultures often partially or totally lose their biosynthetic capacity to gather secondary metabolites. Although some reports are available on the use of co-cultured differentiated tissues (e.g., shoots and roots) for producing secondary metabolites, most efforts are focused on transformed (hairy) roots.

Hairy roots are the result of genetic transformation by *A. rhizogenes*. For secondary metabolite production, these roots have suitable properties. They do not need hormones in the medium and often grow as fast as or faster than plant cell cultures. Hairy roots have the biggest advantage in that their cultures often show almost the same or greater biosynthetic ability for the production of secondary metabolite as their mother plants. A large number of secondary metabolites are synthesized *in vivo* in roots, and often the synthesis is related to root differentiation. Even when secondary metabolites accumulate only in the aerial part of an intact plant, it has been shown that hairy root cultures accumulate the metabolites. For instance, although accumulation of lawsone normally occurs only in the aerial part of a plant, hairy roots of *Lawsonia inermis* grown in half- or full-strength MS medium are found to produce lawsone under dark conditions. It was thought that artemisinin accumulates only in the aerial part of *Artemisia annua*, but many laboratories have demonstrated that hairy roots have the ability to produce artemisinin.

3.3   STRUCTURE AND FUNCTIONS OF TI AND RI PLASMIDS AND THEIR APPLICATION

The vectors that are most commonly used for transferring gene in higher plants are based on the tumour-inducing mechanism of *A. tumefaciens*, which causes crown gall disease. *A. rhizogenes* is a closely related species and it causes hairy root disease. Understanding the molecular basis of these diseases has enabled the use of these bacteria for developing gene transfer systems. The transfer of a DNA segment to the plant nuclear genome from the bacterium has been reported to cause the disease. The transferred DNA segment is called T-DNA and it is part of a large Ti plasmid found in virulent strains of *A. tumefaciens*. Similarly, virulent strains of *A. rhizogenes* contain Ri (root-inducing) megaplasmids.

Ti plasmids contain the following four regions:

1. Region A has the T-DNA. A mutation in this region causes tumours with altered morphology (shooty or rooty mutant galls). Sequences that are homologous to this region are always transferred to plant nuclear genome. Hence, the region is called T-DNA (transferred DNA).
2. Region B is responsible for replication.
3. Region C is responsible for conjugation.
4. Region D is responsible for virulence. As a mutation in this region abolishes virulence, this region is called virulence (*vir*) region. It plays a key role in transferring T-DNA to the plant nuclear genome. Efficient plant transformation vectors have been developed using the components of this Ti plasmid.

T-DNA consists of a region consisting of three genes: two genes *tms1* and *tms2* represent 'shooty locus' and one gene *tmr* represents 'rooty locus'. The biosynthesis of two phytohormones—indole acetic acid or IAA (an auxin) and isopentyladenosine 5'-monophosphate (a cytokinin)—occurs in this region.

Almost perfect 25 bp direct repeat sequences flank the T-DNA regions on all Ti and Ri plasmids. These sequences are necessary for T-DNA transfer and act only in the *cis* orientation. Flanked by these 25 bp repeat sequences in the correct orientation, any DNA sequence can be transferred to plant cells. This property has been used successfully for the production of transgenic plants by *Agrobacterium*-mediated gene transfer in higher plants. In addition to 25 bp flanking border sequences (with T-DNA), the *vir* region is also essential for the expression of Ti plasmid gene in vegetative *Agrobacterium* before the plant phenolic compound acetosyringone induces the *vir* gene.

The *vir* region can function even in the *trans* orientation, whereas border sequences function in the *cis* orientation with respect to T-DNA. Therefore, the physical separation of T-DNA and *vir* region on two different plasmids does not affect T-DNA transfer, provided both plasmids are located in the same *Agrobacterium* cell. This property has been utilized in designing vectors for transferring genes in higher plants. Of the six operons of the *vir* region, four operons (except *virA* and *virG*) are polycistronic. Genes *virA*, *virB*, *virD*, and *virG* are crucial for virulence; the genes *virC* and *virE* are required for the formation of tumour.

Under all conditions, the *virA* locus is constitutively expressed. In vegetative cells, although the *virG* locus is expressed at low levels, it is quickly induced to higher expression levels by exudates found in wounded plant tissue. The expression of other *vir* loci is regulated by the *virA* and *virG* gene products. The *virA* product (VirA) is present on the inner membrane of *Agrobacterium* cells. It is probably a chemoreceptor that senses the presence of phenolic compounds (found in exudates of wounded plant tissue), such as acetosyringone and b-hydroxyaceto-syringone. Transduction of signal proceeds via VirG (product of gene *virG*) activation (possibly phosphorylation), which in turn induces the expression of other vir genes.

3.4 *AGROBACTERIUM*-MEDIATED GENE TRANSFER, *AGROBACTERIUM*-BASED VECTORS, AND THEIR APPLICATION

The only organisms known to be capable of inter-kingdom gene transfer are *Agrobacterium* species. This soilborne Gram-negative bacterium induces formation of tumours in most dicotyledonous and some monocotyledonous species. These tumours do not require the continuous presence of bacteria for proliferation, which indicates that plant cells are genetically transformed. The bacterium can be utilized to obtain functional products using engineered DNA segments of interest. In this process, the desired DNA segment is cloned into the T-DNA region of 'disarmed' plasmid, which is then introduced into *Agrobacterium* and subsequently transferred to plants. From these 'disarmed' plasmids, genes responsible for tumour formation are removed. This ensures that the transformed cells are regenerated into fertile plants that transmit the desired DNA fragment to their progeny. *Agrobacterium* can be used for the transformation of recalcitrant plants, fungi, and human cells under laboratory conditions. *Agrobacterium*-mediated transformation is regarded as a valuable model system for the study of host–pathogen recognition and delivery of macromolecules to target cells.

The interaction of *Agrobacterium* with plant cells involves the following steps: recognition, expression of virulence (*vir*) gene, attachment to the host cell, targeting of Vir factors and T-DNA into the host cell, and chromosomal T-DNA integration (Figure 3.1). Expression of *vir* gene is induced in *Agrobacterium* after its exposure to phenolic compounds from plants. This chemical recognition, after physical interaction of bacterium and plant cells, induces the formation of bacterial transfer machinery. The whole machinery facilitates the transfer of T-DNA strand into the host plant cell along with a number of Vir factors. After entering the host plant cell, the T-DNA is translocated into the nucleus, which on expression reprogrammes plant cells for a tumour phenotype.

3.4.1 Recognition of Plant Cells as Host by *Agrobacterium*

The presence of Ti plasmid is a selective advantage of *Agrobacterium* species; some strains lack this plasmid and are able to survive independent of plant host. However, because plant transformation is a complex process and energetically demanding, *vir* gene expression must be carefully regulated. Identification of *vir* genes, which are required for virulence but lie outside the T-DNA, was a major step towards understanding the transformation process. With the exception of *virA* and *virG*, the *vir* genes were found to be essentially silent unless the bacteria are cultured with plant cells. Induction of *vir* gene depends on molecules exuded by the plant. For transformation, attachment to plant cells is necessary and it is mediated by *Agrobacterium* genes. Thus, the host recognition by *Agrobacterium* resulting in transformation involves two independent processes: *vir* gene activation and attachment to the host cell.

3.4.2 *Agrobacterium vir* Gene Expression

For the *vir* gene activation by plant factors, the genes *virA* and *virG* are needed. These two genes are constitutively expressed at a basal level, but they can be highly induced in a feed-forward manner. The *virA* and *virG* genes encode a two-component phospho-relay system in which VirA is a membrane-bound sensor and VirG is the intracellular response regulator. On signal sensing, VirA, a histidine kinase, transfers its phosphate to a particular aspartate of VirG, thereby activating VirG to function as a transcription factor. Then

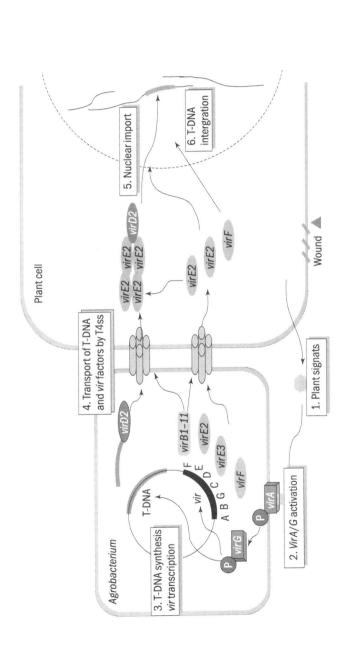

Fig. 3.1 *Overview of the Agrobacterium–plant interaction. Plant signals (1) induce VirA/G activation (2), and as a result there is T-DNA synthesis and vir gene expression in Agrobacterium (3). (4) Through a bacterial-type IV secretion system (T4SS), T-DNA and Vir proteins are transferred into the plant cell to assemble a T-DNA/Vir protein complex. (5) The T-DNA complex is imported into the host cell nucleus. (6) T-DNA becomes integrated into the host chromosomes by an illegitimate recombination.*

phosphorylated VirG binds at specific 12 bp DNA sequences of *vir* gene promoters (*vir* boxes) and activates transcription (Brencic and Winans 2005). The signals perceived by VirA are phenols, low pH, aldose monosaccharides, and low phosphate. For *vir* gene induction, phenols are indispensable. Other signals sensitize VirA to phenol perception. For example, sugars allow induction of the VirA/VirG system at much lower phenol concentrations and increase the response by several folds. Phenols, such as acetosyringone, were identified as the inducers of *vir* gene expression through analysis of root exudates and leaf protoplasts. Now acetosyringone is regularly being used for increasing plant transformation efficiency mediated by *Agrobacterium*. The ability of the VirA/VirG system to detect phenols and sugars is the reason for the broad host range exhibited by *Agrobacterium*.

3.4.3 Plant Entry Sites for *Agrobacterium*

Wounded plant sites mainly act as entry points for *Agrobacterium*. Although a wound site may be a portal of entry, other processes that occur at these sites can facilitate transformation. For example, compounds released by the wound, such as phenols and sugars, induce *vir* gene expression. Sugars act as chemotactic attractants of *Agrobacterium* for *vir* gene activation. Thus, wound-specific features, such as low pH, a high activity of the phenylpropanoid pathway, and sugars associated with cell wall synthesis/wound repair, correlate with enhanced transformation frequency and efficiency. Even though transformation can take place in unwounded plants (with *Agrobacterium* cultures grown in pre-induction medium), it seems that *Agrobacterium* has optimized the VirA/VirG system for responding to signals from wound sites. Cell division at wound sites is equally essential for transformation. However, cells located in the root elongation zone have been found to be highly transformable. Cells in this non-meristematic zone do not undergo a normal cell cycle, but they undergo endoreduplication.

3.4.4 Host Cell Attachment by *Agrobacterium*

After the *vir* gene activation, an association between pathogen and host cell is a prerequisite for both the transfer of T-DNA and proteins and for the transformation to take place. Quantitative binding assays have revealed an easily removed non-specific interaction and a specific interaction. The specific attachment of *A. tumefaciens* to plant cells does not depend on the Ti plasmid. Rather, this attachment is facilitated by the chromosomally encoded bacterial genes *chvA*, *chvB*, and *pscA* (*exoC*). These genes are involved in the synthesis and localization of periplasmic b-1,2 glucan.

From early studies, it can be found that when *A. tumefaciens* cells are exposed to soluble pectic cell wall fractions of the plant, both specific binding of *Agrobacterium* to plant cells and tumour induction frequencies decrease. This indicates the presence of the elusive *Agrobacterium* receptor-like components. BTI domain proteins are possible candidates, which have been separated from a screen for VirB2-interacting proteins. A direct contact of BTI1 with *Agrobacterium* T-pilus is proposed in the initial interaction of *Agrobacterium* with plant cells because of its (BTI1) transient increase just after *Agrobacterium* infection and its preferential localization to the periphery of root cells.

New insights into possible plant molecules involved in the attachment process have emerged from genomic studies. Many *Arabidopsis* mutants that are recalcitrant to *Agrobacterium* transformation (rat mutants) have been isolated. That *Agrobacterium* can no longer efficiently bind to some rat mutants has been proved. A well-characterized mutant is affected in the gene that encodes a cell wall arabinogalactan protein to which bacteria bind poorly. The process of recognition and the physical interaction between *Agrobacterium* and host cells are revealed when the mutants are analysed.

3.4.5 *Agrobacterium* Secretion of T-DNA and Vir Proteins into Plant Cells

Following the activation of *vir* gene and the attachment of *Agrobacterium* to plant cells, a VirB–VirD4 transporter complex is formed that helps Vir proteins and T-DNA to cross the peptidoglycan layer, the outer membrane, and the inner bacterial membrane, as well as the plant host cell wall and membrane. The VirB complex belongs to the class of T4SS, which are found in a broad range of Gram-negative bacteria and are involved both in the conjugative transfer of plasmids between bacteria and in the translocation of Vir factors from pathogens to host cells during infection. The VirB complex is essential for virulence and consists of at least 12 proteins (VirB1–VirB11 and VirD4). The proteins associate with the cell envelope and form a multi-subunit envelope-spanning structure. VirD2–T-DNA, VirE2, VirE3, VirF, and

VirD5 are the bacterial factors transported into host cells by the VirB complex. VirD2 nicks the T-DNA at 25-nucleotide-long repeats that border T-DNA and VirD2 and then becomes covalently bound to the 50-end of T-DNA. VirD2 is transported with the T-strand into the plant cell, where it is involved in nuclear import and integration of T-DNA into the host genome. VirE2 coats the T-strand along its entire length *in vitro* and is a single-stranded DNA-binding protein. It possibly interacts with T-DNA in the cytoplasm of the plant cell and also plays a role in nuclear import and integration. To plants or other *Agrobacterium* species, the VirB/D4 complex can transport not only the Ti derived T-DNA but also the broad-host-range plasmid RSF1010, which indicates that conjugative intermediates should also be the substrates.

3.4.6 Import of *Agrobacterium* Vir Factors into Host Cells

Substrates are transported across the bacterial cell envelope by *A. tumefaciens* virB encoded T4SS. For exporting targeted substrates, certain C-terminal motifs are required. These export signals mediate the interaction between substrates and T4SS. To mediate the transport of fusion proteins to plants, the C-termini of VirF, VirE2, and VirE3 are enough. The minimum size of VirF required for directing protein translocation to plants is the C-terminal 10 amino acids. From this, the minimal consensus sequence R-X(7)-R-X-R-X-R, which the VirB complex required for substrate secretion, could be derived. C-terminal fusions of VirE2 block its translocation to host cells. Accordingly, the protein is rendered non-functional in *A. tumefaciens* with the insertion of a FLAG tag at the C-terminus of VirE2 or the truncation of the C-terminal 18 amino acids of VirE2. This does not affect the capability of the protein to bind single-stranded DNA. However, when such VirE2 C-terminal mutant derivatives are overexpressed in transgenic plants, susceptibility to transformation by an *A. tumefaciens* virE2-deficient strain increases. This indicates that the mutations disrupt a part of amino acids that is required for translocation, such as a secretion signal.

In an experimental approach, transport of these proteins in the absence of T-DNA has been studied using fusion proteins of VirE2 or VirF to the Cre recombinase. In the experiments designed, the transport of Cre-VirE2 or Cre-VirF fusions into host plant cells led to recombination, conferring kanamycin resistance to host tissues. This shows the dependence of transportation of bacterial factors on the VirB/VirD4 complex. Besides, it was found that VirE3 and the C-terminus of VirD5 mediate in targeting of substrates into host cells. VirE3 may play a role in nuclear localization of VirE2, and VirD5 may act as a transcription factor in plant cells.

3.4.7   Host Cell Entry of *Agrobacterium* Factors

The mechanism involved in the movement of Vir proteins and T-DNA–protein complex into the host cell wall and membrane barriers is still not clear. In the transfer of a plasmid mediated by T4SS, the interaction between donor and recipient is facilitated by pilus, which is then followed by fusion of outer membranes in a mating junction. It is not known how the transferred conjugal intermediate traverses the bacterial wall and the inner membrane. Also, little information is available about the transfer across host cell barriers mediated by VirB. The wide host range transformed by *Agrobacterium* indicates that the specificity of host–pathogen interaction may not be necessary to break the cell wall of the host and membrane barriers.

3.4.8   Targeting of *Agrobacterium* T-DNA into the Host Cell Nucleus

Once T-DNA enters a plant cell, it finds its way into the nucleus. In this process, a number of plant proteins and several *Agrobacterium* Vir proteins are likely involved. The proteins VirD2 and VirE2 contain plant active nuclear localization signal (NLS) sequences. VirD2 is covalently linked to the 50-end of T-DNA and contains two NLS regions. Both the NLS regions can direct chimeric proteins to the nucleus. Steric considerations indicate that for nuclear targeting of T-DNA complex, the bipartite NLS in the C-terminus of VirD2 is biologically important. VirE2 protein contains two separate bipartite NLS regions that can target linked reporter proteins to plant nuclei. Fluorescently labelled single-stranded DNA coated with VirE2 and microinjected into plant cells accumulates in the nucleus, and naked single-stranded DNA remains in the cytoplasm.

Deletion of the VirD2 bipartite NLS leads to the complete loss of transformation, which suggests that VirE2 NLS domains cannot compensate for the loss of VirD2 NLS. For nuclear targeting of T-complexes synthesized *in vitro* in permeabilized HeLa cells, VirD2 and VirE2 proteins were demonstrated to be vital. The C-terminal NLS of VirD2 is found to be essential for virulence, but it is not necessary for the production of intrabacterial T-strand. This NLS does not play any role in targeting of the T-DNA to the nucleus. However, deletion of VirD2 NLS was found to alter the VirD2 protein structure in such a way that it could still nick the T-DNA border, but it was unable to pass through nuclear pore or T4SS. In the absence of VirD2 NLS, VirE2 might provide nuclear targeting. Gelvin (1998) showed that an *A. tumefaciens* virE2 virD2DNLS double mutant could form tumours on VirE2-producing transgenic tobacco, but not on wild-type tobacco. Gelvin suggested that the NLS of VirE2 might play a role in directing T-DNA to nucleus. Several studies have found that VirE2 plays an additional role as a transmembrane DNA transporter. It can insert itself into artificial membranes and form channels, which facilitate the efficient transport of ssDNA (single-stranded DNA) through membranes. VirE2 appears to actively pull ssDNA into the host, as demonstrated by biophysical experiments and particle-bombarded tobacco cells expressing VirE2 fusion protein transiently.

Although VirE2 protein plays a crucial role in the transformation process, certain strains of *A. rhizogenes* that lack this protein can still efficiently transfer T-DNA. This is made possible through GALLS proteins. Despite being dissimilar to VirE2, GALLS protein can restore pathogenicity to the virE2 mutant *A. tumefaciens*. The GALLS gene encodes for two proteins: full-length GALLS and a C-terminal domain, which initiates at an internal in-frame start codon. In the full-length GALLS protein are domains for nuclear localization, ATP binding, and type IV secretion. In plant cells, interaction between GALLS-FL and VirD2 has been observed. These findings, as well as the nuclear localization of GALLS-FL and its predicted helicase activity, suggest that GALSS-FL may pull T-strands into the nucleus. VirF has been shown to degrade host cell factors during infection. In the host nucleus, it is believed that VirF mediates the degradation of the T-DNA complex with the help of host proteasome machinery. This facilitates the release of the T-DNA and its subsequent integration into the chromosome.

3.4.9 Functions of Vir Proteins in T-DNA Integration

There is very limited knowledge about the exact mechanism of integration of T-DNA into the plant genome or the function of specific proteins in this process. Illegitimate recombination or non-homologous end joining is the major means by which the foreign DNA is integrated into plants. Using a similar mechanism, T-DNA integrates into plant chromosomes. According to an alternative model, for integration, it is essential to have the initial invasion of plant DNA by T-DNA. Besides, the analysis of T-DNA integration sites indicates that microhomologies are involved in the integration process. The fact that most T-DNA is not stably integrated into chromosomal DNA is revealed by the measurements of the relative amounts of transient versus stable expression of reporter genes in plant cells infected by *Agrobacterium*. T-DNA enters the nucleus as a single-stranded molecule, but much of it becomes double stranded. This is because for the conversion to a transcriptionally competent form, the synthesis of a complementary DNA to the T-strand is required (Narasimhulu, Deng, Sarria, *et al.* 1996). It is not yet known whether T-DNA integrates by the single-strand invasion of the locally denatured plant DNA or the substrate for integration is the extrachromosomal double-stranded T-DNA. Since VirD2 is linked covalently to the T-DNA strand, it probably has some role in the integration process: VirD2 is capable of mediating site-specific cleavage and reversal of it, although it is only site specific. Host proteins mediate the ligation of T-DNA to plant DNA. VirD2 could not facilitate the *in vitro* ligation–integration reactions in which T-DNA was ligated to a model target sequence; however, this reaction occurred in the presence of plant extracts. There is generally precise integration of the 5′ end of T-strand into the plant DNA; just some 5′ nucleotides are deleted by the integration of T-DNA into the plant genome. This may be because VirD2 protects the capped 5′ T-strand end from exonucleases as mutations in VirD2 cause imprecise ligation and deletion of the 50-end of T-DNA to plant chromosomal DNA. Furthermore, the o domain of VirD2, which is a conserved region outside the NLS, is essential for tumorigenesis, but it has insignificant impact on transient T-DNA expression. In virulence, the virE2 mutants are extremely attenuated (Stachel and Nester 1986). VirE2 acts as a gated pore for the passage of ssDNA. Thus, a defect in the nuclear host transport might explain the severe attenuation of virE2 mutant *Agrobacterium* strains. However, integrated T-DNA molecules that are transferred from the virE2 mutant *Agrobacterium* strains also show considerable deletions corresponding to the 3′ ends of the T-strand. This suggests that VirE2 also plays a major role in nucleolytic protection of T-strand.

3.4.10 Plant Factors and Defence Responses Involved in *Agrobacterium* Tumour Formation

The host plant inadvertently participates in *Agrobacterium* transformation. This assistance takes place at several stages, such as import of Vir protein/T-DNA, dissociation of Vir/T-DNA complex, integration of T-DNA, and reprogramming of gene expression for development of tumour. Many host factors exploited by *Agrobacterium* for achieving transformation have been identified. For identifying host proteins that interact with *Agrobacterium* Vir proteins, significant progress has been achieved with yeast two-hybrid (Y2H) screen. Another progress was made through large plant mutant screens.

3.4.11 Functions of Plant Proteins in T-DNA Nuclear Import

It has been reported that VirD2 interacts *in vitro* with many members of the *Arabidopsis* importin-α and cyclophilin families and in the Y2H system. Importin-α, the NLS-binding protein of the nuclear import machinery, specifically interacts with the bipartite NLS region of VirD2, facilitating its nuclear import. Deng, Chen, Wood, *et al.* (1998) found an *Arabidopsis* cyclophilin that interacts with a central domain of VirD2 in Y2H experiments. Since a few cyclophilins show the peptidyl-prolyl isomerase activity, the authors conjectured that this protein might act as a chaperonin to hold VirD2 in a transfer-competent conformation during the trafficking of T-strand through the plant cell. However, a recent study found the cyclophilin-binding domain of VirD2 is not necessary for virulence, putting this hypothesis into question.

Moreover, VirD2 can undergo post-translational modification. The phosphorylation of a serine residue close to the bipartite NLS apparently controls the nuclear targeting of VirD2 (Tao, Bhattacharjee, and Gelvin 2004). Through the alanine substitution of this residue, the cytoplasmic localization of a β-glucuronidase (GUS)–VirD2 NLS fusion protein occurs. In addition, Tao, Bhattacharjee, and Gelvin (2004) discovered DIG3, which is a type 2C serine/threonine protein phosphatase that negatively impacts the nuclear import of a GUS–VirD2 NLS fusion protein. A biochemical approach helped in the identification of CAK2M, a VirD2-interacting kinase. VirD2 is phosphorylated by CAK2M *in vivo*. It is likely that CAK2M corresponds to the kinase that governs the nuclear import of VirD2.

Another substrate of CAK2M is RNA polymerase II large subunit (RNApolII CTD). This substrate has the ability to recruit TATA box-binding proteins (TBPs) to actively transcribed regions. A preferential integration of the T-DNA into promoters of transcribed chromatin domains has been shown by the comparative sequence analysis of target sites and insert junctions. Furthermore, experiments with *Arabidopsis* cells transformed by *Agrobacterium* revealed an association between VirD2 and TBP, one of the most conserved nuclear proteins in eukaryotic cells. It was suggested that TBP or CAK2M might target VirD2 to the CTD, thereby controlling the integration of T-DNA. However, an alternative explanation for the interaction between TBP and VirD2 might have to be found.

The earlier concept of preferential T-DNA integration into transcriptionally active regions is not supported by a genome-wide analysis of T-DNA integration sites in *Arabidopsis* carried out under non-selective conditions. Instead, it is found that the integration of T-DNA happens randomly. Moreover, the selection pressure applied for the recovery of T-DNA insertions is responsible for the earlier reported enrichment of such integration sites in gene-rich or transcriptionally active regions of chromatin. However, studies could not verify the results leading to the non-selected T-DNA integration events. This was because only sequence analysis was presented and transgenic plants could not be recovered owing to the special experimental design. A novel PCR-based method was developed for high-throughput sequencing of the T-DNA/genomic DNA junction of 150,000 T-DNA insertion mutants. Researchers have isolated and characterized two VirE2-interacting proteins VIP1 and VIP2. Most data related to VIP1 originate from experiments with tobacco cells.

The elevated levels of VIP1 promote susceptibility and transformation efficiency in *Agrobacterium*, as shown by the larger calli formed in the infected plants that overexpress VIP1. More experiments on interaction suggest that VIP1 serves as a bridge between VirE2 and plant importin α-1 (IMPa-1 IMPα-1), thereby aiding in the nuclear import of VirE2 and its associated T-DNA.

3.4.12   Nuclear Transport of T-DNA Complex

VIP1 is responsible for the shuttling of T-DNA complex into the nucleus of the host cell. However, *Agrobacterium* VirE3 can partially complement this function. Similar to VIP1, *Agrobacterium* VirE3 has been found capable of binding to VirE2 and IMPa-1. With VirE2 and ssDNA, VIP1 forms ternary complexes *in vitro*. By analysing an *Arabidopsis* vip1-1 mutant, which generates a truncated VIP1 protein, researchers found that the protein's C-terminal region is essential for stable transformation but not necessary for transient transformation. VIP1 is a mobile protein and subject to stress-dependent cytoplasmic-nuclear trafficking. VIP2, the other VirE2-interacting protein, exhibits nuclear localization. A NOT domain (negative on TATA-less) is the most characteristic feature of VIP2, which makes VIP2 a putative transcriptional repressor protein. That VIP2 is required for stable (but not transient) transformation has been revealed by the virus-induced gene silencing of VIP2 in *Nicotiana benthamiana* and characterization of the *Arabidopsis* vip2 mutant. Studies of Y2H and bimolecular fluorescence complementation (BiFC) have shown that VIP2 interacts with both VirE2 and VIP1. From the microarray analysis, researchers found a significant decrease in the transcriptional response of vip2 to *Agrobacterium* compared to wild-type plants. Besides, in vip2 mutants, about 52 histone/histone-associated genes are constitutively repressed. These observations prompted the authors to suggest that impaired *Agrobacterium*-responsive gene induction and constitutive histone gene repression were responsible for the recalcitrancy of vip2 mutants to *Agrobacterium* infection and the decreased transformation efficiency.

Information on certain aspects of VirE2 is still controversial. Ballas and Citovsky (1997) have reported specific interaction between VirE2 and *Arabidopsis* importin IMPa-1 and nuclear localization of VirE2, whereas other studies have revealed a predominant cytoplasmic localization of VirE2 and the interaction between VirE2 and several importin isoforms in plants. Furthermore, impa-4 (but not other importin mutants) is recalcitrant to transformation. This deficiency can be overcome by ectopic overexpression of heterologous importin isoforms.

The differences for reported VirE2 localization are likely due to stress-dependent subcellular translocation of the VirE2-interacting protein VIP1. Presumably, by directing VirE2 to nucleus, stress-triggered phosphorylation of VIP1 mediates VIP1 nuclear localization and virulence. Hence, phosphorylated VIP1 may pull cytoplasmic VirE2 into nucleus under stress conditions, such as those occurring in cell bombardment. VIP1 interacts with VirE2 and IMPa-1. Tzfira, Vaidya, and Citovsky (2002) observed interaction between VIP1 and IMPa-1, but they did not find any interaction between VirE2 and

IMPa-1. Therefore, the authors suggested that VIP1 might function as an adaptor molecule and aid in the import of VirE2-bound T-strands into nucleus. In the study of Y2H and BiFC interaction, Bhattacharjee, Lee, Oltmanns, *et al.* (2008) found that VirE2 could directly bind all tested isoforms of importin (IMPα-1IMPa-1, IMPa-2, IMPa-3, IMPa-4, IMPa-7, and IMPa-9). They combined their own results with those of Tzfira, Vaidya, and Citovsky (2002) and proposed that VirE2 might use various cellular mechanisms for nuclear import, thus creating additional scope for the entry of T-complex into the nucleus. This is a plausible explanation, particularly when one considers that MPK3, which is the VIP1-nuclear-targeting kinase, is transiently (5–15 min) activated on contact with *Agrobacterium*. Therefore, for securing the T-DNA complex entry over a prolonged period, VIP1-independent nuclear translocation of VirE2 might be required. In addition, since VIP1 can bind IMPa-1 and VirE2 (irrespective of its phosphorylation status), it might still indirectly facilitate VirE2 nuclear translocation by guiding VirE2 to the nuclear periphery. Direct VirE2–IMPa-4 interaction would then achieve nuclear VirE2 import. This assumption is supported by the study of Lee, Fang, Kuang, *et al.* (2008). In the particle bombardment of onion cells, the presence of VirE2–IMPa-1 protein complexes was observed around the nucleus, but VirE2–IMPa-4 complexes were found exclusively within the nucleus.

3.4.13 Functions of Plant Proteins in T-DNA Integration

Altering chromatin conformation would probably affect T-DNA integration since T-DNA must interact with chromatin to integrate into plant chromosomal DNA. To find out which chromatin proteins are essential for transformation, researchers adopted forward and reverse genetic approaches. In this manner, they identified mutants in or near various histone genes, histone deacetylase (HADC) genes, and histone acetyltransferase genes as rat mutants. Furthermore, they screened 340 stable RNAi mutant lines in *Arabidopsis* for rat phenotypes. These lines consist of 109 chromatin genes of 15 gene families, including chromatin remodelling complexes, bromodomain and chromodomain proteins, global transcription factors, DNA methyltransferases, HADCs, histone H1, histone acetyltransferases, methyl binding domain proteins, MAR-binding filament-like proteins, SET domain proteins, and nucleosome assembly factors. Some level of decreased transformation susceptibility reproducibly results from the silencing of 24 chromatin genes reproducibly. Since T-DNA integrates into the plant genome through an illegitimate recombination, plants that lack DNA repair and recombination might also be deficient in T-DNA integration. Such DNA metabolism mutants are probably hypersensitive to DNA damaging agents such as UV radiation and bleomycin. For transient and stable *Agrobacterium*-mediated transformation, Sonti, Chiurazzi, Wong, *et al.* (1995) studied many *Arabidopsis* mutants that were sensitive to radiation. Among these, they found uvh1 and rad5 mutants resistant to stable transformation (but not to transient transformation). This was assessed by the formation of kanamycin-resistant calli. But a detailed analysis confirmed stable transformation deficiency only for rad5, but not for uvh1 mutants. Tumour growth on non-selective medium and stable phosphinothricin (PPT) resistance were similar to that of wild type. Besides rad5, a few radiation-sensitive *Arabidopsis* ecotypes are also recalcitrant to transformation mediated by *Agrobacterium*.

The study of almost 40 *Arabidopsis* ecotypes for susceptibility to root transformation by *Agrobacterium* has revealed that ecotype UE-1 is slightly radiation hypersensitive and transformation deficient. High transient but low stable transformation efficiency was observed in rat mutants when they were further tested. Out of the initial 21 mutants tested, five were found to be integration deficient. One of these mutants rat5 has an insertion in the 3′ untranslated region of a histone H2A gene. The rat5 mutant is efficiently transformed by flower vacuum infiltration although it is highly recalcitrant to stable transformation by root inoculation. These results indicate that the factors required for efficient transformation are present in the female gametophyte but absent in root somatic tissue. The rat5 mutant is haplo-insufficient (dosage dependent) since rat5 plants can be complemented to transformation proficiency with the wild-type RAT5 histone H2A gene. Recent studies have further emphasized the role of histones in T-DNA transformation. These studies assessed the effect of overexpression of several histones on transgene expression and *Arabidopsis* transformation. It was found that transgene DNA accumulated more quickly after transfection in plants overexpressing histone HTA1. It has been suggested that the increased transformation mediated by *Agrobacterium* through histone overexpression is owing to the protection of incoming transgene DNA during the early stages of transformation.

As the overexpression of *Arabidopsis* HTA1 can enhance the efficiency of transformation in both *Arabidopsis* and rice, the mechanism by which histones confer susceptibility seems to be conserved. Mutations in fas1 and fas2, which encode two subunits of the chromatin assembly factor CAF1, show enhanced frequencies of homologous recombination and T-DNA integration. Research on the *Arabidopsis* protein KU80 further highlights that the host's repair machinery actively participates in T-DNA integration. An essential protein in the non-homologous end-joining complex is KU80 that directly binds to double-stranded T-DNA intermediates, which are rapidly converted from T-strands early in the infection process and are necessary intermediates of T-DNA integration. Ku80 mutants are defective in T-DNA integration but not in transient T-DNA expression, whereas overexpression of KU80 leads to increased susceptibility to *Agrobacterium* infection and enhanced resistance to DNA damaging agents. However, the role of KU80 in the transformation of germline cells is still unknown. It has been reported that KU80 is both required and dispensable for T-DNA integration.

3.4.14 *Agrobacterium* and Plant Innate Immune Response

As other pathogens, when *Agrobacterium* attacks a plant, it is perceived as an invader and results in triggering of the 'innate immune response' of the plant, which is characterized by the expression of defence genes and the accumulation of reactive oxygen species. This reaction is accomplished through the perception of pathogen-associated molecular patterns (PAMPs) by specific receptors. So far, only a few receptors have been identified although many PAMPs have been isolated. Putative plant receptors for *Agrobacterium* include a vitronectin-like protein, a rhicadhesin-binding protein, and several VirB2-interacting proteins. A recent study of vitronectins has suggested that this group of proteins is unlikely to act as receptors for site-specific *Agrobacterium* attachment. Flagellin, a highly conserved bacterial protein, is the most intensively studied PAMP. In *Arabidopsis*, it is perceived by the receptor protein FLS2, a leucine-rich repeat receptor kinase (LRR-RLK). FLS2 becomes activated and starts phospho-relay-based signal transduction through the MAPK cascade MEKK1–MKK1/2–MPK4 after the perception of flagellin or the highly conserved 22-amino acid peptide (flg22) derived from it. Consequently, WRKY33, the MPK4-activated transcription factor, contributes to the defence-related transcriptional reprogramming. The flagellin proteins of *Agrobacterium* and symbiotic bacteria (rhizobia) are not recognized and they do not stimulate a defence response. This indicates that other PAMP–receptor pairs play a role in the recognition of these organisms. Researchers have identified a prominent *Agrobacterium* PAMP: the elongation factor Tu (EF-Tu). *Agrobacterium* EF-Tu is fully active as an elicitor although it is highly conserved in all prokaryotes. Interestingly, flg22 and EF-Tu share many characteristics despite their chemical dissimilarity. While acting with no apparent synergy, both PAMPs impede the growth of seedlings and initiate a common set of signalling events and defence responses. Among these responses are mitogen activating protein kinases (MAPK) activation, alkalization of the medium, and an oxidative burst. In addition, a microarray analysis of the *Arabidopsis* response to flg22 and the EF-Tu derived peptide elf18 found a distinct correlation of differential gene expression, although no apparent flg22- or elf18-specific subsets of genes were identified.

These PAMPs were found to rapidly induce a high number of RLK encoding genes (100 of 610). A targeted reverse genetics approach has made possible the identification of a receptor kinase EFR1, which is necessary for the perception of EF-Tu. EFR1, like FLS2, is a member of the LRR-RLK protein family. For root nodule symbiosis, the important role of LRR-RLKs in plant–microbe sensing is already evident. Mutants of *Lotus japonicus* affected in the LRR-RLK symbiosis receptor kinase (SYMRK) do not engage in symbiosis. Similarly, *Arabidopsis* mutants that lack fls2 do not respond to flg22 treatment. The efr mutants are not sensitive to elf18, while they show a normal flg22 response and an otherwise normal phenotype. The high susceptibility to *Agrobacterium* infection is the major characteristic of efr mutants, as evident by the heightened expression of a T-DNA harboured reporter gene (GUS) on the transient transformation of seedlings. On pre-treatment with flg22, receptor-binding sites for EF-Tu increased. In both wild-type and efr mutants, co-injection of flg22 terminated GUS expression, whereas co-injection of elf18 terminated GUS expression only in wild type and not in efr. An easily transformed plant species is *N. benthamiana*, which lacks an EF-Tu recognition system. However, transgenic *N. benthamiana* plants that express *Arabidopsis* EFR can induce elf18-triggered defence responses. From these observations, it can be stated that EFR-mediated EF-Tu perception inhibits *Agrobacterium*-mediated transformation.

A recent study highlights the molecular mechanism that associates the activation of the EFR receptor to intracellular signal transduction and stresses the differences between flg22- and elf18-triggered restrictions of *Agrobacterium* transformation. *Arabidopsis* mutants affected in the G protein β subunit (agb1-2) exhibit significantly decreased ROS production on flg22 or elf18 treatment, but there is no apparent effect on the stress-triggered MAPK activation (analysed by immunoblotting with an antibody recognizing active MPK3 and MPK6). Also, these mutants are impaired in the elf18-triggered immunity against *Agrobacterium*, but not in the flg22-triggered immunity. Thus, AGB1 plays the role of a positive regulator integrating flagellin and EF-Tu perception into ROS production, particularly in EF-Tu signalling, to restrict *Agrobacterium* transformation.

3.4.15 Gene Expression Reprogramming in Response to *Agrobacterium*

An *Agrobacterium* attack results in a major reprogramming of gene expression in plants. In the pre-microarray era, expression study (e.g., cDNA-AFLP) found that many *Agrobacterium*-induced genes are associated with plant defence and general stress responses. The transfer of T-DNA and Vir proteins can modulate the plant gene expression in tobacco cell culture by the use of different *Agrobacterium* strains. T-DNA and Vir protein transfer plays the role of suppressors of the defence response. An attachment-deficient *Agrobacterium* mutant was found to hyper-induce defence-related genes in cell culture of *Ageratum conyzoides*. It is interesting to note that non-pathogenic *E. coli* also triggered such hyper-induction. It was concluded that *Agrobacterium* can reduce plant defence in an attachment-dependent way. These results correlate with the finding that co-injection of plants with *Agrobacterium* plus flg22 or elf18 elicitors abolishes transformation. Salicylic acid (SA) is a factor that might explain the negative correlation between transformation efficiency and stress status. In regulating the plant response to pathogens, this plant hormone is an important signal. It accumulates in local and systemic tissues of plants exposed to stress and induces the expression of genes involved in pathogenesis. Salicyclic acid was found to be involved in the repression of the Vir regulon, attenuation of the role of the

VirA kinase, and degradation of an *Agrobacterium* quormone (Yuan, Edlind, Liu, *et al.* 2007). Accordingly, plant mutants that overproduce SA are recalcitrant to the formation of tumours. To summarize, a minimally stressed state of host cells seems desirable for maximal transformation efficiency from the bacterial point of view. Plants may escape *Agrobacterium* infection if their defence system is in an alerted state.

The study of the response of tobacco BY2 cell cultures to various *Agrobacterium* strains, including T-DNA and Vir protein transfer-incompetent strains, allowed the analysis of the stress versus transformation efficiency ambivalence. It also made possible to distinguish general transcriptional responses resulting from attachment or proximity of *Agrobacterium* near BY2 cells from transformation-specific responses induced by the transfer of T-DNA and Vir proteins into plant cells. *Agrobacterium* initiates an early (3–12 h) induction of stress genes. The expression of these stress genes is subsequently repressed by transfer-competent strains. The transfer of T-DNA and Vir protein occurs as a result of the decline of this general defence. On the contrary, Vir transfer-deficient strains initiate a second wave of defence gene expression (after 24 h) and fail to transfer T-DNA. Along with the observed constitutive repression of several histone genes in the vip2 mutant, which are impacted in stable but not in transient transformation, these findings highlight the significance of histones in *Agrobacterium* transformation and suggest that an enhanced pool of histones is required for enabling the integration of T-DNA into the host genome. Researchers are yet to discover the mechanisms of transcriptional reprogramming of tissue infected by *Agrobacterium* and the extent to which bacterial factors contribute to this reprogramming. Based on the findings that VirE3 exhibits transactivation activity in yeast, localizes to plant nuclei, and binds pBrp, a general plant-specific transcription factor, a function of VirE3 as a potential plant transcriptional activator mediating the expression of tumour development-specific genes, has been proposed.

3.4.16 Defence Versus Transformation—How to Evade Innate Immune Response

It is obvious from the above sections that the defence response triggered in the host cell is a major barrier for *Agrobacterium* infection. Much higher transformation efficiencies can be achieved once plant defence is blocked, reduced, or evaded. Moreover, *Agrobacterium* have learnt to turn the tables in that they can even benefit from being recognized as a pathogen in at least one aspect. Like other microbes, *Agrobacterium* activates MAPKs, primarily MPK3, MPK4, and MPK6. *In vivo* interaction, kinase assays, and Y2H experiments have revealed that VIP1 interacts with and becomes specifically phosphorylated by MPK3. Phosphorylation of VIP1 triggered by stress leads to the rapid translocation of this protein from cytoplasm to nucleus. Unlike VirD2, which enters the host nucleus on its own, VirE2 requires the help of both importin α and host protein VIP1. Thus, *Agrobacterium* not only forces VIP1 to deliver VirE2/T-DNA into the host cell nucleus, but also actively manipulates the subcellular localization of this plant protein. Furthermore, when the nuclear transfer of the T-DNA complex occurs, the integration of T-DNA into the plant genome is likely to be achieved through another manipulation of the host cell machinery.

Since both *Agrobacterium* and flg22 induce activation of MPK3 and rapid nuclear accumulation of VIP1, question arises on the role of the stress-dependent cytoplasmic-nuclear translocation of VIP1 in plant's response to other stresses. In the promoter regions of stress-responsive genes, VREs are over-represented. Under conditions that activate the MPK3 pathway, VIP1 binds *in vivo* to the promoter of a stress-responsive transcription factor MYB44. In stress-triggered induction of VIP1 target genes, the MPK3 mutants are impaired, whereas plants overexpressing VIP1 exhibit constitutively enhanced transcript levels. From these observations, researchers derived the function of VIP1 as a mediator of MPK3-mediated stress gene modulation. The question arises, how does the role of VIP1 as an activator of the defence responses correspond with the finding that VIP1 overexpression leads to increased transformation efficiency? A reasonable explanation is that the elevated transformation rate is obtained through a more efficient nuclear transfer of the T-DNA complex since VIP1 can 'piggypack' more VirE2–T-DNA molecules. The degradation of the T-DNA complex and VIP1 due to the action of VirF and plant proteasome releases T-DNA and prevents the expression of defence gene induced by VIP1. According to this model, VIP1 phosphorylation/localization and the presumably directly correlated nuclear import of VirE2/T-DNA are the limiting factors in transformation and not the VirF, VirD2, or fidelity of the plant proteasomal machinery. VIP1 has the dual function as a stress-responsive transcription factor and as a T-DNA/

VirE2 shuttle. Therefore, it might be one of the factors that determine between successful transformation and failure of transformation because of increased basal stress levels. Since MPK3 is especially sensitive to various stresses, its target protein VIP1 can be potentially phosphorylated and translocated to the nucleus and can trigger stress gene expression. Therefore, *Agrobacterium* and the plant 'compete' for one of the two VIP1 functions.

3.5 BINARY VECTORS

To produce transgenic plants, for a long time *Agrobacterium*-mediated genetic transformation have been carried out to generate transgenic plants. Initially, complex microbial genetic methodologies were employed to introduce genes of interest (goi) into *Agrobacterium*. Through the use of these methodologies goi were inserted into the T-DNA region of large Ti plasmids. It was later discovered that T-DNA transfer could still be achieved if the T-DNA region and *vir* genes required for T-DNA transfer and processing were split into two replicons. This binary system allowed facile manipulation of *Agrobacterium*.

Agrobacterium transfers T-DNA, which constitutes a small (5%– 10%) portion of a resident Ti or Ri plasmid, to numerous plant species. The bacterium can also be manipulated in the laboratory to transfer T-DNA to fungal and even animal cells. Transfer requires three major components: (i) T-DNA border repeat sequences (25 bp) flanking the T-DNA in direct orientation and delineating the region to be processed from Ti/Ri plasmid, (ii) *vir* genes located on Ti/Ri plasmid, and (iii) various genes (chromosomal virulence (chv) and other genes) present on bacterial chromosomes. These chromosomal genes are responsible for bacterial exopolysaccharide synthesis, secretion, and maturation. However, a few chromosomal genes critical for virulence probably mediate the bacterial response to the environment.

The *vir* region consists of approximately 10 operons (depending on the Ti or Ri plasmid) that perform the following functions:

1. **Sensing plant phenolic compounds and transducing this signal to induce expression of *vir* genes (*virA* and *virG*):** VirA and VirG constitute a two-component system that responds to particular phenolic compounds produced by injured plant cells. Since wounding is critical for efficient plant transformation, *Agrobacterium*

can detect an injured potential host by sensing these phenolic compounds. When these phenolic inducers activate VirA, there is initiation of a phospho-relay that finally results in phosphorylation and activation of the VirG protein. Activated VirG binds to *vir* box sequences preceding each *vir* gene operon, which allows enhanced expression of each of these operons. Many sugars serve as co-inducers besides the induction of the *vir* genes by phenolics. These sugars are perceived by the protein ChvE encoded by a gene on the *Agrobacterium* chromosome. In the presence of these sugars, *vir* genes are fully induced at lower phenolic concentrations.

2. **Processing T-DNA from the parental Ti or Ri plasmid (*virD1* and *virD2*):** At the 25 bp T-DNA border repeat sequences, VirD1 (a helicase) and VirD2 (an endonuclease) together bind to and nick the DNA. The VirD2 protein covalently links to the 5′ end of processed single-stranded DNA (the T-strand) and leads it out of the bacterium, then to the plant cell and finally to the plant nucleus.

3. **Secreting T-DNA and Vir proteins from the bacterium via a T4SS (*virB* operon and *virD4*):** There are 11 genes in the *Agrobacterium* *virB* operon, and most of these genes form a pore through the bacterial membrane for transferring Vir proteins. At present, five such proteins are known to be secreted through this apparatus: VirD2 (unattached or attached to T-strand), VirD5, VirE2, VirE3, and VirF. VirD4 is a coupling factor that links VirD2–T-strand to the type IV secretion apparatus.

4. **Participating in events within the host cell that involves T-DNA cytoplasmic trafficking, nuclear targeting, and integration into the host genome (*virD2, virD5, virE2, virE3,* and *virF*):** In targeting the T-strand to the nucleus, VirD2 and VirE2 might play certain roles. Moreover, in the plant cell, it is likely that VirE2 protects T-strands from nucleolytic degradation. VirF might be responsible for stripping proteins off the T-strand before T-DNA integration. Within the T-DNA, no gene is necessary, for transferring the T-DNA even though *vir* genes were first defined genetically because of their importance in virulence. The ability to delete wild-type oncogenes and opine synthase genes from within T-DNA and replace them with genes encoding selectable markers and other goi has contributed to the advancement of the plant genetic engineering.

3.5.1 Development of Binary Vector Systems

In the beginning, to introduce goi into T-DNA for subsequent transfer to plants, there used to be complicated genetic manipulations to recombine these genes into the T-DNA region of Ti plasmids. This was because of the following constraints with Ti plasmids:

(i) Ti/Ri plasmids are very large and have a low copy number in *Agrobacterium*.

(ii) They are difficult to isolate and manipulate *in vitro*.

(iii) They are not replicable in *E. coli*, the favoured host for genetic manipulation.

T-DNA regions from wild-type Ti plasmids are mostly large and do not possess unique restriction endonuclease sites suitable for cloning a goi. Besides, for regenerating normal plants, scientists wanted to remove oncogenes from T-DNA. In constructions designed to deliver goi to plants, opine synthase genes were generally considered superfluous.

In 1983, two groups made a breakthrough which enabled laboratories not specializing in microbial genetics to employ *Agrobacterium* for gene transfer. They also found that *vir* and T-DNA regions of Ti plasmids could be split into two separate replicons. Proteins encoded by *vir* genes could act upon T-DNA in *trans* to mediate its processing and export to the plant as long as both these replicons are located within the same *Agrobacterium* cell. Systems in which T-DNA and *vir* genes are situated on separate replicons were termed T-DNA binary systems (Figure 3.2).

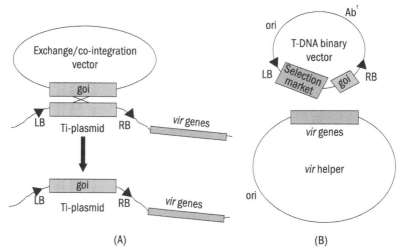

Fig. 3.2 *Plasmids and binary vector systems*

T-DNA is located on the binary vector. The replicon containing *vir* genes are known as the *vir* helper. When the strains harbouring this replicon and a T-DNA do not contain oncogenes that could be transferred to a plant, they are considered disarmed.

It soon became apparent that binary systems were useful for reducing genetic manipulation. It was realized that to introduce a goi into the T-region of a Ti plasmid, the complicated microbial genetic technologies were no longer necessary. Instead, goi could easily be cloned into small T-DNA regions within binary vectors specifically suited for this purpose. The T-DNA binary vector could be mobilized (by bacterial conjugation or transformation) into an appropriate *Agrobacterium* strain containing a *vir* helper region after characterization and verification of the construction in *E. coli*. T-DNA binary vectors and disarmed *Agrobacterium* strains harbouring *vir* helper plasmids have become more sophisticated over the past years and they are now suited for specialized purposes. Commonly used T-DNA binary vectors and vector series are given in Table 3.1.

3.5.2 Properties of Binary Vectors

For genetic engineering experiments, T-DNA binary vectors contain a number of important features. Some of the features are as follows:

Table 3.1 *Agrobacterium* T-DNA binary vectors

Vector series name	Vector ori/ incompatibility group	Important features	Gateway compatible	Bacterial selection marker	Plant selection marker
pBIN	IncPa	mcs with blue/white selection	No	Kan	Kan
pGA	IncPa	cos site ColE1 ori	No	Kan	Kan
SEV	IncPa	Reconstitutes a missing T-DNA border; not a binary vector	No	Kan	Kan/Nos
pEND4K	IncPa	cos site, mcs with blue/white selection	No	Kan/Tet	Kan
pBI	IncPa	Promoterless *gusA* gene for promoter studies	No	Kan	Kan

1. **Left-border (LB) and right-border (RB) of 25 bp direct repeat sequences as the defined limit of T-DNA:** The border repeats of T-DNA are highly conserved sequences across the range of all Ti and Ri plasmids. VirD1/VirD2 endonuclease forms a nick on T-DNA between third and fourth nucleotides, leaving the first three nucleotides with RB and the rest of the nucleotides with LB. However, within the plant, the T-strand is frequently chewed back, which is likely by exonucleases. As VirD2 protects the 5′ end of T-strand and linked to it, there is negligible loss of nucleotides at this end, which is a few nucleotides at most. There is frequent loss of nucleotides from the unprotected 3′ end and it is usually more extensive. Also, deletions to the extent of several hundred nucleotides are not uncommon. In the early T-DNA binary vectors, the plant antibiotic selection marker gene was located near the 5′ end of T-DNA (RB) and goi were placed near the 3′ end. However, huge DNA loss from the 3′ end owing to the nucleolytic degradation could result in antibiotic-resistant transgenic plants with deletions in goi. This difficulty was overcome by inserting the selection marker gene near LB and the goi near RB. Extensive deletion of T-DNA from the 3′ end would lead to selection marker removal and hinder recovery of these plants. Thus, deletion of goi was generally avoided. Sequences near RB (called overdrive sequences) can enhance the T-DNA transmission. These sequences are often incorporated into the regions of T-DNA binary vector RB.

2. **A plant-active selectable marker gene (usually for antibiotic or herbicide resistance):** The most commonly used selection systems use aminoglycoside antibiotics (e.g., kanamycin or hygromycin), herbicides (e.g., PPT or gluphosinate), and herbicide formulations (e.g., Basta or Bialophos). Phospho-mannose isomerase, which employs metabolic markers, is an example of other selection systems. The concentration of kanamycin antibiotic should be carefully determined depending on the tolerance level of plant species to antibiotics so that non-transformed plant tissues are completely killed. These selectable markers were placed close to the RB of T-DNA in early binary vectors, as already mentioned. However, because of the polarity of T-DNA transfer, recent vectors contain the selectable marker near the LB to ensure transfer of goi.

3. **Multiple cloning sites (MCS) within T-DNA for the insertion of goi:** In the early binary vectors (e.g., pBIN19), there were a few

restriction endonuclease cloning sites in a *lacZ*, a complementation fragment, allowing blue/white screening for the presence of transgene insertion. These sites are flanked by promoters and polyA addition signals in many vectors. Recent binary vectors contain multiple rare-cutting restriction endonuclease or homing endonuclease sites. The pSAT vectors contain polyA addition signal, promoter, multiple restriction endonuclease cloning sites (expression cassettes) flanked by rare-cutting or homing endonuclease sites. For protein localization or protein–protein interaction studies, some of these vectors have incorporated tags into these expression cassettes to produce fluorescent fusion proteins. For simultaneous introduction of multiple genes into plants, multiple expression cassettes from pSAT vectors can be loaded into cognate rare-cutting sites in binary vectors. Many recent binary vectors contain Gateway sites, which help in the insertion of genes or exchange of gene cassettes with other vectors. Moreover, to clone large inserts of more than 100 kb, several BAC binary vectors have been developed.

4. **Origin(s) of replication to allow maintenance in *E. coli* and *Agrobacterium*:** If several plasmids are to coexist in the bacterium, the incompatibility group of the plasmid, with the function related to the specific origin of replication, can be essential. Therefore, these plasmids should be part of different incompatibility groups. In some cases, origins of replication can be functional in both *Agrobacterium* and *E. coli* (in which initial constructions are commonly made). These replication origins, which have a broad range of hosts, include those from RK2 (e.g., pBIN19 and derivatives), pSa (e.g., pUCD plasmid derivatives), and pVS1 (e.g., pPZP derivatives). Other replication origins that function in *Agrobacterium* do not function in *E. coli*. These include those from Ri plasmid (e.g., pCGN vectors). Therefore, a ColE1 origin (such as the one used in pUC and pBluescript plasmids) is added to the vector. In *Agrobacterium*, different origins of replication replicate to different degrees. The pSa origin replicates to 2–4 copies per cell, the RK2 and pVS1 origins replicate to 7–10 copies per cell, and the pRi origin replicates to 15–20 copies per cell.

5. **Antibiotic resistance genes within both chromosome and backbone sequences for selection of the binary vector in *E. coli* and *Agrobacterium*:** As a result of chromosomal mutation, a number of *Agrobacterium* strains become resistant to rifampicin.

In addition, sucrose can be a sole carbon source for commonly used *Agrobacterium* strains. Most *E. coli* laboratory strains K12 cannot use sucrose as the source of carbon. Therefore, *E. coli* from *Agrobacterium* cultures is eliminated when it is grown on minimal medium containing rifampicin and sucrose. This is a useful selection approach following the introduction of binary vector into *Agrobacterium*. The matching of binary vectors with specific *vir* helper *Agrobacterium* strains should be carefully made. Many of the strains express genes for resistance to carbenicillin, erythromycin, kanamycin, or gentamicin. Thus, it is not always possible to use binary vectors with the same selection marker in these strains. For instance, several T-DNA binary vectors based on pBIN19 utilize kanamycin resistance as the bacterial selection marker. Since *A. tumefaciens* EHA101 is resistant to kanamycin, it cannot be easily used with these pBIN19 derivatives. But these binary vectors can be utilized in *A. tumefaciens* EHA105, the near-isogenic kanamycin-sensitive strain. Besides, some *Agrobacterium* strains show resistance towards low quantities of spectinomycin, which is an antibiotic used together with pPZP plasmids and their derivatives. To ensure effective killing, various concentrations of spectinomycin should be tested with the *vir* helper strain which lacks the binary vector. Extra care is necessary if a binary vector has a tetracycline resistance gene. Since *A. tumefaciens* C58 possesses a tetracycline resistance determinant, it is resistant to low levels of this antibiotic.

Some *Agrobacterium* strains or binary vectors might have a β-lactamase gene that confers resistance to carbenicillin. However, it is still relatively easy to kill these bacteria ever if they have infected plants. Augmentin and timentin are the β-lactam antibiotics that contain clavulanate, which inhibits β-lactamases. At concentrations of 100–150 mg/L, timentin will completely eliminate the growth of *Agrobacterium* C58-based strains that harbour a β-lactamase gene. Since *Agrobacterium* Ach5-based strains (such as LBA4404) fail to express β-lactamase activity well, even lower concentrations of carbenicillin or timentin can kill these strains.

3.5.3 Alternative T-DNA Binary Systems

T-DNA binary vector systems generally always consist of T-DNA and *vir* regions localized on plasmids. However, it is not necessary that they function this way. Replicons containing *vir* genes or T-DNA do not need to be plasmids. It has been demonstrated by many laboratories that T-DNA can be integrated into an *Agrobacterium* chromosome and launched from this replicon. Specialized vectors have been produced to enable integration of DNA into a specific neutral (i.e., not associated with virulence) region of the chromosome of *A. tumefaciens* C58 (Lee, Fang, Kuang, *et al.* 2008). Although there can be lower transformation frequencies on launching T-DNA from the *Agrobacterium* chromosome, this technique can reduce integrated transgene copy number and do away with the integration of vector backbone sequences into the plant genome.

3.6 USES OF 35S PROMOTER

Promoter is the molecular switch that every gene possesses to turn it on or off in every organism. A promoter is positioned in front of a gene (upstream) and is made up of DNA. A promotor is usually gene specific and activated by molecular signals given by the organism. Thus, it regulates the formation of a gene product (e.g., enzymes or hormones) according to the requirement of the organism.

3.6.1 Cauliflower Mosaic Virus

The cauliflower mosaic virus (CaMV) (a double-stranded DNA virus) infects a variety of crucifers, such as cauliflowers, cabbages, oilseed rape, and mustard. To replicate itself in a plant cell, the virus uses two promoters, that is, 35S and 19S. The two promoters function overrides the regulatory system of the plant cell. These promoters are constantly switched on and cannot be regulated or switched off by the plant.

Bacteria and fungi can uptake DNA from various sources in their surroundings, such as water, soil, or gut. They can use this DNA either as a food source or as genetic information by integrating it into their own genome. The ongoing evolutionary process contributed by this is a kind of horizontal gene transfer. The most common form of horizontal gene transfer in bacteria is the exchange of genetic information with each other, for example, antibiotic or herbicide resistance gene. Thus, there is a possibility of transfer of genes from genetically modified (GM) crops to bacterial population present in the gut or soil. This has already been observed under various laboratory conditions. For a sound risk assessment, the crucial question is whether a transferred GM gene could be used or activated by bacteria.

Thus, after the identification of the risk associated with such gene transfer by the British Medical Association (BMA) and the Cartagena Protocol on Biosafety, the use of antibiotic resistance genes as markers has been withdrawn. However, other genetically engineered genes, such as the genes responsible for herbicide tolerance or insect resistance, did not receive much attention. Generally, horizontal gene transfer is not considered a factor of risk that should be assessed for the approval of GM crops and is denoted as 'negligible to zero' without any evidence. It has not been recognized as the route through which bacteria and yeast may utilize the newly introduced genes through genetic engineering and whether these genes (antibiotic resistance) may provide a selective advantage and shift the balance of microorganisms in gut or soil.

Assessments are based on the assumed lack of both risk and evidence. For the CaMV 35S promoter, any question of risk was immediately ignored, as this promoter was considered plant specific and was not active in other organisms; for example, fungi, bacteria, or human cells. However, this assumption is faulty and puts the current process of risk assessment and the accuracy of safety data into doubt.

Studies in 1990 showed that the 35S promoter of CaMV is active, both in plants and in the gut bacterium *E. coli,* yeast, and extracts of human cancer cell lines.

3.7 REPORTER GENES

An integral part of plant transformation strategies is the dominant stable markers. Genes conferring resistance toxic levels of amino acids, antibiotics, antimetabolites, herbicide resistance genes, and hormone biosynthetic genes are the selectable marker genes that have become available. The properties of the selection agent, the resistance gene, and the plant species determine the requirement of a particular resistance. The effect of the dying, untransformed cells should be minimal over transformed cells. The selection agent should fully inhibit the growth of untransformed plant cells. Therefore, the lowest concentration of the selection agent is generally used, which suppresses the growth of untransformed cells. The sensitivity of plant cells to the selection agent is dependent on the explant type, genotype, tissue culture conditions, and development stage. Therefore, it should be determined under the actual conditions of the transformation and regeneration process. Finally, the transcriptional and translational control signals to which the resistance gene is fused also influence the level of resistance. Hence, it might be essential to test several gene constructs (Herrera-Estrella, De Block, Messens, *et al.* 1983).

3.7.1 Neomycin Phosphotransferase II (NPTII)

The most widely used selectable marker for the transformation of plants is aminoglycoside 3'-phosphotransferase II [APH(3')-II] or NPTII. Aminoglycoside antibiotics such as geneticin, neomycin, kanamycin, and paromomycin are inactivated by the enzyme through phosphorylation. The enzyme is encoded by the *nptII* (or *neo*) gene, derived from the transposon Tn5. The toxicity of geneticin is more than that of kanamycin, but the most widely used selective agent is kanamycin at a concentration of 50–500 mg/L. Kanamycin is ineffective as a selection marker for several legumes and gramineae, although kanamycin resistance is a very useful marker for a diverse range of plant species.

The *nptII* gene has been used to generate useful selectable markers owing to its ability to fuse with many promoters. A mutated form of the *nptII* gene present in some vectors encodes a protein with a reduced enzymatic activity. The *nptII* gene can be employed to study gene expression and regulation both as a selection marker and as a reporter gene. The ability of construction of N-terminal fusion that retains the enzymatic activity is one of the major advantages of NPTII in this respect. A non-destructive callus induction test can be employed to check kanamycin resistance in putative transformants and their progeny. To detect the NPTII protein quantitatively or semi-quantitatively, several *in vitro* assays are available. The activity of NPTII in plant cell extracts is detectable by an enzymatic assay where substrates used are kanamycin and $[\gamma\text{-}^{32}p]$ ATP. Thin-layer chromatography, less elaborate dot-blot analysis, or polyacrylamide gel electrophoresis and immunological assay are used to detect phosphorylated aminoglycosides. Finally, to follow the segregation of the *nptII* gene in the progeny of primary transformants, a seed germination assay on a kanamycin-containing medium can be used, and kanamycin solution can also be applied locally or sprayed on soil-grown plants. The NPTII protein has been found to be safe for human consumption.

3.7.2 Hygromycin Phosphotransferase

Hygromycin B is an aminocyclitol antibiotic that interferes with protein synthesis. A hygromycin phosphotransferase gene (*hpt*) is originally derived from *E. coli*. With wide application as a resistance gene for plant transformation, the bacterial *hpt*-coding sequence has been modified for expression in plant cells. Deleting or substituting amino acid residues near the carboxyl terminus of the enzyme increased the level of resistance in tobacco protoplasts. However, the hydrophilic carboxyl-terminal end itself was preserved and might be necessary for strong HPT activity. Being more toxic than kanamycin, hygromycin B usually kills sensitive cells more quickly. Selection, *in vitro*, is applied at concentrations ranging from 20 mg/L for *A. thaliana* to 200 mg/L for tall fescue. By the use of a non-destructive callus induction test, hygromycin resistance can be checked. The *hpt* gene in the progeny of transgenic plants can be segregated with a seed germination assay. For detecting HPT, an enzymatic assay has been described. Careful working procedures are recommended because of the human toxicity of hygromycin B.

3.8 MATRIX ATTACHMENT REGION

Also known as scaffold/matrix attachment regions (S/MARs), the matrix attachment regions (MARs) are sequences in the DNA of eukaryotic chromosomes in which the nuclear matrix attaches. The structural organization of the chromatin within the nucleus is mediated by S/MARs. For the chromatin scaffold, these elements constitute anchor points of the DNA and organize the chromatin into structural domains. Studies of individual genes revealed that the dynamic and complex organization of the chromatin mediated by S/MAR elements has a significant role in the regulation of gene expression. Among the many factors that might affect the expression at different genomic sites, the higher order structure of chromatin is prominent. For instance, the transcriptional ability of large regions is believed to depend on the extent to which the 30 nm chromatin fibre is unwound to the level of 11 nm nucleosome fibre. This permits access to DNA by RNA polymerase and transcriptional regulatory proteins. In each of the different chromosomal domains, unwinding may be regulated independently as demonstrated by the current models of the eukaryotic chromatin structure. It is believed that these domains correspond to both the domains classically defined by nuclease sensitivity mapping and the 'loop domains' visualized in electron microscope studies of partially deproteinized metaphase chromosomes and interphase nuclei.

Electron micrograph studies also showed a proteinaceous matrix or 'scaffold' to which DNA is attached at intervals to make a series of loops that vary from 40 kb to several hundred kilobases. Studies have revealed that scaffold attachment points act as domain boundaries and play important roles in regulating gene expression. Observations in animal systems suggest that scaffold attachment can insulate transgenes from the effect of the surrounding chromatin. It has also been shown that certain DNA sequences, called scaffold attachment regions (SARs), that bind to the nuclear scaffold *in vitro* reduce the position effect variation *in vivo* when they are inserted on both sides of globin or lysozyme. This has prompted to conclude that SARs define independent domains in which the higher order chromatin structure is determined independently from that of the surrounding chromatin. Besides, SARs and other types of sequences might reduce the impact of enhancers and other *cis*-acting regulatory elements near a given integration site.

3.8.1   Scaffold Attachment Models

The simplest models that explain the SAR effect presume that SARs mediate *in vivo* by binding to the nuclear scaffold. In the loop domain model, SARs are believed to define the ends of chromatin loops that correspond to the independent regulatory domains. As per this model, the scaffold attachment insulates the introduced DNA from the influence of factors in the chromatin surrounding the site of integration. It is believed that such influences are predominantly negative and blocking them should tend to enhance gene activity because most of the chromatin is in an inactive conformation at any given time. Since multicopy transformants probably have tandemly arrayed copies of introduced constructs, they should also contain a series of closely spaced SARs. Some or all these SARs, as per the loop domain model, might associate with the nuclear scaffold *in vivo* to create a series of domains insulated from the effects of *cis* elements and structural effects from outside the transgenic locus. Moreover, this arrangement would create small-loop domains that bring the cis-regulatory elements and the scaffold closer. This corresponds to the observation that genes in small-loop domains are expressed at a higher level than the genes in large-loop domains. Alternatively, segments of chromatin with multiple points of attachment with scaffold might physically resist compaction into heterochromatin, which allows continued access by polymerases and transcription factors. This model would be particularly appealing if it is proved that chromatin compaction is involved in co-suppression. This is because it would give a mechanistic explanation for the observation that SAR-containing constructs are more resistant to co-suppression.

For gene expression studies, the selection of a suitable reporter for the experiment is crucial. Each reporter provides a different amount of information about the expression of a gene. There are three general types of reporter gene constructs: transcriptional reporters, translational reporters, and 'smg-1-based' transcriptional reporters (Figure 3.3).

A promoter fragment from a gene of interest driving green fluorescent protein (GFP) is a constituent of transcriptional reporters (Figure 3.3A). The cis-regulatory information, which is required to provide a tentative expression pattern of the endogenous gene under study, is present a few kilobases immediately upstream of the start codon of the promoter.

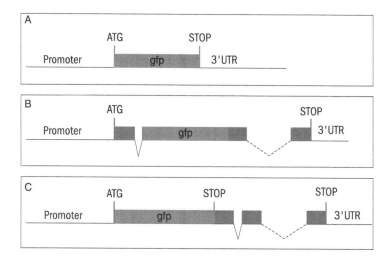

Fig. 3.3 *Translational reporters*

Translational reporters are in-frame gene fusions between GFP and a gene of interest (Figure 3.3B). A translational reporter ideally contains the whole genomic locus of a gene (including 5' upstream region, exons, introns, and 3'-UTR). Although GFP can be inserted at any point in the ORF, it is preferable at a site that does not disrupt protein function and topology.

Transcriptional reporters based on *smg-1* include all *cis*-regulatory information of a gene. GFP, along with its stop codon, is inserted between the promoter and the first exon (Figure 3.3C). Hence, *cis*-regulatory sequences found upstream controls transcription of the transgene, intronically, or downstream of the gene of interest, yet only GFP is translated. The produced mRNA for nonsense-mediated mRNA decay is targeted by the inclusion of the GFP stop codon. Hence, these reporters are required to be injected into a genetic background deficient of nonsense-mediated mRNA decay, such as *smg-1*.

3.8.2 Sequences to be Included in Reporter Constructs

The genomic information of *Caenorhabditis elegans* is very compact. Most *cis*-regulatory information is located within several kilobases immediately upstream of a gene. However, many examples of crucial *cis*-regulatory sequences found within introns or the 3'-UTR are available. Yet, some studies have found *cis*-regulatory sequences at unusually long distances (beyond upstream or downstream flanking genes) from the gene of interest. The commonest reporter gene constructs simply contain a large portion of the 5' intergenic sequence (transcriptional reporter). Reporters should ideally contain as much *cis*-regulatory sequence (from upstream to downstream gene) as possible.

3.9 BIOLISTIC GENE TRANSFER

For transforming plants, various methods are employed for physical delivering of DNA. These include microinjection, PEG-mediated transformation of protoplasts, electroporation of protoplasts and tissues, and microprojectile bombardment of tissues. Among these, microprojectile bombardment is the most widely used method for genotype-independent transformation of plants. Initially, this method was reported to be used for the delivery of DNA and RNA into the epidermal cells of *Allium cepa*. Since then, this technique has been used for the transformation of algae, cereals, pulses, and yeast and filamentous fungi. Biolistic method has been used for the transformation of chloroplasts in *Chlamydomonas* and mitochondria in yeast and *Chlamydomonas*. Two systems are basically used with various modifications for particle bombardment. The electric discharge particle acceleration device (ACCELTM) accelerates DNA-coated gold particles using an instrument at a velocity that can be changed by regulating the input voltage. When powered by a burst of helium gas, Biolistic PDS-1000/He device accelerates the macrocarrier, upon which DNA-coated gold particles (microprojectiles) are uniformly bombarded over target cells.

The particle bombardment system has the following advantages:

- Plants uninfected by *Agrobacterium* can be transformed.
- Specialized vectors may not be required for DNA transfer.
- Multiple DNA fragments/plasmids can be introduced by co-bombardment; thus, the requirement for the construction of a single large plasmid with multiple transforming sequences is eliminated.
- Reporter gene expression in *Agrobacterium* leading to false positive results can be avoided.
- Plants lacking good regeneration systems can be aided by the transformation protocols.
- Transformation of organelle is achieved only by particle bombardment.

Although transgenics have been successfully created by using biolistic gene delivery system, certain disadvantages have been observed in this technique, for example, high copy number and rearrangements of transgene(s), which result in gene silencing or genomic rearrangements. However, the problems related to gene silencing and genetic integrity have been addressed with various improvements. Osmotic conditioning of cells has helped in the improvement of the overall transformation efficiency of biolistics.

3.10   TRANSGENE STABILITY AND GENE SILENCING

Researchers involved in generating transgenic plants with an increased level of endogenous gene expression by using heterologous strong promoters or a foreign gene of some utility face the problem of silencing transgenes in a vast population of transgenic plants. The silencing of a gene at the transcription level is referred to as transcriptional gene silencing (TGS). It involves in the inhibition of transcription, which may result by the methylation of the promoter region. Multiple copies of transgenes and inverted repeats (IRs) in transgenes have been found to be associated with TGS that can act in both *cis* and *trans*, suggesting a homology-dependent mechanism of gene silencing. TGS in plants might have evolved to act as a defence mechanism against the activities of endogenous transposable elements present in multiple copies.

Gene silencing at the post-transcriptional level is referred to as post-transcriptional gene silencing (PTGS). In this case, although the genes are transcribed, their mRNA is degraded. Methylation of the coding region of transgenes is associated with PTGS. However, PTGS has been observed even when unrelated foreign genes have been transferred to plants. PTGS has been associated with the strength of the promoter, high copy number of transgenes, or stability of the transcripts. A threshold level of transcript accumulation seems to exist for the activation of PTGS, and viruses that infect plants accumulate their transcripts at a very high level. Thus, PTGS can be a possible mechanism to stop viral gene transcription. Studies propose that some plants use PTGS to recover from viral infection and avoid from future infections. Again, some plant viruses are found to have evolved to counter or suppress PTGS for their successful survival.

Though DNA methylation is involved in both TGS and PTGS, it is not very clear whether this is the cause of gene silencing. However, efforts have been made to find an answer to this question. Methylation at asymmetric sites has also been observed although CG and CNG symmetric sequences are common sites of methylation. A mutant form of 35S promoter that lacked all symmetrical methylation acceptor sites was synthesized to find the involvement of DNA methylation in gene silencing. This promoter was utilized to drive the *bar* gene. A considerable decrease in the number of PPT-resistant progeny was noticed when resistant lines obtained were crossed with plants carrying the 271-silencer locus (which possessed the ability of silencing other 35S promoters). At only asymmetric methylation acceptor sites, the silenced lines were found to be methylated. PPT resistance was restored when the 271-silencer locus was outcrossed. Experiments on similar lines, but using wild-type 35S promoter, revealed that the silenced state was maintained even after outcrossing the silencer locus. These studies indicate that methylation at symmetric methylation sites is not needed for the initiation of silencing. But this is necessary for the maintenance of silencing through generations. Mutation in *MET* gene, which encodes a DNA methyltransferase, results in a drastic reduction in DNA methylation and causes the release of silencing at most loci, but some loci still remain silenced, which might be due to the presence of other functional methyltransferases. The mutation in *DDM1* locus, which encodes a plant homologue of SW12/SNF2 protein, a component of yeast chromatin remodelling complex, also causes loss of methylation and release of silencing. This study proposes that for the maintenance of methylation and gene silencing, a normal chromatin structure is required.

Studies with mammalian systems have indicated that MeCP2, a methylated DNA-binding protein that can bind to even a single methylated CpG, recruits a repressor complex that includes HADC. In the case of a similar situation in plants, it should be possible for a mutation in the HADC gene to relieve silencing without impacting the methylation status. It has been found that transgenic inhibition of HADC in *Arabidopsis* is associated with pleiotropic effects that are similar to those noticed in the case of *MET/DDMI* mutation, without any considerable reduction in DNA methylation. Also, silencing is released without impacting DNA methylation by the mutation of *MOM* locus, which stands for 'Morpheus' molecule'. This gene might be another component of the chromatin remodelling complex. Are TGS and PTGS linked? Although TGS and PTGS operate differently, both involve DNA methylation, albeit of different regions. It has been established that expressed RNA can trigger the methylation of the coding region and PTGS. Upon infection, the potato spindle tuber viroid, which does not code for any protein and contains a small RNA genome, can direct methylation of its transgene integrated into plant genome.

A recent report indicates that the generation of double-stranded RNA complementary to the coding sequence from a single transcript, which can pair to itself, may induce PTGS of genes having a homologous sequence. Such double-stranded RNA can also be generated *in vivo* from promoter sequences having IRs or other complex structures by read-through transcription from nearby promoters. Recent experimental evidence supports this possibility. Mette, Aufsatz, van der Winden, *et al.* (2000) expressed double-stranded RNA corresponding to the promoter sequence and observed TGS. These studies indicate that TGS and PTGS may have some common features. It has been shown that aberrant RNAs are synthesized during PTGS. Using these RNAs as substrates, RNA-dependent RNA polymerase synthesizes antisense RNAs. The pairing of both sense and antisense RNAs results in the formation of double-stranded RNAs, which are targeted for degradation. Experimental evidence has been obtained in tomato for the presence of RNA-dependent RNA polymerase. During PTGS96, degradation products of about 25 bases corresponding to the silenced transgene have been found. In a recent study, Bernstein, Caudy, Hammond, *et al.* (2001) found an RNase from *Drosophila* that they termed 'Dicer' because it chops RNA into small pieces of uniform size. The RNA is first chopped into 22-nucleotide fragments, which are then incorporated into a multi-subunit complex and function as a guide RNA for targeting the degradation of single-stranded mRNA. Such a mechanism might be functional in plants.

Researchers have discarded plants that show silencing of the desired phenotype and utilized plants that do not show silencing as transgene silencing is an unpredictable phenomenon. Efforts to overexpress endogenous genes have often silenced both transgenes and endogenous genes, a phenomenon known as co-suppression. Attempts to overexpress useful genes have suffered because RNA expression above a threshold has been found to lead to PTGS. A few researchers have successfully achieved high expression levels of genes of interest, but others have faced problems. De Neve, De Buck, Jacobs, *et al.* (1997) generated five different homozygous lines of *Arabidopsis* for producing antibodies. They found occurrence of gene silencing in all the lines. Gene silencing has been linked to the methylation of promoter alone, promoter and coding sequence, coding sequence alone, and coding region and downstream sequences. Though the presence of duplicated sequences, methylation acceptor sites, high copy number of foreign genes, and strength of promoter with silencing are all correlated, efforts to overcome silencing by addressing these features have not been well documented. Common occurrence of multiple copies of transgenes in the case of the biolistic method of transformation has made the *Agrobacterium*-mediated method of transformation well known. However, silencing of even single copy genes has been reported. To avoid the silencing of transgenes, some attempts have been made, and one of them is the use of MAR in transformation vectors for preventing the effect of heterochromatin on integrated genes.

To protect transgenes from the influence of nearby *cis*-acting elements, insulation elements have been used in vertebrate systems. Plant viruses have evolved strategies for avoiding PTGS that plants utilize to avert viral infections. In tobacco plants infected by these viruses, 2b protein of cucumber mosaic virus (CMV) and HC-protease (HC-Pro) of potato virus Y (PVY) have been found to interfere with PTGS. While HC-Pro blocks the maintenance of gene silencing in tissues in which silencing has already taken place, 2b protein prevents the initiation of gene silencing at growing points of plants. Recently, researchers have identified a calmodulin-like protein named rgs-CaM in tobacco that can suppress PTGS. These proteins can be utilized to stop gene silencing. However, it is yet to be investigated whether the expression of such proteins would make these plants vulnerable to viral infections. For avoiding transgene silencing, specified guidelines should be followed while designing a transgene. Depending on the correlation of various features of vectors and gene silencing, attempts should be made to avoid the use of potential methylation acceptor sites. By using different promoters and termination signals to drive the expression of different genes of the vector, DNA repeats can be avoided. Vectors can be designed to incorporate insulators or MARs and even genes for suppression of gene.

REFERENCES

Ballas, N. and V. Citovsky. 1997. Nuclear localization signal binding protein from Arabidopsis mediates under nuclear import of Agrobacterium VirD2 protein. *PNAS, USA.* 94: 10723–28

Berlin, J., H. Beier, L. Fecker, E. Forche, W. Noe, F. Sasse, O. Schiel, and V. Wray. 1985. Conventional and new approaches to increase the alkaloid production of plant cell cultures. In *Primary and Secondary Metabolism of Plant Cell Cultures*, K. H. Neumann, W. Barz, and E. Reinhard (eds). Berlin: Springer–Verlag

Bhattacharjee, S., L. Y. Lee, H. Oltmanns, H. Cao, J. H. Veena, J. Cuperus, and S. B. Gelvin. 2008. IMPa-4, an Arabidopsis importin alpha isoform, is preferentially involved in agrobacterium-mediated plant transformation. *Plant Cell* 20: 2661–80

Brencic, A. and S. C. Winans. 2005. Detection of and response to signals involved in host-microbe interactions by plant associated bacteria. *Microbiol Mol Biol Rev* 69: 155–194

Bernstein, E., A. A. Caudy, S. M. Hammond, and G. J. Hannon. 2001. Role for a bidentate ribonuclease in the initiation step of RNA interference. *Nature* 409(6818): 363–66

Cangelosi, G. A., L. Hung, V. Puvanesarajah, G. Stacey, D. A. Ozga, J. A. Leigh, and E. W. Nester. 1987. Common loci for *Agrobacterium tumefaciens* and *Rhyzobium meliloti* exopolysaccharide synthesis and their role in plant interaction. *Journal of Bacteriology* 169: 2086–91

Cangelosi, G. A., G. Martinetti, C. C. Leigh Lee, C. Theines, and E. W. Nester. 1989. Role of *Agrobacterium tumefaciens* chvA protein in export of β-1,2-glucan. *Journal of Bacteriology* 171: 1609–1615

Cangelosi, G. A., E. A. Best, C. Martinetti, and E. W. Nester. 1991. Genetic analysis of *Agrobacterium tumefaciens*. *Methods in Enzymology* 145: 177–81

De Neve, M., S. De Buck, A. Jacobs, M. Van Montagu, and A. Depicker. 1997. T-DNA integration patterns in co-transformed plant cells suggest that T-DNA repeats originate from co-integration of separate T-DNAs. *Plant J.* 11: 15–29

Deng, W., L. Chen, D. W. Wood, T. Metcalfe, X. Liang, M. P. Gordon, L. Comal, and E. W. Nester. 1998. Agrobacterium virD2 protein interacts with plant host cyclophilins. *PNAS USA*. 95: 7040–45

Gelvin, S. B. 1998. Agrobacterium virE2 proteins can form a complex with T strands in the plant cytoplasm. *J. of Bacteriol.* 180: 4300–02

Herrera-Estrella, L., M. De Block, E. Messens, J. P. Hernalsteens, M. Van Montagu, and J. Schell. 1983. Chimeric genes as dominant selectable markers in plant cells. *EMBO Journal* 2: 987–96

Hookyaas, P. J. J. 1998. Agrobacterium molecular genetics. In *Plant Molecular Biology Manual*, edited by S. B. Gelvin and R. A. Schilperoort, pp. A4–A13. Dordrecht, The Netherlands: Kluwer Academic Publishers

Lee, L.-Y., M.-J. Fang, L.-Y. Kuang, and S. B. Gelvin. 2008. Vectors for multi-color biomolecular fluorescence complementation to investigate protein–protein interactions in living plant cells. *Plant Methods* 4: 24

Mette, M. F., W. Aufsatz, J. van der Winden, M. A. Matzke, and A. J. M. Matzke. 2000. Transcriptional silencing and promoter methylation triggered by double-stranded RNA. *EMBO J.* 19(19): 5194–5201

Narasimhulu, S. B., X. B. Deng, R. Sarria, and S. B. Gelvin. 1996. Early transcription of Agrobacterium T-DNA genes in tobacco and maize. *Plant Cell* 8: 873–86

Sonti, R. V., M. Chiurazzi, D. Wong, C. S. Davies, G. R. Harlow, D. W. Mount, and E. R. Singer. 1995. *Arabidopsis* mutants deficient in T-DNA integration. *Proceedings of the National Academy of Sciences of the United States of America* 92: 11786–90

Stachel, S. E. and E. W. Nester. 1986. The genetic and transcriptional organization of the vir region of the A6 Ti plasmid of *Agrobactrium tumefaciens. EMBO J.* 5: 1445–54

Tao, Y., S. Bhattacharjee, and S. B. Gelvin. 2004. Expression of plant protein phosphatase 2C interferes with nuclear import of the *Agrobacterium* T-complex protein VirD2. *Proceedings of the National Academy of Sciences of the United States of America* 101: 5164–69

Tzfira, T., M. Vaidya, and V. Citovsky. 2002. Increasing plant susceptibility to Agrobacterium infection by overexpression of the Arabidopsis nuclear protein VIP1. *Proc Natl Acad Sci USA* 99: 10435–40

Yuan, Z. C., M. P. Edlind, P. Liu, P. Saenkham, L. M. Banta, A. A. Wise E. Ranzone, A. N. Binns, K. Kerr, and E. W. Nester. 2007. The plant signal salicylic acid shuts down expression of the *vir* regulon and activites quormone-quenching genes in *Agrobacterium. Proceedings of the National Academy of Sciences of the United States of America* 104: 11790–95

Plant Genetic Engineering for Productivity and Performance

4.1 RESISTANCE TO HERBICIDES

The inaccessibility of herbicide molecules at the site of action develops resistance in plants. Herbicide molecules cannot concentrate at the target site in the right lethal amount. Thus weeds do not die and they develop resistance to the herbicide. Herbicides can get excluded from the action site due to various reasons. Morphological barriers on plants, such as hairy epidermis, low foliage number and size, and enhanced waxy coating on the cuticle lead to differential uptake of herbicides. Differential translocation also excludes herbicides from the target site. Apoplastic or symplastic path prevents or delays the transfer of appropriate concentration of herbicide to the target site. Herbicides can also be sequestered at a number of locations before they reach the site of action. For instance, some lipophilic herbicides can be immobilized by partitioning into lipid-rich glands or oil bodies.

Detoxification of herbicides is another reason for their exclusion from the site of action. The detoxification reaction can be oxidation, reduction, hydrolysis, or conjugation. Resistance genes are influenced by enzymes because of metabolic detoxification. For example, in the velvetleaf weed *Abutilon theophrasti*, increase in the enzyme glutathione S-transferase causes detoxification of atrazine herbicide, and hence the weed gains resistance. Similarly, in *Echinochloa colona*, increased content of aryl-acylamidase detoxifies propanil herbicide. Herbicide metabolism can also be increased due to fast-acting cytochrome P450 monooxygenase, with target enzymes such as acetyl coenzyme A carboxylase (ACC), acetolactate synthase (ALS), and photosystem II (PSII).

4.1.1 Site of Action Resistance

The site of action is altered in such a way that it is no longer vulnerable to the toxicity of the herbicide. Many species have been reported around the world to have developed resistance against sulphonylurea. In *Lactuca sativa* biotype, the site of action of sulphonylurea, that is, the ALS enzyme, is modified in such a way that the herbicide cannot bind to its site. Hence, the enzyme is not harmed and the weed biotype is not killed by the application of herbicide.

In the gene encoding the protein, which is mostly an enzyme to which the herbicide binds, a single-nucleotide change or mutation occurs. The amino acid sequence of the protein is altered by the change of one nucleotide, which destroys the herbicide's ability to interact with the protein and also the herbicide fails to incapacitate the enzyme's normal functioning. Other unrelated physiological pathways are indirectly hampered by mutations, leading to herbicide resistance. This affects the growth and development of the resistant biotype. Mutated resistance against sulphonylurea in *Kochia scoparia* diminishes the acetolactate enzyme (ALA) sensitivity to normal patterns of feedback inhibition. As a result, the availability of amino acids for cell division is increased and growth and development are accelerated. Consequently, the biotype of *K. scoparia* resistant to sulphonylurea dominates germination compared to its counterpart susceptible biotype. In some cases, the site of action is enlarged, resulting in the dilution effect of the herbicide. Thus, applying the herbicide at the normal rate fails to inactivate the total amount of enzyme protein produced. Hence, the additional amount of enzyme produced by the plant biotype enables it to carry out its normal metabolic activities, overcoming the lethal effects of the herbicide.

4.1.2 Assessing Herbicide Resistance

It is not necessary that herbicide treatment has failed because of weed resistance. The parameters discussed next should be taken into account before attributing the failure of a herbicide to weed resistance.

4.1.2.1 *Visual diagnosis*

• The product label contains important information about the herbicide's potential effects. So check whether other listed weeds on the label have been controlled satisfactorily. Maybe only one weed species is resistant to the herbicide in field trials. Thus, factors other than herbicide resistance may be the reason for lack of weed control.

- If the uncontrolled weeds exist in patches and each patch contains different species, herbicide resistance might not be the case because all species might not develop resistance. There could be several other reasons.
- If the herbicide has been repeatedly used in the same field and its mode of action is same, the weed might be developing resistant biotype.
- The area should be surveyed for any previous case of herbicide resistance. Farmers might have observed resistant to the same herbicide. So enquire if the herbicide failed in the same area of the field in the previous year.
- Also, there might be a decline in the level of weed control in the last few years.

In case of any of the aforementioned cases, there are chances that the weed species is developing herbicide resistance.

4.1.2.2 *Bioassay*

If it is found from the diagnostic survey that herbicide resistance has probably been evolved in a certain biotype; confirmation tests should be carried out. For this purpose, an adequate sample of seed or plant material should be collected from the suspected population.

4.1.2.2.1 Plant assay/Seed collection

Healthy plants and stage of seed harvest are important factors for reliable germination and proper growth of plants during experiments. The best time to collect ripened seeds is when a plant has shed at least 20% of its seeds. From an average area of 50–100 m, an adequate amount of seeds should be collected by gentle rubbing. The name of the species, date, and location should be labelled on the bag.

4.1.2.2.2 Greenhouse/Plant–pot assay

Greenhouse or plant–pot assay is a preliminary bioassay on mortality and plant vigour check. The discriminatory spraying effect of the herbicide is revealed by this test. However, it must include a treatment of susceptible set of pots for reference and not for evaluation of herbicide sensitivity or insensitivity. Appropriate design and replications can be made by seeking statistical advice.

4.1.2.2.3   Dose–response experiments

A standard response curve can be formulated by using a range of doses. By calculating the dose ratio required to achieve the same effect in both resistant and susceptible populations, the curve enables the quantification of resistance. Usually, the dose required for a 50%–70% reduction in the measured parameter (foliage weight or the number of surviving plants) relative to the untreated control is determined (Figure 4.1). Ratios of these estimates (termed ED_{50}, GR_{50}, LD_{50}, or 150) with respect to that of a susceptible population give the resistance index (RI), which describes the degree of resistance in simple terms. A good estimate of ED_{50} can be obtained by keeping the dose range relatively wide; preferably six doses should be used. Moreover, best results are obtained when each dose is twice the preceding dose (e.g., 10, 20, 40, 80, 160, and 320 g a.i./ha). Since herbicides are more active under greenhouse conditions, there should be doses both below and above the field recommended rate in the range.

4.1.2.2.4   Single-dose resistance assay

After obtaining the dose–response information from the preceding experiment, more populations can be tested by using a single or 2–3 discriminating doses for the future screening assays.

4.1.2.2.5   Specific discreet tests

Many other detailed confirmatory analytical tests, such as floating leaf disc assay, enzyme specificity/sensitivity assay, Petri dish germination assay, and chlorophyll fluorescence assay, can be used.

4.1.3   Herbicide Resistance Management (Prevention and Delaying Resistance)

A weed management strategy applied to minimize selection pressure for resistance will block the development of resistance. All or a combination of some of the strategies discussed in the following sections can prevent or delay the evolution of resistance in weeds.

4.1.3.1   Herbicide rotation

The use of the same herbicide or a different herbicide with a similar mode of action for consecutive years increases the evolution of resistance. However, a process called herbicide rotation can reverse

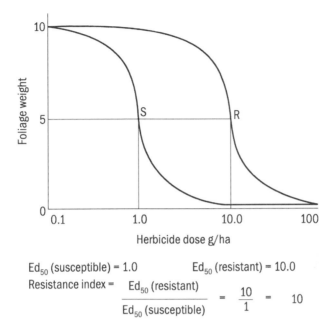

Ed$_{50}$ (susceptible) = 1.0   Ed$_{50}$ (resistant) = 10.0

Resistance index = $\dfrac{Ed_{50} \text{ (resistant)}}{Ed_{50} \text{ (susceptible)}}$ = $\dfrac{10}{1}$ = 10

Fig. 4.1 *Dose–response curves for a susceptible and resistant population*

the action. In this process, different herbicides with different modes of action are used in subsequent seasons. Herbicide-resistant transgenic crops, such as Roundup Ready soya beans, are gradually gaining popularity beyond North America. Single use of a herbicide, for example, glyphosate in soya bean or corn, is necessary. This will enhance weed-resistant culture. In addition, two or more herbicides with different modes of action can be mixed, called tank mix application, to prevent or postpone resistance pressure. For example, in Australia, the continuous use of amitrol and atrazine for 10 years led to the evolution of resistance in ryegrass grown along railway lines. A tank mix of glyphosate and sulphometuron methyl was applied to effectively control this resistant biotype. However, this combination has to be further rotated to delay the evolution of resistance.

Furthermore, the potential threat of resistance evolution can be reduced by including non-chemical control options such as integrated pest management (IPM). Employing the least persistent herbicide reduces the resistance risk. If the weed control spectrum of two herbicides in a tank mix is the same, each herbicide should be separately used in seasonal rotation. The herbicide should not be sprayed in one season because susceptible seeds will grow vigorously from the geo-seed-bank reservoir. As a result, the fewer resistant biotype will be out-competed by the non-resistant biotype. It will also face strong competition and threat to survival due to natural rejection.

4.1.3.2 Crop rotation

The same ecological culture results in the use of same inputs, including a herbicide, when the same crop is grown every season. This can be prevented by following the technique of crop rotation. It helps in manipulating cultivation techniques, planting time, spectrum of weed infestation, as well as choice of herbicides with different modes of action, different stages, and different ways of application. The resistant biotypes can be eradicated by using various combinations of weed control strategies.

4.1.3.3 Post-treatment monitoring

Any pattern of uncontrolled weed patches should be carefully watched. Certain patches that are neglected during spraying should be differentiated from resistant biotypes. Those patches that are not spray neglects should be eliminated manually or by the superimposed herbicide application.

4.1.3.4 Integrated cultural practices

All possible non-chemical control methods can be combined to effectively prevent evolution of weed resistance. Cultivation practice, in contrast to no-tillage farming, buries the emerged weed seedlings and solarizes the soil. Weeding by hand eliminates 90%–100% weed plants. Mulching with organic matter will result in reduced weed population with crop stand.

4.1.3.5 Herbicide resistance management (Post-evolution)

Integrated weed management approaches should be adopted to control herbicide-resistant biotypes. The resistant population can be reduced or eradicated by following crop rotation or, preferably, fallow tillage, followed by close cultivation. Extensive manual weeding can achieve 100% eradication and ensure fewer weeds in the next season. Clean and certified seeds should be planted each season, and farm machinery should be cleaned before use. Infested seeds of the resistant biotype can be removed from the machinery by using a power washer.

Farmers can be encouraged to follow the aforementioned diverse weed control strategies. Farmers who rely heavily on selective and non-selective herbicides for controlling weed and crop intensively on the same fields have a much greater risk of developing weed resistance in their fields. These farmers should be alerted to the possibility of resistance appearing under these conditions and should be encouraged to modify their cropping practices.

4.1.3.6   *Advanced services for resistance management*

The scenario of mono-cropping pattern in which single-action, broad-spectrum, ready herbicides are used with transgenic crop seeds has aggravated the problem of herbicide resistance. The following measures should be adopted to overcome the problem of herbicide resistance:

- Research on DNA fingerprinting test for resistant and susceptible biotypes should be encouraged.
- Researchers should be provided with electronic database based on region, crop, or herbicide-resistant species.
- Genetic resistance (single gene or additive gene effect) must be investigated.
- Technologies for remote aerial sensing of fields for clump areas should be developed.

Resistance to herbicides should be timely detected and resistance management strategies should be implemented before the situation aggravates. Otherwise, higher cost and human endeavour would be required for undertaking any extreme strategy. Currently, herbicide producers and policymakers (the state or federal governments) do not seem to take precautionary approach with regard to resistance management. The marketing plans of herbicide manufacturers lack recognition of the resistance concern. Thus, all stakeholders, including farmers, scientists, and herbicide makers, must jointly address this issue for herbicides to remain as an effective agricultural input. There has to be concerted efforts to invest in resistance research. As professional strategies can restrict the extent of resistance, it is recommended that farming practices be assessed as discussed in the previous sections. Hence, professional vigilance for probable resistance, followed by laboratory testing, should be monitored.

4.2   INSECT RESISTANCE

In the 21st century, genetic engineering and molecular biology have emerged as powerful tools to develop and manipulate crop genotypes for safe and sustainable agriculture. Safer alternatives should replace harmful technologies and chemical inputs to control insect pests in agricultural ecosystems. Many insecticidal proteins and molecules are available in nature that are effective against pests and also beneficial to mammals, insects, and other organisms. For example, the insecticidal proteins present in *Bacillus thuringiensis* (Bt) have shown positive results. These proteins have been successfully expressed in many crop species over the past five decades.

Large-scale production of Bt crops has increased concerns about the development of resistance in insects. However, many strategies have been formulated to prevent or delay the development of resistance (Ranjekar, Patankar, Gupta, *et al.* 2003). These strategies should be seriously considered in India as they have released the first Bt crop resistant to insects for commercial cultivation. One of the strategies could be combining various insect-resistant proteinase inhibitors present in many plants as a viable alternative to Bt to control pests such as *Helicoverpa armigera*. Besides plants, many bacteria such as *Photorhabdus luminescens* contain novel insecticidal proteins. Hence, insects can be prevented from developing resistance when there is judicious expression of multiple insecticidal proteins that differ in their toxicity mechanism. Finally, following IPM strategies during the cultivation of transgenic crops will ensure durable insect resistance.

The menace of insect pests is a major factor destabilizing productivity in agricultural ecosystems. Insect pests such as lepidopterans and orthopterans cause extensive damage to crops and stored seeds. The predominantly tropical and subtropical climate in India favours a perpetual onslaught of insects on the biodiversity of agricultural, horticultural, and forest species. Breeding for resistance to insect pests is the top priority for many important crops. Table 4.1 lists some of the important pests in major crops of India.

Table 4.1 Important pests on major crops of India

Crop	Insect pest	Family
Rice	Yellow stem borer (*Scirpophaga incertulas*)	Lepidoptera
	Brown plant hopper (*Nilaparvata lugens*)	
Mustard	Mustard aphid (*Lipaphis erysimi*)	Hemiptera
Pigeon pea	Gram pod borer (*Helicoverpa armigera*)	Lepidoptera
Sugarcane	Cotton boll worm (*H. armigera*)	Lepidoptera
Tomato	Top borer (*Scirpophaga nivella*)	Lepidoptera
	Leaf miner (*Stomopterix nertaria*)	Lepidoptera
Brinjal	Tuber moth (*Phthorimaea operculella*)	Lepidoptera
Cauliflower and cabbage	Fruit borer (*H. armigera*)	Lepidoptera
	Shoot and fruit borer (*Leucinodes orbonalis*)	Lepidoptera
	Diamondback moth (*Plutella xylostella*)	Lepidoptera

Two factors have enhanced pest incidence: introduction of high-yielding varieties that are more responsive to nitrogen and lack of proper crop rotation practices. Though chemical pesticides have provided considerable protection to crops over the last few decades, but their extensive and indiscriminate use has degraded the environment and resulted in adverse effects on human health and other organisms. The application of these pesticides has also eradicated beneficial insects and developed resistant insects. As the objective of the new millennium is to achieve higher and stable crop productivity, it is necessary to apply safe and environment-friendly strategies to attain these goals.

There are a large number of effective insecticidal molecules that are innocuous to humans and other organisms. These molecules should be harnessed and utilized with the help of molecular biology and genetic engineering techniques in a safe and sustainable manner.

4.2.1 Insecticidal Proteins of *Bacillus Thuringiensis*

Bacillus thuringiensis is a Gram-positive, sporulating bacterium that synthesizes crystalline proteins during sporulation. These proteins are highly non-toxic to mammals and other organisms. Bt strains and their insecticidal crystal proteins (ICPs) have been used all over the world as eco-friendly biopesticides and for mosquito control for the past few decades (Schnepf 1995). The advent of molecular biology and genetic engineering has made it possible to use Bt more effectively and rationally in crop plants (Kumar, Sharma, and Malik 1996).

A range of lepidopteron insects recognized worldwide as major agricultural pests are affected by Bt strains and ICPs. The host range has subsequently expanded by the discovery of new strains. Now strains toxic to coleopterans, dipterans, lice, mites, and even nematodes are available. Most Lepidoptera families contain species susceptible to Cry1 and Cry2 crystal proteins produced particularly by Bt serotypes *kurstaki* and *aizawai*. The crystal toxins are at present classified based on their amino acid sequence homology. ICPs are categorized under 40 different classes, and some of them exhibit specificity to multiple insect orders (Crickmore, Zeigler, Feitelson, *et al.* 1998). The toxicity of these ICPs towards different pests has been studied. Because of the high commercial value, Bt ICPs are undergoing extensive screening programmes. Molecular structures of at least three different ICPs have been studied. When the insect larva ingests these crystals, they are solubilized in the highly alkaline midgut into individual protoxins. These protoxins vary in molecular weight from 133 kDa to 138 kDa, depending on their type. The midgut proteases act on the protoxins and cleave them into two halves. The N-terminal half is the toxin protein and usually of 65–68 kDa. The toxin protein fragment can be divided into three domains: the first is involved in pore formation, the second determines receptor binding, and the third protects the toxin from proteases. The toxin protein binds to specific receptors in the midgut epithelial membranes. Upon receptor binding, domain I inserts itself into the membrane and, as a result, a pore is formed. The insect dies due to disturbances in osmotic equilibrium and cell lysis.

Bacillus thuringiensis ICPs are delivered through spray formulations, engineered Bt, and other bacteria, but delivery through these means has some limitations. The biopesticide sprays have short half-life, can be removed by wind and rain, and cannot reach burrowing insects. Engineered bacteria often do not proliferate at the rate and quantity required to kill the target insects. However, these drawbacks can be overcome by expressing ICPs in plant cells at levels sufficient enough to kill the larvae.

In 1987, the first transgenic plants were developed that used *cry* genes. Tobacco plants were engineered with truncated genes encoding

Cry1Aa and Cry1Ab toxins and were found to be resistant to the larvae of tobacco hornworm (Jouanin, Bonade-Bottino, Girard, *et al.* 1998). However, the levels of cry protein expression in the plant tissues were not very high. In 1990, researchers at the Monsanto Company made a significant breakthrough when they modified the *cry* genes (*cry1Ab* and *cry1Ac*) for better expression in plant cells. They altered the codon usage of prokaryotic genes of Bt to resemble that of higher plants. Besides, many properties such as presence of putative poly(A)-type signals and splice sites that destabilize Bt mRNAs in plant cells were removed without altering the amino acid sequence of the ICPs. Considerable protection against lepidopteran and coleopteran pests was achieved by the expression of *cry1Ac* gene in cotton and *cry3Aa* gene in potato, respectively. Consequently, rice, maize, peanut, soya bean, canola, tomato, cabbage, and other crop plants were transformed with various modified *cry* genes (de Maagd, Bosch, and Stiekema 1999).

Selvapandian, Reddy, Kumar, *et al.* (1998) gave an example of native gene (*cry1IA5*) expression resulting in significant resistance to *H. armigera* in transgenic tobacco. Another landmark was the introduction of a native *cry1Ac* gene into the chloroplast genome of tobacco. This expressed the cry protein to a very high level, about 3%–5% of leaf soluble protein (McBride, Svab, Schaaf, *et al.* 1995). Besides providing high foreign protein expression, chloroplast transformation also ensures maternal transmission of the foreign gene, which prevents the spread of transgene through pollen. This strategy can prove very useful in the future if used in important crop plants such as cotton and rice. However, it remains to be seen whether transformed chloroplast genomes can protect the reproductive and fruit bearing parts, which are often targeted by insects. There are currently more than 30 plant species that have been transformed with Bt *cry* genes (Schuler, Poppy, and Denholm 1998) (Table 4.2).

Bacillus thuringiensis crops were first commercialized in 1996 in the United States when the bollworm-resistant cotton Bollgard was introduced. Later, potato and maize crops were also commercialized (Krattiger 1997). Intensive efforts are under way in India to introduce *cry* genes in rice, potato, cotton, sorghum, and vegetables. Investigations were carried out on different ICPs for their relative toxicity to various target pests. Transgenic crop species carrying different *cry* genes are at various stages of development. The Tata Energy Research

Institute (now The Energy and Resources Institute) developed the first transgenic plants of tobacco by using modified *cry1Ab* and *cry1C* (obtained from Dr Bert Visser, CPRO-DLO, the Netherlands). These plants showed considerable protection against tobacco caterpillar (*Spodoptera litura*) in limited field trials at the Central Tobacco Research Institute. For resistance to yellow stem borer (YSB), researchers at the Bose Institute (Kolkata) introduced a modified *cry1Ac* gene in rice (IR64). Under the control of ubiquitin promoter, a synthetic *cry1Ac* gene was introduced in rice (Pusa Basmati 1, Karnal Local, and IR64). Transgenic lines exhibiting total protection against neonate larvae of YSB were identified. Field evaluation of these transgenic varieties was carried out in 2002, and lines resistant to YSB were identified. To confer resistance to fruit borers, crops of vegetables such as brinjal and tomato were transformed by synthetic or modified *cry1Ab* and *cry1Ac* genes, respectively. Limited field trials of Bt brinjal and Bt tomato were carried out for three and two growing seasons, respectively. The degrees of insect protection found in brinjal and tomato were 75% and 94%, respectively. To achieve considerable protection against tuber moth and *H. armigera*, four genotypes of potato were transformed by modified *cry1Ab*. Many public and private organizations are also carrying out research to develop insect-resistant varieties of castor, cotton, groundnut, rice, sorghum, sunflower, and tobacco. In collaboration with Monsanto, Mahyco introduced a modified variety of cotton into parental lines of hybrids that were bred specially for Indian agronomic conditions. These transfers required four backcrosses and

Table 4.2 Plant species transformed with Bt *cry* genes

Crop	Gene	Target pest
Cotton	*cry1Ab/cry1Ac*	Bollworms
Corn	*cry1Ab*	European corn borer
Potato	*cry3a*	Colorado potato beetle
Rice	*cry1Ab/cry1Ac*	Stem borers and leaf folders
Tomato	*cry1Ac*	Fruit borers
Potato	*cry1Ab*	Tuber moth
Eggplant	*cry1Ab/cry1B*	Shoot and fruit borer
Canola	*cry1Ac*	Diamondback moth
Soya bean	*cry1Ac*	Soya bean looper
Corn	*cry1H/cry9C*	European corn borer

two selfed generations. These hybrids were tried on field at various places, and various experiments on gene flow and effects of pollen and plants on non-target organisms were carried out. It was found that Bt cotton required no or minimal pesticide sprays, whereas non-transgenic plants required 9–12 sprays to control bollworms. The Government of India approved the commercial release of Bt cotton in March 2002.

Nunhems Proagro Seeds Pvt. Ltd is awaiting results of field trials of Bt vegetables carrying modified *cry* genes. The insecticidal efficacy of Bt ICPs to pests such as *S. litura*, *Earias insulana*, *Chilo partellus*, *Spilosoma obliqua*, and *Maruca testulalis* should be systematically evaluated as has been done for *H. armigera* and *L. orbonalis*. Valuable information can be obtained from the biochemical analysis of receptor binding in relation to δ-endotoxins for designing suitable toxin combinations to be expressed in transgenic plants.

4.2.2 Polyunsaturated Fatty Acids

When mutant strains of *Arabidopsis* were analysed, it was found that the size of the chloroplast and the formation of its membranes were influenced at low temperatures by polyunsaturated fatty acids in the chloroplast membranes. The activity of a chloroplast ω-9 fatty acid desaturase is absent in the fad5 mutant, which accumulates high levels of palmitic acid (16:0). On the contrary, the fad6 mutant lacks the activity of the chloroplast ω-6 fatty acid desaturase, and it accumulates high levels of fatty acids (16:1 and 18:1). Correspondingly reduced levels of polyunsaturated fatty acids were found in the chloroplast galactolipids of both mutants. For three weeks, young seedlings of these mutants and the wild type (grown for seven days at 22°C) were subjected to chilling treatment (5°C). The mutant strains exhibited more chlorosis than the wild type. There was also a decrease in the amount of chlorophyll to approximately half of that in the wild-type strain. The chloroplast membrane (thylakoid membrane) content in the mutant strains decreased and the chloroplast became smaller.

However, more mature plants showed no difference when they were subjected to similar chilling treatment. There were no variations observed between wild-type and mutant strains regarding the membrane content, chloroplast size, or chlorophyll. On the contrary, long-term culture (42 days) at 6°C inhibited growth in the fad2 mutant, and the plants eventually withered and died. The fad2 mutant lacks the activity of 18:1 fatty acid desaturase, and thus there is a marked decrease in the amount of polyunsaturated fatty acids in the extrachloroplast membrane lipids. Therefore, it can be concluded from these observations that polyunsaturated fatty acids are necessary for surviving in low temperature. Moreover, under chilling stress, such fatty acids in chloroplast membrane lipids are found to contribute more to the regular formation of chloroplast membranes than to maintaining their stability when they are mature.

The cyanobacterium gene with a broad specificity for the 9 desaturases (des9) was introduced into tobacco (D9-1 line). The 16:1c7 fatty acids found in higher plants are bound to the sn-2 position of the chloroplastic lipid monogalactosyldiacylglycerol (MGD). They exist only in minute quantities as intermediates in the production of 16:3. The characteristics of the 16:1t3 fatty acid bound to the sn-2 position of PG are similar to those of the saturated 16:0 fatty acids. The 16:0 fatty acids that are bound to the sn-1 position of PG and all other lipid species are not desaturated.

The 16:1c9 fatty acids produced by the D9-1 line are not found naturally in higher plants. These acids accounted for more than 10% of all fatty acids esterified to individual membrane lipids. The wild-type plants grown at 25°C underwent chlorosis when chilled at 1°C for 11 days. However, this effect was not observed in the D9-1 line subjected to the same treatment. When the seeds of both wild-type and D9-1 plants were germinated at 10°C and grown for 52 days, the wild-type seedlings became pale green and showed damage from chilling, but the growth was maintained in D9-1 seedlings and grew similar to their response at 25°C. It should be noted that the chilling resistance increased when the glycerol-3-phosphate-acyl transferase gene was introduced and the increase was limited only to the process of recovery from chilling damage. Yet the D9-1 line seedlings showed resistance to stress from direct exposure to low temperatures and reduced the lower temperature limit for plant growth.

4.2.3   Trienoic Fatty Acids

Fatty acids present in plant membrane lipids have the highest percentage of trienoic fatty acids (TAs). TAs are polyunsaturated fatty acids having three *cis* double bonds. Their content varies according to environmental conditions and plant species. TAs are formed from dienoic fatty acids (DAs) (having two *cis* double bonds) through the activity of ω-3 fatty acid desaturase, which appears to be deeply embedded within biomembranes. It is difficult to characterize it by conventional biochemical approaches. Hence, its genes are cloned using genetic techniques such as map-based cloning methods that use mutant strains of *Arabidopsis*. The cloned ω-3 fatty acid desaturase genes are divided into two types—one localized in the chloroplasts (*FAD7* and *FAD8*) and the other localized in the ER (*FAD3*). In the leaf tissue, DAs (16:2+18:2) were reduced and TAs (16:3+18:3) were increased by overexpressing the *Arabidopsis FAD7* gene in tobacco. When the low-temperature tolerance of transgenic and wild-type plants was evaluated, no discernible difference in the performance of mature plants was noticed. However, very young seedlings showed differences in low-temperature tolerance. Growth inhibition and chlorosis of young leaves were observed when there was transfer of wild-type plants from 25°C to 1°C for 7 days without going through an acclimation process and subsequent return to the original temperature environment. But transgenic tobacco in which the level of TAs in the leaf tissue had increased did not show this kind of damage. No significant difference was observed between the wild-type and transgenic tobacco plants with respect to their resistance to chilling and freezing when the levels of TAs in phospholipids, which are the main constituents of extrachloroplastic membranes, were increased in tobacco by overexpressing the ER-localized ω-3 fatty acid desaturase gene (*NtFAD3*).

On the contrary, rice transgenic lines showed different results with increased levels of TAs in the extrachloroplastic membranes. In rice, due to increased TA levels, chilling injury was alleviated at the seedling stage and the germination rate under a low-temperature environment improved. TAs enhance tolerance to low temperatures was proved by the results obtained after the analysis of transgenic plants into which *FAD7* and *FAD3* genes were introduced. However, the effect of TAs was relatively insignificant and might be limited to specific plant species and their tissues or growth processes.

The increase in TAs did not result in the expected improvement in low-temperature tolerance. So the reverse concept was examined, that is, whether a decrease in the level of TAs in the organism increases the high-temperature tolerance of plants. To increase the amount of enzymes produced within the plant, the ω-3 fatty acid desaturase gene was linked to a potent expression promoter such as the cauliflower mosaic virus 35S promoter. However, transgenic lines were found in which the expression of the intrinsic ω-3 fatty acid desaturase gene was co-suppressed by gene silencing in parts of the transgenic tobacco plants that were created for this purpose.

An analysis of the correlation between the TA content of the biomembranes and the ability of the plant to tolerate high temperatures was done. For this purpose, transgenic tobacco lines in which the activity of chloroplast-localized ω-3 fatty acid desaturase was suppressed by gene silencing were used. Although the level of TAs in the chloroplast membrane lipids of transgenic tobaccos was extremely low, the increase in the level of DAs was in correspondence to the decrease in the level of TAs. Moreover, a few changes were detected in the lipid molecular species of biomembranes.

When plants were grown at temperatures ranging between 15°C (cool) and 25°C (a more suitable growth temperature), no differences between the growth of two transgenic tobacco lines (T15 and T23) and that of the wild type were observed. The fresh weights of the aerial parts of T15, T23, and wild-type plants after germinating and cultivating for 45 days were 489 ± 71 mg, 513 ± 88 mg, and 497 ± 43 mg at 25°C, and 6.2 ± 1.4 mg, 6.9 ± 1.2 mg, and 6.6 ± 0.9 mg at 15°C, respectively. These results showed by tobacco plants are in accordance with the growth of *Arabidopsis* fad7 and fad8 mutants in the normal cultivation temperature range of 12–28°C. Furthermore, the growth of the two transgenic lines and the wild type was similarly suppressed at temperatures below 10°C. Even though there was no difference in growth between the transgenic tobacco lines and the wild type from low to normal growth temperature, there were clear differences in growth at high temperatures. For instance, for T15, T23, and wild-type plants, the fresh weights of above-ground portions of plants, grown for 45 days at 30°C, were 492 ± 81 mg, 445 ± 62 mg, and 399 ± 69 mg, respectively. At a higher temperature (36°C), there were significant differences in the growth of the transgenic tobacco lines and the wild type. After cultivating plants for 45 days at 36°C, the fresh weights of the aerial parts of the T15 and T23 lines and the wild-type plants were taken and they were found to be 124 ± 49 mg, 123 ± 23 mg, and 13 ± 6 mg, respectively. The growth of the transgenic lines continued to be uninhibited beyond 45 days at 36°C. Hence, the resistance to high temperature was not temporary and, therefore, different from the protection conferred by induction of a heat shock protein.

On exposing plants to a considerably higher temperature (47°C), the leaves of wild-type plants withered within two days and the plants showed chlorosis after three days, which finally resulted in their death. On the other hand, the high temperature did not damage the transgenic plants, although it inhibited their growth. When returned to a more suitable temperature (25°C), the plants continued to grow. A measurement of photosynthetic activity was done to find out the mechanism by which there was improvement in high-temperature tolerance. The photosynthetic activity in wild-type tobacco plants significantly decreased at temperatures of 40°C and above, but the decrease was mild in transgenic tobacco plants with decreased levels of TAs. An analysis of the chloroplast membrane by differential scanning calorimetry revealed an endothermic peak in the range of 40–45°C accompanying the thermal denaturation of proteins in wild type, but not in the transgenic lines with decreased TA levels. Transgenic lines may have improved thermal stability in those proteins or functions that are associated with chloroplast membrane lipids. Such protection could work primarily on proteins constituting the photosynthetic machinery. Plants can also have high-temperature tolerance when there is an increase in the level of fatty acid molecular species accompanied by a higher degree of saturation than that of DAs. It has been reported that the saturation of thylakoid membrane lipids due to mutations in fatty acid desaturation or catalytic hydrogenation increases the thermal stability of the membranes. However, the increased saturation could raise the temperature at which lipids, such as MGD, phase separate into non-bilayer structures, disrupting membrane organization. Hence, there might be an increase in the sensitivity of such plants to low temperatures.

4.2.4 Regulation of Polyunsaturated Fatty Acid Levels by Temperature and Other Stresses

No significant difference was observed between the growth rates of wild-type plants and transgenic tobacco in which the TA level was altered. This result was obtained at normal cultivation temperatures and even at the lowest temperature (15°C). Thus, it can be suggested that plants develop high-temperature tolerance without compromising on their tolerance to low temperatures owing to reduction in the TA level. However, the plant's ability to withstand stress might be affected by other factors, such as growth inhibition, decrease in photosynthetic activity, and chlorosis, within lower temperature ranges, especially near the freezing point. Thus, tolerance to either low or high temperatures can be increased by altering the composition of membrane lipids. Therefore, to apply this technique for basic research on temperature stress tolerance in plants, it is necessary to clarify how the TA level is regulated in conjunction with the environmental change.

In response to a shift in ambient temperature, the expression of the chloroplast FAD8 ω-3 desaturase gene changes, but temperature has no effect on the expression of FAD7, a second chloroplast ω-3 desaturase. In the case of the *Arabidopsis* fad7 mutant, deficient in *Fad7* desaturase activity, the only activity of chloroplast ω-3 fatty acid desaturase which can be measured is that of the Fad8 desaturase. This helps in clearly monitoring the temperature dependency of the enzyme expression. It is surprising to note that the expression of the Fad8 desaturase is switched on and off by a difference of only a few degrees Celsius on either side of 25°C. Thus, it can be concluded that this temperature regulation operates through a mechanism quite different from that governing the expression of temperature-dependent genes, such as the hereditary spastic paraplegia (*HSP*) genes, found so far. Recently, a series of chimeric genes were created from both *FAD7* and *FAD8* genes, which encode the isozymes of chloroplast ω-3 desaturase, and introduced into the *Arabidopsis fad7 fad8* double mutant. It was found that the temperature-dependent expression of the *FAD8* gene in transgenic plants was not because of the five flanking regions, including the promoter region and the untranslated region, but it was due to the exon/intron structure inherent to the *FAD8* gene. Therefore, it is not likely that expression of *FAD8* gene is regulated at the transcriptional levels, as seen in the bacterial desaturase genes. The levels of 18:3 in the root tissues of wheat increase considerably at low temperatures. On the contrary, the mRNA level of the desaturase gene *TaFAD3* demonstrates a minor change based on the increased accumulation of the ER ω-3 fatty acid desaturase protein. Thus, the increased level of 18:3 at low temperatures might be regulated at the translational or post-translational level. In response to changes in the environment or other factors inducing stress, there appears to be a complex process of regulation of expressions of each of the ω-3 fatty acid desaturase genes—the chloroplast-localized (*FAD7*) and ER-localized (*FAD3*) types. For instance, environmental stimuli such as wounding, salt stress, and pathogen invasion rapidly increase the defence-related signal molecule jasmonate. There might be an important role of TAs, especially α-linolenic fatty acids, as a precursor to jasmonate. The stimulus results in a rapid induction of the expression of the chloroplast ω-3 desaturase genes (*FAD7* and *FAD8*). Conversely, synergistic and antagonistic interactions of plant hormones such as auxin, cytokinin, and abscisic acid (ABA) regulate the expression of the ER ω-3 desaturase gene (*FAD3*). Also, the tissue specificity of the expression of this gene is further modified based on the growth stage in plant development. There seems to be a relationship between the regulation of fatty acid desaturation of membrane lipids and the wide range of mechanisms that enable plants to adapt to their environment throughout development.

4.2.5 Future Approaches

Most approaches to molecular breeding have attempted to introduce a single gene to improve host resistance to temperature stress, which altered only a single character. However, it is generally believed that multiple characters are involved in providing tolerance over a wide range of temperatures. As such, attempts have been made to simultaneously alter multiple related characters to engineer better acclimation to freezing and chilling temperatures. *Arabidopsis* and tobacco underwent transformation to express the transcription factors (activators) of those genes that play a role in acclimation. Intrinsic low-temperature-inducible genes are expressed selectively in such transgenic plants, and the tolerance to cold or freezing are enhanced.

The effects of various stresses that plants receive from their surrounding environment are complex. For instance, on being subjected to abiotic environmental stresses such as high or low temperature, intense light, or drought, a plant's sensitivity to biotic stresses such as viruses and bacteria could also intensify. In the previous studies on plant resistance to temperature stress, temperature extremes were regarded as the sole source of stress and the multiple stresses that affect populations under natural conditions were ignored. For example, even if plants are conferred with tolerance to drought and salt stresses or freezing, this remains within only a single stress category because these stresses are mainly caused by a single effect of dehydration on cells. Therefore, it remains a challenge for the future to develop plants that can resist a wide range of stresses. For this, key strategies that plants use to deal with complex stresses (both biotic and abiotic) should be identified. For instance, there is close relationship between the degree of unsaturation of fatty acids in the membrane lipids of plant cells and the plant's temperature tolerance. At the same time, the degree of unsaturation also influences the replication of viruses that infect the plant. Therefore, whether a viral infection can occur or not can be possibly determined by the degree of unsaturation of lipid fatty acids. In the future, researchers should emphasize on cases in which a single factor regulates the tolerance to seemingly unrelated physical stresses and viral propagation. From the rapid developments of functional genomics in recent years, scientists will be able to understand the route by which signals that activate resistance to various stresses are transmitted. Researchers, by inferring the origins or intersections of such signals, might be able to discover intrinsic resistance factors that are vital for plants to cope with a wide range of stresses.

4.3 ABIOTIC STRESS—SALT

Salinity of soil is a major hurdle in food production because it affects crop yield and limits use of previously uncultivated land. According to an estimate of the United Nations Environment Programme, salt stress is found in approximately 20% of agricultural land and 50% of cropland in the world. The caloric and nutritional potentials of agricultural production are also limited by natural boundaries imposed by soil salinity. These constraints are most acute in areas where insufficient infrastructure or political instability limits food distribution. Agricultural production on soils affected by salinity is possible by water and soil management practices, but achieving additional gain by these approaches is not so easy. There are many crop improvement strategies based on the use of molecular marker techniques and biotechnology that can be used alongside traditional breeding efforts. DNA markers enhance the recovery rate of the isogenic recurrent genome after hybridization. They also facilitate the introgression of quantitative trait loci necessary to increase stress tolerance. Alleles of interest from wild relatives were successfully transferred into commercial cultivars by molecular marker techniques. The basic resources for biotechnology are genetic determinants of salt tolerance and yield stability.

This goal can be achieved through biotechnology strategies by carrying out substantial research on identifying salt tolerance effectors and the regulatory components that control these during the stress. Further knowledge of these stress tolerance determinants will provide additional information for examining the plant response to salinity. This will reveal the mechanisms plants use to sense salt stress, transduce signals to mediate a defensive response, and define the signal pathway outputs or effectors that are necessary for stress survival and steady-state growth in the saline environment. Molecular genetics and advances in plant transformation have made it possible to evaluate biotechnological strategies based on activated signal cascades and engineered biosynthetic pathways. Targeted gene or protein expression and alteration of the natural stress responsiveness of genes for the development of salt-tolerant crops are also important factors to consider. Now there is knowledge about the molecular identities of key ion transport systems fundamental to plant salt tolerance. Recently, it has been identified that the salt overly sensitive (SOS) stress signalling pathway has a pivotal regulatory function in salt tolerance, fundamental of which is the control of ion homeostasis. This review will summarize research on plant ion homeostasis in saline environments and present a model integrating the current understanding of sensing salt stress. This will lead to the activation of the SOS pathway and the regulation of ion transport systems that facilitate ion homeostasis.

4.3.1 Genetic Diversity for Salt Tolerance in Plants

The extensive genetic diversity for salt tolerance found in plant taxa is distributed over many genera. Most plants are sensitive or hypersensitive to salt (glycophytes). These are different from halophytes, which are native to saline environments. Some halophytes can tolerate extreme salinity owing to their special anatomical and morphological adaptations or avoidance mechanisms. However, these are unique characteristics for which it is not likely that the genes can be easily introgressed into crop plants.

In recent decades, researchers have found that most halophytes and glycophytes apply similar strategies to tolerate salinity by using analogous tactical processes. Cytotoxic ions in saline environments, mostly Na^+ and Cl^-, are compartmentalized into the vacuole and used as osmotic solutes. Hence, many molecular entities that facilitate ion homeostasis and salt stress signalling have been found to be similar in all plants. In fact, the paradigm for ion homeostasis facilitating plant salt tolerance resembles that described for yeast. For the dissection of the plant salt stress response, it is possible to use model genetic organismal systems because cellular ion homeostasis is controlled and affected by common molecular entities. Studies of the plant genetic model *Arabidopsis* have greatly increased the understanding of the integration and coordination of cellular salt tolerance mechanisms in an organismal context. As *Arabidopsis* is a glycophyte, a salt-tolerant genetic model will be required to determine whether salt tolerance is affected by the form or function of genes or by differences in the expression of common genes because of transcriptional or post-transcriptional control.

4.3.2 Cellular Mechanisms of Salt Stress Survival, Recovery, and Growth

Hyperosmotic stress and ion disequilibrium are caused by high salinity resulting in secondary effects or pathologies. Plants normally survive by avoiding or tolerating salt stress. They either remain dormant during the salt episode or undergo cellular adjustment. Tolerance mechanisms reduce osmotic stress and ion disequilibrium or lessen the consequent secondary effects caused by these stresses. The saline solution's chemical potential initially establishes an imbalance in water potential between the apoplast and the symplast. This leads to decrease in turgor, which can cause growth retardation if it is severe. Cessation of growth occurs when turgor is lowered below the yield threshold of the cell wall. Dehydration of cells starts when the difference in water potential is greater than what can be compensated by turgor loss. Cellular response to turgor reduction is an osmotic adjustment. The cytosolic and organellar mechanisms of glycophytes and halophytes are equivalently sensitive to Na^+ and Cl^-; therefore, osmotic adjustment can be achieved in these compartments by accumulation of compatible osmolytes and osmoprotectants. However, Na^+ and Cl^- are effective osmolytes for osmotic adjustment and are compartmentalized into the vacuole to lessen cytotoxicity. Cell growth in plants occurs primarily because of directional expansion mediated by an increase in vacuolar volume. Hence, compartmentalization of Na^+ and Cl^- helps in osmotic adjustment, which is essential for development of cells. Ions may move directly into the vacuole from the apoplast through membrane vesiculation or indirectly through a cytological process that juxtaposes the plasma membrane to the tonoplast. Compartmentalization can then be achieved with minimal or no exposure of the cytosol to toxic ions. However, it is still not clear to what extent such processes contribute to vacuolar ion compartmentalization. The majority of Na^+ and Cl^- movement from the apoplast to the vacuole is probably mediated through ion transport systems located in the plasma membrane and tonoplast. Presumably, to control the net influx across the plasma membrane and vacuolar compartmentalization, strong coordinated regulation of these ion transport systems is necessary. For some key transport systems required for ion homeostasis, the SOS signal pathway is an essential regulator.

4.3.3 Osmolytes and Osmoprotectants

As discussed earlier, for achieving salt tolerance, compatible solutes have to accumulate in the cytosol and organelles and function in osmotic adjustment and osmoprotection. Although some compatible osmolytes are essential ions, such as K^+, the majority of them are organic solutes. Accumulation of compatible solutes as a response to osmotic stress is a ubiquitous process in organisms ranging from bacteria to plants and animals. However, the solutes accumulate in varying amounts depending on the organism and even plant species. Organic osmotic solutes consist of simple sugars (mainly fructose and glucose), sugar alcohols (glycerol and methylated inositols), and complex sugars (trehalose, raffinose, and fructans). Others constituents include quaternary amino acid derivatives (proline, glycine betaine, α-alanine betaine, and proline betaine), tertiary amines (1, 4, 5, 6-tetrahydro-2-methyl-4-carboxyl pyrimidine), and sulphonium compounds (choline sulphate and dimethyl sulphonium propionate). Many organic osmolytes are supposed to be osmoprotectants because they accumulate at levels insufficient to facilitate osmotic adjustment. Thylakoid and plasma membrane integrity is preserved by glycine betaine when it is exposed to saline solutions or to freezing or high temperatures. Moreover, many osmoprotectants increase stress tolerance of plants on being expressed as transgene products. Osmoprotectants have an adaptive biochemical function of scavenging reactive oxygen species that are by-products of hyperosmotic and ionic stresses. They then cause membrane dysfunction and cell death.

Compatible solutes have a common feature. These compounds can accumulate in high concentrations without affecting intracellular biochemistry. They can preserve the activity of enzymes in saline solutions. The effects of these compounds on the pH or charge balance of cytosol or luminal compartments of organelles are minimal. Basic intermediary metabolites can be diverted into unique biochemical reactions to synthesize compatible osmolytes. Often, stress triggers this metabolic diversion. For instance, glycine betaine is synthesized in higher plants from choline with the help of two reactions that are catalysed in sequence by choline monooxygenase (CMO) and betaine aldehyde dehydrogenase (BADH). Pinitol is synthesized from *myo*-inositol by the sequential catalysis of inositol *O*-methyltransferase and ononitol epimerase.

4.3.4   Ion Homeostasis—Transport Determinants and Their Regulation

Sodium chloride (NaCl) being the principal soil salinity stress, researchers have focused on the transport systems that utilize Na^+ as an osmotic solute. More than 30 years of research has established that intracellular Na^+ homeostasis and salt tolerance are modulated by Ca^{2+} and that high $[Na^+]_{ext}$ negatively affects acquisition of K^+. For uptake through common transport systems, Na^+ competes effectively with K^+ because $[Na^+]_{ext}$ is considerably greater than $[K^+]_{ext}$ in saline environments. Ca^{2+} increases K^+/Na^+ selective intracellular accumulation. Studies carried out in the last decade have defined many molecular entities that mediate Na^+ and K^+ homeostasis. These studies have also given insights into the role of Ca^{2+} in the regulation of transport systems. Of late, researchers have identified the SOS stress signalling pathway as the primary regulator of plant ion homeostasis and salt tolerance. This signalling pathway is functionally similar to the yeast calcineurin cascade that controls influx and efflux of Na^+

across the plasma membrane. When an activated form of calcineurin is expressed in yeast or plants, salt tolerance is enhanced further. This implicates the functional similarity between calcineurin and SOS pathways. There is little information available about the mechanistic entities responsible for Cl^- transport or Cl^- homeostasis regulation.

4.3.5 Ion Transport Systems that Mediate Na^+ Homeostasis

4.3.5.1 H^+ pumps

In the plasma membrane and tonoplast, H^+ pumps energize transport of solute necessary to compartmentalize cytotoxic ions away from the cytoplasm. They facilitate the function of ions as signal determinants and provide the driving force (H^+ electrochemical potential) for secondary active transport. These pumps function to establish membrane potential gradients that facilitate electrophoretic ion flux (Figure 4.2). The plasma membrane-localized H^+ pump is a P-type ATPase. It is generally responsible for the large pH and membrane potential gradient across this membrane. The pH and membrane potential across the tonoplast are generated by a vacuolar-type H^+-ATPase and a vacuolar pyrophosphatase. The activity of H^+ pumps is enhanced by salt treatment. The induced gene expression may account for some of the upregulation. Based on the analyses of phenotypes caused by the semi-dominant aha4–1 mutation, plasma membrane H^+-ATPase was confirmed as a salt tolerance determinant. The growth of roots and shoots (relative to wild type) of plants grown on a medium supplemented with 75 mM NaCl was reduced by the mutation of AHA4, which is expressed predominantly in the roots. The decreased root length of salt treated aha4–1 plants results from reduced cell length. In a medium supplemented by NaCl, the leaves of aha4–1 plants accumulate more Na^+ and less K^+ than those of wild type. It is assumed that AHA4 plays the role in the control of Na^+ flux across the endodermis.

4.3.5.2 Na^+ influx and efflux across the plasma membrane

Researchers have recently gained much insight into Na^+ transport systems involved in net flux of the cation across the plasma membrane. It is presumed that transport systems with greater selectivity for K^+

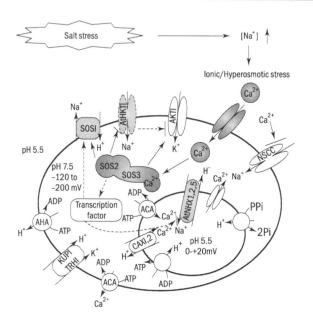

Fig. 4.2   *Salt stress-induced, Ca²⁺-dependent signalling that mediates Na⁺
homeostasis and salt tolerance. SOS1, plasma membrane Na⁺/H⁺
antiporter; SOS2, serine/threonine kinase; SOS3, Ca²⁺ binding protein;
AHA, plasma membrane H⁺-ATPase; ACA, plasma/tonoplast membrane
Ca²⁺-ATPase; KUP1/TRH1, high-affinity K⁺–H⁺ co-transporter; AKT1,
K⁺ in channel; NSCC, non-selective cation channel; CAX1 or 2, Ca²⁺/H⁺
antiporter; NHX1, 2 or 5, endomembrane Na⁺/H⁺ antiporter*

facilitate Na⁺ 'leakage' into cells. Na⁺ is a competitor for uptake
through plasma membrane K⁺ inward rectifying channels, such as
those that are in the Shaker-type family (e.g., AKT1). Na⁺ influx might
also be facilitated by K⁺ outward rectifying channels. For the Na⁺
influx across the plasma membrane, the high-affinity K transporter
(HKT) from wheat and low-affinity cation transporter (LCT) may
also be responsible. Both HKT and LCT confirm their role in Na⁺
uptake by transporting Na⁺ when they are expressed in heterologous
systems. Wheat HKT1 is an Na⁺–H⁺-dependent K⁺ transporter. Na⁺
influx is reduced and salt tolerance is enhanced by modifications to
HKT1 that also enhance K⁺ transport. This further establishes that
Na⁺ conductance occurs through this protein. Physiological data also

implicate non-selective cation (NSC) channels in Na$^+$ influx. Scientists have recently characterized the properties of HKT proteins from *Arabidopsis*, rice, and eucalyptus. AtHKT1 is the only member of the *Arabidopsis* gene family, whereas both rice and eucalyptus have at least two genes. AtHKT1 expression enhances the NaCl sensitivity of a yeast strain deleted for the plasma membrane Na$^+$ efflux system (ena1–4Δ). However, it does not suppress the K$^+$ deficiency of trk1 trk2 mutant cells that have lessened the uptake of this necessary cation. The K$^+$-deficient phenotype of an *Escherichia coli* mutant is suppressed by the AtHKT1 expression, for which acquisition of the cation is disrupted. It is indicated from electrophysiological data that AtHKT1 expressed in *Xenopus oocytes* specifically transports Na+ and that conductance is not dependent on K$^+$, H$^+$, and voltage. It has been confirmed recently that AtHKT1 functions in plants as an effector of Na$^+$ influx. T-DNA insertion and deletion mutations of AtHKT1 were identified in a screen for suppressors of NaCl sensitivity of the sos3–1 mutant. A correlation exists between the suppression of sos3–1 NaCl sensitivity and the reduced cellular accumulation of Na$^+$ and the capacity to maintain [K$^+$]$_{int}$. These results establish that Na$^+$ influx into plants is controlled by AtHKT1. Although AtHKT1 is an Na$^+$ influx system, it cannot be precluded that it functions as a regulator of Na$^+$ and K$^+$ influx systems.

As the transcript is expressed chiefly in the roots, AtHKT1 most likely controls the Na$^+$ entry into the xylem for export to the shoot. On the basis of sequence similarity with wheat HKT, researchers identified rice (*Oryza sativa L. indica*) OsHKT1 and OsHKT2 transcripts. These transcripts accumulate in response to low K$^+$, but treatment with 30 mM NaCl reduces their steady-state abundance. Data from yeast complementation and *Xenopus oocyte* expression indicate that OsHKT1 acts as an Na$^+$ influx system like AtHKT1, but OsHKT2 is an Na$^+$/K$^+$ symporter. The transport characteristics of the two *Eucalyptus camaldulensis* HKT1 (EcHKT1 and EcHKT2) orthologs resemble that of the wheat protein. Interestingly, hypotonic treatment activates these proteins, which indicates an osmosensing capacity. Recently, sos3–1 hkt1 double mutations were identified in *Arabidopsis*, which confirmed the existence of an Na$^+$ entry system different from HKT1 functioning in planta. Decrease in [Ca]$_{ext}$ to micromolar concentrations nullifies the capacity for an hkt1 knockout mutation to suppress Na$^+$ sensitivity of sos3–1. The presence of a Ca^{2+}-inhibited Na^+ influx system is indicated by these results. After several years of physiological research, the presence of Ca^{2+}-insensitive and Ca^{2+}-sensitive Na^+ conductance in the analyses of plant cell patches was found. Recently, the uptake of Ca^{2+}-sensitive component of Na^+ was attributed to NSC channels. The combined data from electrophysiological and in planta mutant dissection show that there are two Na^+ influx systems. The first system is one whose activity is directly inhibited by Ca^{2+}. The other system is HKT, for which genetic evidence indicates that it may be regulated negatively by the SOS signal pathway (Figure 4.3). The secondary active Na^+/H^+ antiporter SOS1 mediates the energy-dependent transport of Na^+ across the plasma membrane of plant cells. SOS1 is phylogenetically similar to SOD2 of *Schizosaccharomyces pombe*, NHA1 of *Saccharomyces cerevisiae*, and NhaA and NhaP of *Pseudomonas aeruginosa*. A large C-terminal domain (1162 amino acid) is present in the SOS1 protein, which distinguishes it from other members of the phylogenetic family. The SOS1 antiporter has distant relationship with the plant endomembrane NHX type of antiporters. NaCl stress upregulates the expression of SOS1 gene, and this is dependent on other components of the SOS signal pathway.

4.3.5.3 Na^+ vacuolar compartmentalization

The vacuolar compartmentalization of Na^+ is facilitated by an Na^+/H^+ antiporter, which is energized by the ΔpH across the tonoplast. The *Arabidopsis* AtNHX1 was isolated by functional genetic complementation of a yeast mutant defective for the endosomal Na^+/H^+ antiporter yeast (ScNHX1) and had sequence similarity to mammalian NHE transporters. Tomato plants and transgenic *Arabidopsis*, which overexpress AtNHX1, accumulate abundant quantities of the transporter in the tonoplast and show enhanced salt tolerance. These results demonstrate the crucial role of the AtNHX family in vacuolar compartmentalization of Na^+. Six loci in *Arabidopsis* were categorized as AtNHX gene family members based on the predicted amino acid sequence and topological similarities to AtNHX1. Phylogenetically, the proteins can be divided into two subgroups—one containing four (AtNHX1–4) members and the other containing two (AtNHX5 and AtNHX6) members. The expression of AtNHX2 and AtNHX5 suppresses the Na^+/Li^+-sensitive phenotype of a salt-sensitive yeast mutant (ena1–4 nha1 nhx1Δ), which indicates that both AtNHX2 and AtNHX5 are orthologous to yeast ScNHX1 and AtNHX1. AtNHX2 suppresses the yeast mutant's Na^+/Li^+-sensitive phenotype to a greater extent than AtNHX1. AtNHX1 and AtNHX2 are expressed constitutively in shoot and roots. Transcript abundance of AtNHX1 and AtNHX2 is induced by hyperosmotic stress (NaCl, sorbitol), and this osmotic response depends on the hormone ABA. Transcript abundance of AtNHX5 is induced by NaCl and not ABA. SOS mutants have greater steady-state transcript abundance of AtNHX1, AtNHX2, and AtNHX5 than wild-type Col-0 gl1. This indicates that the SOS pathway regulates negatively transcriptional expression of these Na^+/H^+ antiporter genes. Data from yeast complementation and expression profiling show that AtNHX2 and AtNHX5, like AtNHX1, are functional determinants of salt tolerance. The expressions of AtNHX1 and AtNHX2 are regulated by a common hyperosmotic stress signalling pathway, but the AtNHX5 expression is controlled by a different cascade. Post-transcriptional mechanisms that control AtNHX antiporter activation are still unknown.

4.3.5.4 *Ca^{2+} signalling and the activation of the SOS signal transduction pathway*

Three genetically linked *Arabidopsis* loci SOS1, SOS2, and SOS3 are components of a stress signalling pathway that controls ion homeostasis and salt tolerance. From the genetic analysis of Na^+/Li^+ sensitivity, it was established that SOS1 is epistatic to SOS2 and SOS3. The SOS mutants also show a K^+-deficient phenotype in a medium supplemented with μM $[K^+]_{ext}$ and $[Ca^{2+}]_{ext}$. The Na^+ and K^+ deficiency of SOS2 and SOS3 is suppressed with μM $[Ca^{2+}]_{ext}$ (Zhu-Salzman, Shade, Koiwa, *et al.* 1998). SOS1 exhibits hyperosmotic sensitivity unlike SOS3 and SOS2. It can be observed from these results that the SOS signalling pathway controls Na^+ and K^+ homeostasis and is activated by Ca^{2+}. SOS3 encodes a Ca^{2+}-binding protein with sequence similarity to the regulatory B subunit of calcineurin (protein phosphatase 2B) and neuronal Ca^{2+} sensor. The interaction of SOS3 with the SOS2 kinase and SOS2 activation is dependent on Ca^{2+}. The in planta function of SOS3 as a determinant of salt tolerance is dependent on Ca^{2+} binding and *N*-myristoylation. The SOS2 serine/

threonine kinase (446 amino acids) has a 267-amino acid N-terminal catalytic domain similar in sequence to yeast SNF1 (sucrose non-fermenting) kinase and the mammalian AMP-activated protein kinase (AMPK). The kinase activity of SOS2 is necessary for its function as a salt tolerance determinant. Autoinhibition is caused when the SOS2 C-terminal regulatory domain interacts with the kinase domain. SOS3 interacts with the kinase at a 21 amino acid motif in the regulatory domain of SOS2, and the motif is the autoinhibitory domain of the kinase. When the SOS3 binds to this motif, autoinhibition of SOS2 kinase activity is blocked. If the autoinhibitory domain is deleted, it results in constitutive SOS2 activation, independent of SOS3. Moreover, a Thr168 to Asp mutation in the activation loop of the kinase domain constitutively activates SOS2. Genetic and biochemical evidence suggests that components of the SOS signal pathway work in the hierarchical sequence as shown in Figure 4.4. Ca^{2+} binds to SOS3, leading to the interaction with SOS2 and activation of kinase.

Among the SOS signal pathway outputs are transport systems that facilitate ion homeostasis. The Na^+/H^+ antiporter SOS1 present in plasma membrane is controlled by the SOS pathway at the transcriptional and post-transcriptional levels. It was recently found that the functional disruption of AtHKT1 suppresses the salt-sensitive phenotype of sos3–1. This indicates that the SOS pathway negatively regulates this Na^+ influx system. The SOS pathway also controls negatively the expression of AtNHX family members implicated as determinants of salt stress response. $[Ca^{2+}]_{ext}$ increases salt tolerance, and salinity stress elicits a transient $[Ca^{2+}]_{cyt}$ increase, from an internal or an external source, that is involved in adaptation. Recent experiments with yeast have given insights into Ca^{2+} activation of salt stress signalling that controls ion homeostasis and tolerance. The hyperosmotic component of high salinity induces a short duration (1 min) rise in $[Ca^{2+}]_{cyt}$ which is due to influx across the plasma membrane through the Cch1p and Mid1p Ca^{2+} transport system. The PP2B phosphatase calcineurin (a key intermediate in salt stress signalling that controls ion homeostasis) is activated by the transient rise in $[Ca^{2+}]_{cyt}$. This leads to the transcription of ENA1, which encodes the P-type ATPase that is primarily responsible for Na^+ efflux across the plasma membrane.

It is proposed in the model that the hyperosmotically induced localized $[Ca^{2+}]_{cyt}$ transient activates calmodulin, which is tethered to Cch1p-Midp. In turn, calmodulin activates signalling through the calcineurin pathway, thus mediating ion homeostasis and salt tolerance. A paradigm for salt-induced Ca^{2+} signalling and the activation of the SOS pathway can be suggested from these results. Constituents of the SOS pathway, SOS3 or upstream elements, may be linked with an osmotically responsive channel through which Ca^{2+} influx could begin signalling through the pathway. A new elevated $[Ca^{2+}]_{cyt}$ steady state is established in yeast cells, which are maintained in a medium supplemented with NaCl, after the hyperosmotic induction of the short duration $[Ca^{2+}]_{cyt}$ transient. The newly established $[Ca^{2+}]_{cyt}$ possibly contributes to cellular capacity for growth in salinity. The principal effectors that regulate the amplitude and duration of the $[Ca^{2+}]_{cyt}$ transient are the vacuolar membrane H^+/Ca^{2+} antiporter Vcx1p and endomembrane-localized Ca^{2+}-ATPases. The $[Ca^{2+}]_{cyt}$ steady state formed in a medium containing salt likely involves coordination of channel activation that facilitates influx from external and internal sources and energy-dependent transport systems that compartmentalize the divalent cation. Thus, it can be reasonably assumed that the salt-induced $[Ca^{2+}]_{cyt}$ transient detected in plant cells and a new $[Ca^{2+}]_{cyt}$ steady state are controlled by the ECA and ACA Ca^{2+}-ATPases and CAX1 and CAX2 transporters, which are orthologs of Vcx1p. Nonetheless, Ca^{2+} performs at least two functions in salt tolerance—a pivotal signalling role in the salt stress response leading to adaptation and a direct inhibitory effect on an Na^+ entry system.

On analysing the database, it was found that there are at least seven additional SOS2 isoforms (PKS: protein kinase S) and six SOS3 isoforms (SCaBPs:SOS3-like calcium-binding proteins) of SOS3. It is yet to be established whether these isoforms are also salt tolerance determinants. One can assume that the signalling intermediate functions of these proteins are similar to that of the prototype proteins but in different cell types or at unique stages of development. These isoforms are probably constituents of signal pathways that respond to different inducers but are still components of the plant response to salt stress. Nevertheless, these proteins likely possess both unique and overlapping functions. Some of these isoforms might also act as negative regulators of SOS signal transduction by physical interaction with the positive effector or competition for substrates needed for signalling. Such positive and negative regulation of signal modulation may constitute a 'fine-tuning' required to achieve the appropriate plant response for adaptation to stress and stability of yield. Additional insights into these suggestions might help in understanding how gene families coordinate plant's response to salt stress. Probably, the control system is even more complex because other SOS signal pathway intermediates and outputs, as well as other signalling cascades necessary for salt tolerance, may exist.

Recent developments in salt stress signalling and effector output determinants that facilitate ion homeostasis have revealed some biotechnology tactics that might produce salt-tolerant crops by increasing yield stability under salinity. There are two feasible strategies: (i) regulation of the salt stress signalling pathway controlling tolerance effectors and (ii) modulation of effector activity or efficacy. Recently, it has been demonstrated that a constitutively active form of SOS2 kinase can be achieved by deletion of the autoinhibitory domain or by site-specific modifications to the protein kinase catalytic domain. This provides an approach to regulate stress signalling that controls ion homeostasis. Salt tolerance is enhanced by the constitutive activation of yeast calcineurin in plants, which predisposes the plants to survive the stress. In addition, the overexpression of AtNHX1 increases salt tolerance by increasing vacuolar Na^+ compartmentalization. This reduces the toxic accumulation of Na^+ in the cytosol and promotes growth in the saline environment. The regulation of the net Na^+ influx across the plasma membrane would perhaps increase the efficacy of salt tolerance, which can be achieved by the overexpression of vacuolar antiporter. The net Na^+ flux across the plasma membrane should be controlled by modulating the activation or expression of SOS1 (Na^+ efflux) and (Na^+ influx) or by expressing more efficacious forms of the Na^+ transport proteins. For instance, mutant variant forms of HKT1 transport more K^+ at the expense of Na^+ and result in greater salt tolerance. The expression of the signal intermediates and effectors can be conditioned by promoters that direct tissue- and inducer-specific regulation of the target genes. Therefore, an effective plant response can be achieved by coordinating the regulation of numerous determinants of salt tolerance. Many of the costs associated with salt tolerance in nature can be reduced by agricultural practices that compensate for some essential evolutionary necessities.

REFERENCES

Crickmore, N., D. Zeigler, J. Feitelson, E. Schnepf, J. Van Rie, D. Lereclus, J. Baum, and D. Dean. 1998. Revision of the nomenclature for the Bacillus thuringiensis pesticidal crystal proteins. *Microbiology and Molecular Biology Reviews* 62: 807–13

de Maagd, R. A., D. Bosch, and W. Stiekema. 1999. Toxin-mediated insect resistance in plants. *Trends in Plant Science* 4: 9–13

Jouanin, L., M. Bonade-Bottino, C. Girard, G. Morrot, and M. Giband. 1998. Transgenic plants for insect resistance. *Plant Science* 131: 1–11

Krattiger, A. F. 1997. Insect resistance in crops: a case study of *Bacillus thuringiensis* (Bt) and its transfer to developing countries. *ISAAA Briefs No. 2*. ISAAA: Ithaca, New York, USA, p. 42

Kumar, P. A., R. P. Sharma, and V. S. Malik. 1996. The insecticidal proteins of Bacillus thuringiensis. *Advances in Applied Microbiology* 42: 1–43

McBride, K. E., Z. Svab, D. J. Schaaf, P. S. Hogan, D. M. Stalker, and P. Maliga. 1995. Amplification of a chimeric Bacillus gene in chloroplasts leads to an extraordinary level of an insecticidal protein in tobacco. *Biotechnology* 13: 362–65

Ranjekar, P. K., A. Patankar, V. Gupta, R. Bhatnagar, J. Bentur, and A. P. Kumar. 2003. Genetic engineering of crop plants for insect resistance. *Current Science* 84(3): 321–28

Schnepf, H. E. 1995. Bacillus thuringiensis toxins: regulation, activities and structural diversity. *Current Opinion in Biotechnology* 6: 305–12

Schuler, T. H., G. M. Poppy, and I. Denholm. 1998. Insect-resistant transgenic plants. *Trends in Biotechnology* 16: 168–75

Selvapandian, A., V. S. Reddy, P. Ananda Kumar, K. K. Tiwari, and R. K. Bhatnagar. 1998. Transformation of Nicotiana tobaccum with a native cry1Ia5 gene confers complete protection against Heliothis armigera. *Molecular Breeding* 4:473–78

Zhu-Salzman, K., R. E. Shade, H. Koiwa, R. A. Salzman, M. Narasimhan, R. A. Bressan, P. M. Hasegava, and L. L. Murdock. 1998. Carbohydrate binding and resistance to proteolysis control insecticidal activity of Griffonia simplicifolia lectin II. *Proceedings of the National Academy of Sciences of the USA* 95: 15123–128

Molecular Marker-aided Breeding

5.1 MOLECULAR MARKERS

In molecular genetics, one of the most significant developments is the creation and use of molecular markers for the detection and exploitation of deoxyribonucleic acid (DNA) polymorphism. Existence of different types of molecular markers makes it necessary to choose a particular marker. Each molecular marker differs on the basis of principle, methodology, and applications and at present, there is no single molecular marker that fulfils all requirements of researchers. Researchers can choose among the variety of molecular techniques available based on the kind of study to be undertaken, as each combines at least some desirable properties.

A plant's genetic material is encoded in its DNA that distinguishes one plant from another. It is packaged in chromosome pairs, one coming from each parent. A plant's characteristics are controlled by its genes, and is located on specific segments of each chromosome. A genome (King and Stansfield 1990) is all the genes carried by a representative of each of all the chromosome pairs, also known as gamete. In some plant species such as *Arabidopsis thaliana* and rice, the whole genome sequence is available (The Rice Genome Sequencing Project 2005). Scientists use genetic markers, an indirect method, to identify specific genes located on a particular chromosome. A genetic marker is a chromosomal landmark or allele that allows for the tracing of a specific region of DNA. It is a specific piece of DNA with a known position on the genome. The DNA is a gene whose phenotypic expression is usually easily discerned, used to identify an individual or a cell that carries it, or used as a probe to mark a nucleus, chromosome, or locus (King and Stansfield 1990).

The markers and the genes they mark stay together in each generation of plants produced. This is because they stay close together on the same chromosome. Scientists are able to create a genetic linkage map when they learn where markers occur on a chromosome and how close they are to specific genes. Genetic maps have various advantages, including detailed analysis of associations between economically important traits and genes or quantitative trait loci (QTLs), and facilitate the introgression of desirable genes or QTLs through marker-assisted selection. Genetic markers are (i) based on visually assessable traits (morphological and agronomic traits), (ii) based on gene product (biochemical markers), and (iii) relying on a DNA assay (molecular markers). Understanding of biological sciences has greatly improved because of the development of electrophoretic assays of isozymes (Markert and Moller 1959) and molecular markers (Nakamura, Leppert, O'Connell, *et al.* 1987; Welsh and McClelland 1990; Jaccoud, Peng, Feinstein, *et al.* 2001). Molecular markers are not normal genes because they do not have any biological effect. They are identifiable DNA sequences, found at specific locations of the genome, and passed on from one generation to the next by the standard laws of inheritance. This chapter deals with the basic principles, requirements, and advantages and disadvantages of the most widely used molecular markers.

5.2 DNA EXTRACTION—BEGINNING OF MOLECULAR MARKERS ANALYSIS

For analysing molecular marker types, the first step involves extracting or isolating the DNA from the sample to be studied. The DNA can be either nuclear, mitochondrial, and/or chloroplast DNA. To obtain good quality DNA, it is better to extract from fresh material, although lyophilized, preserved, or dried samples can also be used. DNA extraction can be done by using one of the many protocols available; however, scientists decide which protocol to use depending upon the quality and quantity of DNA needed, nature of samples, and presence of natural substances that may interfere with the extraction and subsequent analysis. The simple protocols, which are also quick, yield low-quality DNA and the methods that are laborious and time-consuming (Murray and Thompson 1980; Dellaporta, Wood, and Hicks (1983); Saghai-Maroof, Soliman, Jorgensen, *et al.* 1984) produce high quality and quantity of DNA (Figure 5.1). The common way to extract DNA is to break (through grinding) or digest away cell walls and membranes so that cellular constituents get released. Membrane lipids get removed when detergents such as sodium dodecyl sulphate

(SDS), cetyl trimethyl ammonium bromide (CTAB), or mixed alkyl trimethyl ammonium bromide (MTAB) are used. The DNA released needs to be protected from endogenous nucleases; for this ethylene diamine tetra acetic acid (EDTA) is part of the extraction buffer to chelate magnesium ions, which is a necessary cofactor for nucleases. The DNA that is extracted often contains RNA, proteins, polysaccharides, tannins, and pigments. These interfere with the extracted DNA and need to be removed. Of these, polysaccharides are the most difficult to remove. There are removed when NaCl, along with CTAB is added (Murray and Thompson 1980; Paterson, Brubaker, and Wendel 1993). In some protocols NaCl is used instead of KC. In the case of removing protein, a protein-degrading enzyme (proteinase-K) is added to the extract that is then denatured at 65°C, and finally, precipitated using chloroform and isoamyl alcohol. To remove RNAs, RNase A, an RNA-degrading enzyme, is used. By centrifugation the DNA released along with lipids proteins, carbohydrates, and/or phenols are separated from these compounds. The DNA in the aqueous phase is then transferred into new tubes, precipitated in a salt solution (e.g., sodium acetate) or alcohol (100% isopropanol or ethanol), and redissolved in sterile water or buffer. Finally, the concentration of the extracted DNA is measured with the help of either 1% agarose gel electrophoresis or a spectrophotometer. While agarose gel is useful to check whether the DNA is degraded or not (Figure 5.2), estimating DNA concentration by visually comparing band intensities of the extracted DNA with a molecular ladder of known concentration is very subjective. The spectrophotometer, on the other hand, measures the intensity of absorbance of DNA solution at 260 nm wavelength and indicates the presence of protein contaminants. After any DNA extraction, there are three possible results:

(i) No DNA has been extracted.

(ii) If the DNA appears too fragmented or sheared, then it means degradation for different reasons.

(iii) If the DNA appears as whitish thin threads, it indicates good-quality DNA. In the case of brownish thread, it means DNA has been oxidized in the presence of contaminants such as phenolic compounds.

Hence scientists need to choose that particular protocol that works best for the species under investigation.

5.3 TYPES OF MOLECULAR MARKERS

Table 5.1 lists the various types of molecular markers.

Table 5.1 Different types of molecular markers

Molecular marker	Abbreviation
Allele-specific associated primers	ASAP
Allele-specific oligonucleotide	ASO
Allele-specific polymerase chain reaction	AS-PCR
Amplified fragment length polymorphism	AFLP
Anchored microsatellite primed PCR	AMP-PCR
Anchored simple sequence repeats	ASSR
Arbitrarily primed PCR	AP-PCR
Cleaved amplified polymorphic sequence	CAPS
Degenerate oligonucleotide primed PCR	DOP-PCR
Diversity arrays technology	DArT
DNA amplification fingerprinting	DAF
Expressed sequence tags	ESTs
Inter-simple sequence repeat	ISSR
Inverse PCR	IPCR
Inverse sequence-tagged repeats	IST
Microsatellite primed PCR	MP
Multiplexed allele-specific diagnostic assay	MASDA
Random amplified microsatellite polymorphisms	RAMP
Random amplified microsatellites	RAM
Random amplified polymorphic DNA	RAPD
Restriction fragment length polymorphism	RFLP
Selective amplification of microsatellite polymorphic loci	SAMPL
Sequence-characterized amplified regions	SCAR
Sequence-specific amplification polymorphisms	S-SAP
Sequence-tagged microsatellite site	STMS
Sequence-tagged site	STS
Short tandem repeats	STRs
Simple sequence length polymorphism	SSLP
Simple sequence repeats	SS
Single nucleotide polymorphism	SNP
Single primer amplification reactions	SPAR
Single-stranded conformational polymorphism	SSCP
Site-selected insertion PCR	SSI
Strand displacement amplification	SDA
Variable number tandem repeat	VNTR

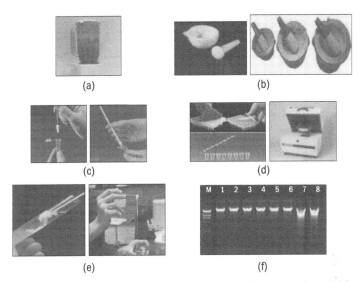

Fig. 5.1 *Different methods of grinding samples for DNA extraction: (a) large-scale DNA extraction using a juice maker; (b) medium-scale DNA extraction using a mortal and a pestle; (c) mini-scale DNA extraction using 1.5 mL/L tubes and small grinding pestle (even a nail or driller); (d) mini-scale DNA extraction using 96-well format microtitre plates or strip tubes (left) and grinding machine (right); (e) DNA after precipitation in 100% isopropanol (note that the DNA is whitish and has tiny thread-like structure, which is an indication of good-quality DNA that can easily be fished using glass hooks); and (f) agarose gel (1%) showing DNA concentration for eight samples and a marker (M) with a known concentration. The DNAs for sample numbers 7 and 8 show partial degradation compared to other samples.*

It should be noted that the marker types:

- ASAP, ASO, and AS-PCR are very similar;
- ISSR, RAMP, RAM, SPAR, AMP-PCR, MP-PCR, and ASSR are synonymous; and
- SSLP, STMS, STR, and SSR are identical.

Still there is a wide range of techniques for researchers to choose from. They need to associate the purpose(s) of a specific project with the various molecular marker types. These molecular markers can be classified into different groups based on:

- mode of transmission (biparental nuclear inheritance, maternal nuclear inheritance, maternal organelle inheritance, or paternal organelle inheritance);
- mode of gene action (dominant or co-dominant markers); and
- method of analysis (hybridization- or PCR-based markers).

5.4 TYPES OF MARKERS

5.4.1 Morphological Markers

These are traditional markers that map the morphological mutant traits in a population, determine the linkage to a desirable or undesirable trait, and carry out indirect selection with the help of physically identifiable mutant for the trait. There are some disadvantages of morphological markers that are as follows:

- Morphological markers are highly dependent on environmental factors. The environment under which a plant grows often influences the expression of these markers and can cause false determination.

- There are undesirable features including dwarfism or albinism in the mutant traits.

- If breeding experiments need to be performed with these markers, then it would be time-consuming and labour intensive. Moreover, in case large populations of plants need to be grown for the experiments, then large plots of land and/or greenhouse space will be required.

5.4.2 Biochemical Markers

In plant breeding, enzymes such as isozymes are used as biochemical markers. Isozymes are expressed in the plant cells, and are extracted and run on denaturing electrophoresis gels. The secondary and tertiary structures of the enzymes are denatured by sodium dodecyl sulphate (SDS). The structures are separated based on their net charge and mass. Polymorphic differences occur on the amino acid level, allowing singular peptide polymorphism to be detected and used as a polymorphic biochemical marker.

In comparison to morphological markers, biochemical markers are better as they are not dependent on environmental growth conditions. A drawback of isozymes in marker assisted selection (MAS) is that most cultivars (commercial breeds of plants) are genetically very similar. Hence, isozymes do not produce a great amount of polymorphism that in the protein primary structure may still cause an alteration in protein function or expression (Figure 5.2).

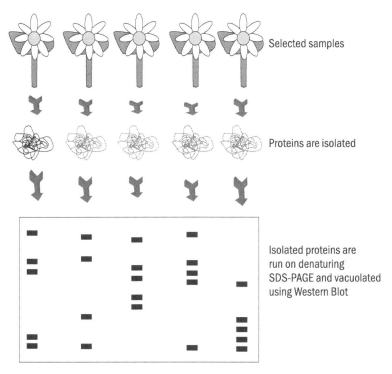

Selected samples

Proteins are isolated

Isolated proteins are
run on denaturing
SDS-PAGE and vacuolated
using Western Blot

Fig. 5.2 *Sequential summary of events occurring in the analysis of isozymes in plant samples. Although useful in some plant varieties, isozymes provide little variation in highly bred cultivars.*

5.4.3 Molecular Markers

These DNA markers are based on naturally occurring polymorphisms in DNA sequences, that is base-pair deletions, substitutions, additions, or patterns. It is used for germplasm characterization, genetic diagnostics, characterization of transformants, study of genome organization, and phylogenic analysis. In this technique, different types of methods are available that help in identifying and magnifying the polymorphisms so that they can be used for breeding analysis. In comparison to other forms of MAS, molecular markers are superior because of the following reasons:

- They are relatively simple to detect as they are abundant throughout the genome even in highly bred cultivars.
- They are not dependent on environmental conditions.
- They can be detected at any stage of plant development.

A suitable molecular marker should be polymorphic; has co-dominant inheritance; should be randomly and frequently distributed throughout the genome; should be easy and cheap to detect; and reproducible.

5.5   RESTRICTION FRAGMENT LENGTH POLYMORPHISM

In 1975, restriction fragment length polymorphism (RFLP) was first used for identifying DNA sequence polymorphisms for genetic mapping of a temperature-sensitive mutation of adenovirus serotypes. Then it was used for human genome mapping and later adopted for plant genomes. In plant breeding, RFLP was the first molecular marker technique developed and used in MAS. The technique is based on restriction enzymes that reveal a pattern difference between DNA fragment sizes in individual organisms. In RFLP genomic, DNAs are digested with restriction enzymes. Isolated from bacteria, these enzymes consistently cut the DNA at specific base-pair sequences. These base-pair sequences are called recognition sites and are not associated with any type of gene. They are distributed randomly throughout the genome. When the restriction enzymes, which are thousands in number and each cuts at a specific sequence, digest the genomic DNA, a series of fragments with varying lengths are produced. Agarose or polyacrylamide gel electrophoresis (PAGE) is used to separate these fragments and yields a characteristic pattern. When run under electrophoresis conditions, the DNA has a uniform charge per unit length. This arises because of the presence of phosphate groups in the backbone. Also the DNA fragments separated via electrophoresis travel on the basis of their molecular weight. The molecular weight is measured by DNA ladders, which are run alongside the DNA in the gel. On the other hand, a series of bands are formed when restriction fragments are separated via agarose gels. Each band corresponds to a restriction fragment of different length. The lighter they are, the farther they travel. The characteristic pattern of an RFLP digest varies based on base-pair deletions, mutations, inversions, translocations, and transpositions. This causes loss or gain of a recognition site that results in a fragment of different lengths and polymorphism. The restriction enzyme will not cut when there is a single base-pair difference in the recognition site. When the base-pair mutation is present in one chromosome and absent in the other, the fragment bands will be present on the gel, and the sample is said to be heterozygous for the marker. Co-dominant markers exhibit this behaviour, which is highly desirable. Dominant markers show a present/absent behaviour that can limit the data available for analysis.

Figure 5.3 summarizes the procedures and principles of RFLP markers. In RFLP, first the DNA is isolated. In this step, from the sample a considerable amount of DNA must be isolated and purified. Purification should be done on a fairly stringent degree because contaminants often interfere with the restriction enzyme and inhibit its ability to digest the DNA. In the second step, restriction enzyme is added. To the purified genomic DNA, restriction enzyme is added under buffered conditions. The enzyme cuts at recognition sites throughout the genome and leaves behind numerous fragments. The third step involves gel electrophoresis. Here the digest is run on a gel and, when visualized, appears as a smear because of the large number of fragments. The fourth step involves transfer to nitrocellulose membrane filter. The fifth step involves probe visualization. As a large number of fragments are extracted, it should be seen that the probes are constructed in such a way that more specific bands can be visualized in the digest. These probes consist of radiolabelled oligonucleotide sequences that will anneal to the fragment sequences so that that they may be visualized on a photographic paper using Southern blot technique. In the sixth and final step, analysis is done. An initial digest needs to be performed on the species of interest in order to develop probes to screen RFLP. Restriction fragments are ligated into a plasmid vector and transformed into bacteria. Positive cultures are then isolated; ligated sequences are removed, amplified by PCR, and radiolabelled for use as RFLP probes.

Restriction enzymes (endonucleases) are bacterial enzymes (e.g., *Mse*I, *Eco*RI, *Pst*I) that recognize specific 4, 6, or 8 bp sequences in

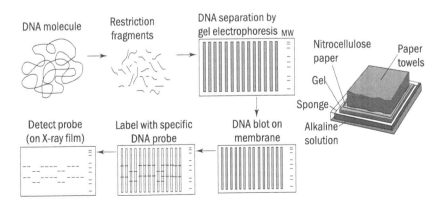

Fig. 5.3 *Different steps of RFLP markers*

DNA, and cleave double-stranded DNA whenever these sequences are encountered. Scientists decide upon the enzymes that can recognize 4, 6, or 8 bp based on the resolution required and the electrophoresis facility available. Enzymes that recognize a 4 bp sequence help in obtaining the greatest resolution. This is because such sites are most abundant in the genome and the fragments produced will be relatively small. This will provide a better chance of identifying single-base alterations. Again, enzymes that identify an 8 bp sequence when used will need fewer probes. The reason for this is that larger fragments of the genome are analysed at one time. As a result, only large alterations of DNA will be visualized and complex electrophoresis system must be used to resolve the fragments into discrete bands. As a compromise, 'six-cutter' enzymes are most often used for RFLP analysis because they are readily available, cheaper, and usually produce fragments in the size range of 200–20,000 bp, which can be separated conveniently on agarose gels. After digestion by restriction enzymes, the DNA is present as a mixture of linear double-stranded molecules of various lengths. Agarose or polyacrylamide gels are then used for separating by electrophoresis. The decision to use either agarose or polyacrylamide is based on the restriction enzymes chosen. Polyacrylamide gels are better equipped to handle fragments produced by four-cutters as they are too small to be resolved by agarose gels. Conversely, agarose gels are used to resolve the fragments produced by six-cutters. Hence, scientists prefer using six-cutter enzymes, as agarose gels are much easier to handle (Potter and Jones 1991). Southern blotting or Southern hybridization technique is used to transfer denatured single strands of DNA fragments separated by gel electrophoresis onto a solid support such as membranes or filter. The basis of this technique is the transfer of the DNA from the gel to a solid support, thus preserving the position of the fragments as they were in the gel, yet enabling hybridization reactions to be performed. Filter-immobilized DNA is then allowed to hybridize to labelled probe DNA. Single locus species-specific probes of about 500–3000 bp in size are preferred for hybridization. Genomic clones (that is fragments of nuclear DNA) and cDNA clones (that is DNA copies of mRNA molecules) are used to identify specific DNA fragments for hybridization. Of the two clones, genomic libraries are easy to construct. The libraries will have many repetitive probes because repetitive sequences constitute the largest proportion of plant genomic DNA. Such probes will hybridize into many fragments on the filters and produce very complex patterns. Despite cDNA libraries being difficult to construct, they contain predominantly unique or low copy number sequences representing expressed genes and usually provide fewer bands on the filter. cDNA probes help in identifying small changes. However, many more probes are used as the proportion of the genome covered by each probe will be relatively small. Appropriate source for RFLP probe is based on the requirement of particular application under consideration. Hybridization of the probe to target sequences occurs when denatured probe solution is left in contact with the filter. The visualization of specific DNA fragments takes place during the hybridization process when the labelled probe binds to complementary DNA on the filter. Traditionally, labelling has been done through radioactive nucleotides. But now non-radioactive methods are also available. Non-specific hybridization must then be washed off under more 'stringent' conditions than those used for initial hybridization. In the case of radioactive probes, the filter is placed against a photographic film and radioactive disintegrations take place resulting in visible bands. An autoradiography of membrane will reveal the set of fragments complementary to the probe. In the case of non-radioactive probes, antibodies against the modified nucleotides and a coupled enzymatic reaction are used to show up the set of fragments directly on the membrane. Digoxigenin is an example of non-radioactive probes. The information obtained with the RFLP technique depends upon both the number of probes and the restriction enzymes used. Each different probe hybridizes with a different set of genomic DNA fragments and each enzyme excises a segment of genomic DNA at different points. An RFLP marker has the following advantages:

- High reproducibility
- Co-dominant inheritance
- Good transferability between laboratories, being a locus-specific marker that allows synteny (conserved order of genes between related organisms) studies.
- Non-requirement of sequence information
- Relatively easy scoring method due to large size difference between fragments.

However, RFLP analysis has several limitations:

- High-quantity and high-quality DNA is required.
- Dependent on development of specific probe libraries for the species.
- Automation is not possible.
- Low level of polymorphism and few loci detected per assay.
- The process is time-consuming, laborious, and expensive.
- Radioactively labelled probes are required.

In the selective mating of individuals of a population or breeding, desired morphological, physiological, or genetic traits such as appearance, yield, and disease resistance are either isolated or combined. This is performed with the assistance of identifiable traits. Molecular biology, in the last 20 years, has revolutionized conventional breeding techniques in all the areas. The duration of breeding programmes have been shortened from years to months to weeks, or even eliminated because of biochemical and molecular techniques. When molecular markers were used in conventional breeding techniques, the accuracy of crosses improved and breeders were able to produce strains with combined traits. All these were impossible before the advent of DNA technology. In a population, if a detectable mutant is identified, the gene that causes the mutation is placed on a genetic map through a series of crosses. This would ascertain its recombination frequency in relation to other genes that had been discovered and mapped earlier. When the mutant gene is in close proximity to the gene for a desired trait, the mutant gene or 'marker' was said to be linked to it because both of them tend to co-segregate. In a breeding cross, this mutant gene could be used to detect whether or not a breeding cross had been successful in transferring the desired trait. If the mutant gene is observed to be expressed in the progeny, the progeny is also most likely to have the desired trait because of its link to the mutant gene. This is the phenomenon of co-inheritance, and the selection of these mutant genes for the tracking of desired traits is called indirect selection. This breeding technique can be used in almost any animal, plant, fungi, or bacteria.

5.6 LINKAGE ANALYSIS

In 1911 D. H. Morgan and his student Alfred H. Sturtevant developed the techniques of genetic mapping and linkage analysis. Today, these techniques are still used much the same way but with far more advanced techniques. Genetic mapping is done as chromosomes cross-over during meiosis; the homologous chromosomes exchange sections of their gene sequence. Two genes that recombine often are termed as far apart on a genetic map, and two that seldom recombine are termed as linked and are very close together on a genetic map. A mapping population is produced to find trait recombination frequencies and form a genetic map. First two genetically divergent parents (that will still produce viable progeny) have to be selected. As the parents need to be the most divergent, a common cultivar and a wild parent are selected. To make sure that the progeny will produce recombinants, these selected divergent parents are screened for polymorphism with the markers that are to be mapped. When the two parents are crossed, it leads to the formation of an F1 hybrid population, which is selfed to produce an F2 population, and hence a mapping population is produced. A uniform, heterozygous population will be produced from the initial cross and each plant will have one chromosome from each parent. During meiosis, there is a possibility that the homologous chromosomes may not cross-over and form recombinants. The expected ratio of phenotypes for the F2 population is the classic Mendelian 1:2:1; divergence from this ratio determines the amount of linkage between genes.

Consider that for parents A and B the two markers X and Y are co-dominant. Both parent A (XX YY) and parent B (xx yy) are homozygous. Gametes (XY) in A and (xy) in B will be produced through meiosis. Hence, a hybrid (Xx Yy) will be produced by F1 cross. Gamete formation in the hybrid occurs when crossing-over becomes important. There are four possibilities of gamete formation XY, xy, Xy, and xY.

The first two are the original or 'parental' genotypes. The next two are 'recombinant' genotypes and called so because they are formed only through crossing-over of the homologous chromosomes. When there are 200 F2 plants, the expected ratio would be 50 (XY XY):100 (XY xy):50 (xy xy) if no recombination was occurring. It would be 13 (XY XY):12 (Xy Xy):25(XY xy):25 (XY Xy):50 (XY xy):25 (xY xy):25 (Xy xy):13 (xY xY):12 (xy xy) in the case of classic, independent assortment. The recombination frequency is determined by dividing the number of recombinant gametes by the total number of gametes. All progenies containing two recombinant gametes are counted twice. Therefore, the recombination frequency for this population is 150/400 = 0.375 and are said to be 37.5 centiMorgans (cM) apart. This is the upper limit of recombination frequency in a selfed F2 population, which implies that alleles X and Y are at opposite ends of the same chromosome; any recombination frequency lower than this indicates that the genes are linked to some degree. Thus, genes are mapped on to a chromosome relative to each other through mounds and mounds of data. This process is now performed almost exclusively by computer programs. Once a genetic linkage map of a chromosome has been established, gene locations that have been mapped in a similar manner can be integrated into the map. The genes can then be assessed for linkage with the closest markers on the map and indirect selection performed using them. For a chromosome map to be useful, it must be saturated with markers of several different types so that a variety of markers can be tried and assessed for their usefulness in detecting that specific trait.

5.6.1 Gel Electrophoresis

Gel electrophoresis is used in the analysis of molecular markers. A common technique, it finds its roots in chromatography. In this technique, molecules with a net charge move through an electric field. The progress of these molecules gets retarded to varying degrees because of the matrix in which they are moving. The migration velocity depends on the following:

- The strength of the electric field
- Net charge on the molecule
- Frictional coefficient of the particle in the matrix

The electrophoresis medium here is almost every time a gel because (i) it is able to suppress small temperature gradients that cause fluctuations in current, (ii) it is not difficult to work with and handle, and (iii) its concentration can be varied to optimize separation. Polyacrylamide and agarose gels are commonly used. Initially, polyacrylamide was used. This is because of its ability to attain high resolution of extremely small molecular weight differences. But as polyacrylamide in its unpolymerized form is a potent neurotoxin, a lot of care is needed in handling it along with being disposed of appropriately. In contrast to polyacrylamide, the resolution power of agarose is not good. Seaweed is boiled at approximately 94°C and after which agarose is extracted. A tight gel matrix forms when it is cooled. It is inexpensive and safe enough to eat. Metaphor agarose is a highly pure form of agarose that, at high concentrations, can achieve highly resolved separations of heavy molecules separated under high electric field conditions.

5.6.2 Polymerase Chain Reaction

The process by which large amounts of DNA are produced from comparatively small amounts, selectively or non-selectively, is known as polymerase chain reaction. Developed by the Cetus Corporation during 1985–86, it is an *in vitro* method of DNA amplification. Here target DNAs are denatured at 95°C, then oligonucleotide primers are annealed to sequences flanking the target DNA for amplification, which allow DNA polymerase to bind and begin synthesizing novel DNA. Repeated over and over again, the cycle each time produces double the amount of DNA present. The final amount of DNA will be 230 times the original amount after 30 cycles. Earlier *Escherichia coli* DNA polymerases were used. However, it was found that *E. coli* were not stable at 95°C and after each denaturation cycle, new polymerase had to be added. The use of *E. coli* was abandoned once *Thermus aquaticus* (Taq) was discovered. These bacteria are found at the opening of thermal vents of the ocean floor and remain stable at the denaturation temperature. In the entire process, these additional bacteria do not need to be added. A PCR is performed with the help of (i) forward and reverse oligonucleotide primers, (ii) amplification buffer (KCl, TrisCl, 1.5 mM $MgCl^2$) that controls the pH drop when incubated at the extension step, (iii) dNTPs at saturation concentration (can be less), (iv) target DNA sequence (purity is not a huge problem as long as pH/Taq is okay) and (v) Taq DNA polymerase.

These components are mixed together with double-distilled/autoclaved water and put into a thermal cycler that adjusts the temperature in the following sequence:

(i)   Denaturation of the template DNA at 94°C

(ii) Annealing of primers to target sequences at 35–65°C

(iii) DNA synthesis from 3' end of each primer by Taq polymerase

The cycle needs to be repeated 30–40 times before completion. As Taq polymerase does not have 5'–3' exonuclease proofreading activity, it can be prone to errors. These errors get propagated after every new cycle. Some manufactures are selling Taq mixed with a thermally stable exonuclease to perform this function.

5.7   RAPD MARKERS

Multiple arbitrary amplicon profiling (MAAP) is a collective term for RAPD, AP-PCR, and DAF (Caetano-Annolles 1994). RAPD, AP-PCR, and DAF were the three techniques that amplified DNA fragments from any species without prior sequences information. The MAAP techniques differ on the basis of the following:

- Modifications in amplification profiles by changing primer sequence and length
- Annealing temperature
- Number of PCR cycles used in a reaction
- Thermostable DNA polymerase used
- Enzymatic digestion of the template DNA or amplification product
- Alternative methods of fragment separation and staining

While RAPD produces quite simple amplification profiles, DAF produces highly complex patterns. The most important innovation with regards to RAPD, AP-PCR, and DAF is that they use single, arbitrary oligonucleotide primer to amplify the template DNA without prior knowledge of the target sequence. Interaction between primer, template annealing sites, and enzymes drives the amplification of nucleic acids with arbitrary primers. They also help in determining the complex kinetic and thermodynamic process. When the single primer binds to sites on opposite strands of the genomic DNA that are within an amplifiable distance (Figure 5.4), generally less than 3000 bp, at an appropriate annealing temperature, a discrete PCR product is produced. In all AP-PCR, DAF, and RAPD, polymorphisms (band presence or absence) result from changes in DNA sequence that inhibit primer binding or interfere with amplification of a particular marker in some individuals; therefore, they can simply be detected as DNA fragments that are amplified from one individual but not from another. The RAPD protocol usually uses a 10 bp arbitrary primer at a constant low annealing temperature (generally 34–37°C). Even though the RAPD primer sequences are chosen randomly, two criteria, however, need to be met:

(1) A minimum of 40% GC content (50%–80% GC content is generally used).

(2) Absence of palindromic sequence (a base sequence that reads exactly the same from right to left as from left to right).

As G–C bond consists of three hydrogen bridges and A–T bond consists of only two, it would be difficult for a primer–DNA hybrid with less than 50% GC to withstand 72°C temperature during which DNA elongation takes place by DNA polymerase. The PCR products which get produced are normally resolved on 1.5%–2.0% agarose gels. They are stained with ethidium bromide (EtBr). Alternatively, polyacrylamide gels are used. They are used in combination with $AgNO_3$ staining (e), radioactivity, or fluorescently-labelled primers or nucleotides. RAPD has become more popular and rapid than AP-PCR and DAF as agarose gel electrophoresis is simple and not expensive. Amplification of one locus causes RAPD fragments, and as a result polymorphism of two kinds occurs. These are: (i) the band may be either present or absent, and (ii) the brightness (intensity) of the band may be different. Band intensity differences may result from copy number or relative sequence abundance (Devos and Gale 1992), and may serve to distinguish homozygote-dominant individuals from heterozygotes, as more bright bands are expected for the former. However, in some cases no relation was found between copy number and band intensity. Fainter bands are generally less robust in RAPD experiment. This means varying degrees of primer mismatch may account for many band intensity differences. Since the source of band intensity differences is uncertain (copy number or primer mismatch), most studies disregard scoring differences in band intensity, although seven-state scale of band intensity is also used (Figure 5.4).

5.7.1 Homology

The reproducibility of RAPD reactions depends on PCR buffer, quality and quantity of template DNA, primer-to-template ratio, annealing temperature, concentration of magnesium chloride, Taq DNA polymerase brand or source, and thermal cycler bra. The reservation regarding RAPD markers' reproducibility can be overcome through the following methods:

- Choosing correct DNA extraction protocol to remove any contaminants
- Optimizing the parameters used
- Testing several oligonucleotide primers
- Scoring only the reproducible DNA fragments
- Using appropriate DNA polymerase brand

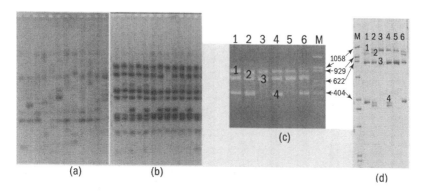

Fig. 5.4 *RAPD amplification products separated on 1.5% agarose gels with EtBr staining (a–c) and 10% polyacrylamide gel with AgNO3 staining (d): (a) poor amplification; (b) good amplification; (c) problem of homology of co-migrating RAPD bands of six samples in agarose gel; and (d) each of the bands, 1 to 4, in the agarose gel appearing as two co-migrating bands in the polyacrylamide gels.*

The reliability of RAPD data is influenced by the presence of artifactual bands (false positives) corresponding to rearranged fragments produced by nested primer-binding sites and intra-strand annealing and interactions during PCR. The reliability of RAPDs gets restricted because of the existence of both false negatives and false positives. All pairwise comparison of RAPD fragments along with samples begins with the assumption that co-migrating bands (i.e., bands that migrate equal distance) represent homologous loci. However, there are some limitations. In polyploid species and also in other studies, based on electrophoretic resolution, it cannot be presumed that equal length equals homology is necessarily true. RAPD bands that were thought of equal length were found to be not homologous on closer examination (see Figure 5.4). Such mistakes can be avoided by using polyacrylamide gels and AgNO$_3$ staining as they help in getting a more accurate resolution of the fragment size (Huff, Peakall, and Smouse 1993). Another limitation of RAPD markers is that most of the alleles segregate as dominant markers, and dominant homozygotes cannot be differentiated from heterozygotes based on this technique. RAPD assays produce fragments from homozygous-dominant or heterozygous alleles. No fragment is produced from homozygous-recessive alleles because amplification is disrupted in both alleles. The original DAF protocol is mainly different from RAPD in that it uses short primers (at least 5 bp), higher primer concentrations, two-temperature cycles instead of three-temperature cycles, and detection of amplification product on $AgNO_3$-stained polyacrylamide gel. The AP-PCR technique differs from RAPD and DAF because in it amplification reaction is divided into three steps, each with different stringencies and concentrations of constituents. In the first cycle, the high primer concentrations are used. Randomly, primers of 20 or more nucleotides, which designed for other purposes, are chosen. Lastly, by radioactivity and autoradiography the detection of amplification products is done (Weising, Nybom, Wolff, *et al.* 1995).

5.8 MICROSATELLITES

Satellite DNAs, minisatellites, and microsatellites are the three types of multiple copies of simple repetitive DNA sequences in genomes of higher organisms. Microsatellites (Litt and Luty 1989) are also called SSRs (Ta), STRs, or SSLPs (McDonald and Potts 1997). They are the smallest class of simple repetitive DNA sequences. Armour, Alegre, Miles, *et al.* (1999) defined microsatellites as 2–8 bp repeats, while according to Goldstein and Pollock (1997) and Schlotterer (1998) they are 1–6 bp or even 1–5 bp repeats. Chambers and MacAvoy (2000) suggested that a strict definition of 2–6 bp repeats should be followed. There are regions where simple repetitive DNA sequence motifs variants are over-represented (Tautz, Trick, and Dover 1986). Microsatellites are born from these regions. Slipped-strand mispairing is the predominant mutation mechanism in microsatellite tracts. According to Eisen (1999), during DNA synthesis, in a microsatellite array slipped-strand mispairing occurs that leads to the gain or loss of one, or more, repeat units. This is dependent on looping out of the newly synthesized DNA chain or the template chain. The relative propensity for either chain to loop out depends on (i) sequences making up the array, and (ii) whether the event occurs on the leading (continuous DNA synthesis) or lagging (discontinuous DNA synthesis) strand. In the microsatellite structure variable numbers of repeat units lead to SSR allelic differences. The repeated sequence is most of the time simple. It consists of two, three, or four nucleotides (di-, tri-, and tetra-nucleotide repeats, respectively). Dinucleotide repeat $(CA)_n$ is an example of a microsatellite. Here n is the total number of repeats that ranges between 10 and 100. These markers represent high levels of inter- and intraspecific polymorphism, especially if the tandem repeat number is 10 or greater (Queller, Strassman, and

Hughes 1993). PCR reactions for SSRs are run in the presence of forward and reverse primers that anneal at the 5' and 3' ends of the template DNA, respectively. Polyacrylamide gels in combination with $AgNO_3$ staining, autoradiography, or fluorescent detection systems are used for PCR fragment separation. In case the differences in allele size is larger than 10 bp when compared to other samples, then agarose gels (usually 3%) in combination with EtBr can be used. It should be kept in mind that building microsatellite primers from scratch for a new species entails a lot of technical challenges. Bruford, Cheesman, Coote, *et al.* (1996); McDonald and Potts (1997); Hammond, Saccheri, Ciofi, *et al.* (1998); Schlotterer (1998) developed many protocols. Zane, Bargelloni, and Patarnello (2002) reviewed the methodologies and explained the technical advances that have helped the development of microsatellite in recent years. They cover a range of methods for obtaining sequences rich in microsatellite repeats and also highlight the availability of companies that will undertake the construction of enriched microsatellite libraries as a commercial service. In developing microsatellite markers many specific steps are undertaken. These involve the following (Roder, Korzun, Wendehake, *et al.* 1998):

- Constructing a microsatellite library
- Identifying unique microsatellite loci
- Finding an appropriate area for primer design
- Attaining a PCR product
- Assessing and interpreting banding patterns
- Evaluating PCR products for polymorphism

When random segments of DNA from the target species are cloned, SSR primers get developed. Then SSR primers are inserted in a cloning vector. The cloning vector is implanted in *E. coli* which then replicates. Colonies get developed, and screened with single or mixed simple sequence oligonucleotide probes that will hybridize to a microsatellite repeat, if present on the DNA segment. If this procedure leads to positive clones for microsatellite, then the DNA is sequenced and PCR primers are chosen from sequences flanking such regions to determine a specific locus. This process involves significant trial and error on the part of researchers, as microsatellite repeat sequences must be predicted and primers that are randomly isolated may not display polymorphism. The best candidate markers should be then selected and amplified in optimal conditions. In order to optimize microsatellite systems, PCR conditions need to be surveyed comprehensively so that the candidate loci are amplified. The main aim is that both the requirements, that is, high specificity and high intensity of amplification products, are balanced completely. It is because of this that signal strength and purity remain the primary focus. The remaining deliberation involves getting products from different loci with non-overlapping ranges of allele sizes, which can be amplified with similar efficiency under a standard set of conditions and can enable multiplexing for high throughput analysis. Some organisms have more microsatellite loci in comparison to others, and screening may produce few useful loci in some species (Cooper 1995). The plenitude of repeats in the target species and the effortlessness with which these repeats can be developed into informative markers are major factors in the efficiency of microsatellite marker development. Around 30% of the sequenced clones get lost when developing plant microsatellites. The reason for this is the absence of unique microsatellites. Even if the sequences contain unique microsatellites, it is possible many of the clones in a library have identical sequences (this would lead to redundancy) and/or chimeric sequences (that is, one of the flanking regions matches that of another clone). At various stages of developing SSR, there is a likelihood that the loci will get lost. Thus, when compared to the original number of clones sequenced, only a fraction of working primer set will remain (Squirrell, Hollingsworth, Woodhead, *et al.* 2003). It is quite arduous to transform microsatellite-containing sequences into useful markers, more so in large genome species. In such species, the high level of repetitive DNA sequences in the genomes makes conversion rates of primer pairs into useful markers low. Some of the reasons for low recovery rate for useful SSR primers are as follows:

• No PCR product is amplified by the primer.
• Complex, weak, or non-specific amplification patterns are produced by the primer.
• The amplification product may not be polymorphic.

Instead of working with dinucleotide arrays, researchers opt for loci containing tri- and tetra-nucleotide repeat arrays as they frequently give fewer 'stutter bands'. Figure 5.5 shows the multiple near-identical 'ladders' of PCR products that are one or two nucleotides shorter or longer than the full-length product. In the case of tri- and tetra-nucleotide repeats, the allele sizing is more accurate to di-nucleotide

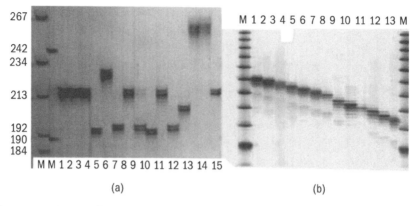

Fig. 5.5 *Microsatellites PCR products separated on polyacrylamide gels with silver nitrate staining. (a) Banding patterns of 15 rice cultivars (number 1 to 15) amplified using RM252 microsatellite primer; M is a molecular size ladder, with the size of each band indicated on the left side of the gel.[1] (b) Gel showing microsatellites allelic diversity for gwm389 primer and 13 hexaploid wheat samples, with each sample possessing a different allele; M is a 10-bp molecular weight ladder and is placed on either side of the gel*

repeats. But di-nucleotide repeat arrays exist much more than tri- or tetra-nucleotide repeat arrays. Running combinatorial screens is easier for them.

Among the markers, researchers prefer SSRs in molecular genetics because they are highly polymorphic even between closely related lines (Figure 5.5), the DNA amount needed is low, high throughput screening can easily be automated, can be exchanged between laboratories, and are highly transferable between populations (Gupta, Varshney, Sharma, *et al.* 1999). One of the reasons why SSR markers from genomic libraries are used in only few of the agriculturally important crops is that developing these markers involves high cost apart from efforts involved in obtaining working primers for a given study species. When the loci become transferrable from one species to another, it would be easier to use genomic SSRs. Recently, researchers have used a new alternative source for developing SSRs from EST databases. The availability of ESTs and other DNA sequence data in plenty has made the development of EST-based SSR markers through data mining quicker, efficient, and inexpensive compared to genomic SSRs (Gupta, Rustgi, Sharma, *et al.* 2003). This is because this approach does not require the lengthy and costly processes of generating genomic libraries and sequencing of innumerable clones for finding

---

[1] Details available at http://www.gramene.org

SSR-containing DNA regions. Moreover, the EST-SSR markers possess lower rate of polymorphism than the SSR markers derived from the genomic libraries. Resolving the differences in SSR allele size is quite hard on agarose gels. High resolutions can be obtained by using polyacrylamide gels with $AgNO_3$ staining. Polyacrylamide gels cost more than agarose gels and are slower than the latter. Researchers in developing countries cannot afford the cost of establishing and running an automatic DNA sequencer.

Microsatellites pose another technical problem: comparing data from different laboratories is not always possible because of the inconsistencies in allele size calling. Such inconsistencies arise from the use of a variety of automatic sequencing machines, each of which provides different fluorescent dyes, gel migration, PCR reactions, and allele calling software. The enzyme used for DNA synthesis (Taq DNA polymerase) acts as a catalyst in adding an extra base (usually an adenine) to the end of the PCR product. Although this problem can be solved by biochemical treatments after PCR or modification of PCR primers; however, they are hardly utilized.

All genes should be transcribed into messenger RNAs (mRNAs), which will serve as templates for synthesizing protein. The resulting mRNAs guide protein synthesis through the process of translation. Since it is unstable outside a cell, an mRNA is converted into a complementary DNA (cDNA) by using an enzyme called reverse transcriptase. A cDNA is produced in a process that is reverse of the process of transcription in cells because the procedure uses mRNA as a template instead of DNA. cDNA is a considerably stable compound. Since it is generated from mRNA, which represents exons by excising (splicing) introns, cDNA represents only the expressed DNA sequence. Once researchers isolate a cDNA that represents an expressed gene, they can sequence a few hundred nucleotides from the 5' or the 3' end and create 5'ESTs and 3'ESTs, respectively (Jongeneel 2000). A 5'EST is created from the part of a transcript (exons) that usually codes for a protein. These regions are conserved across species and do not vary considerably within a gene family. Since the 3'ESTs might fall within non-coding (introns) or untranslated regions (UTRs), they exhibit less cross-species conservation compared to coding sequences. The challenge of identifying genes from genomic sequences varies from organism to organism and depends on the size of genome and the presence or absence of introns, which are intervening DNA sequences that interrupt a gene's protein-coding sequence. The construction of cDNA libraries starts the EST production. In recent years, ESTs have been identified rapidly, and now more than 6 million ESTs are available in computerized databases. Originally meant for identifying gene transcripts, ESTs have also been used for gene discovery, obtaining data on gene expression and regulation, determining sequence, and developing significant molecular markers such as EST-based RFLPs, SSRs, SNPs, and CAPS.

To determine gene expression, probes for DNA microarrays have been designed using ESTs. Single or low-copy RFLP markers also have been developed efficiently using ESTs. These EST-based RFLP markers (EST-RFLP) facilitate the construct ion of high-density genetic linkage maps (Harushima, Yano, Shomura, *et al.* 1998; Davis, Mcmullen, Baysdorfer, *et al.* 1999) and physical maps. Comparative mapping across different species is possible with these markers because of highly conserved sequences in the coding regions. Hence, data available from one species help in marker development and map-based cloning in other species. A computational approach to the development of SSR and SNP markers is also possible using ESTs (Cho, Ishii, Temnykh, *et al.* 2000; Eujayl, Sorrells, Baum, *et al.* 2001). The earlier developed strategies were costly. SSRs in ESTs can be identified with pattern-finding programmes. Using the available sequence information, primer pairs can be designed for screening cultivars of interest for length polymorphisms. In various plant species, around 1%–5% of ESTs have SSRs of suitable length (20 bp or more) for marker development (Kantety, Rota, Matthews, *et al.* 2002). A large number of these SSRs can be found in an organism for which a large number of ESTs have been generated. For instance, Kantety, Rota, Matthews, *et al.* (2002) found that 3.2% ESTs contained SSRs (di-, tri-, and tetra-nucleotide motifs with a minimum repeat length of 18 bp) in 262,631 ESTs in five different grass (rice, maize, wheat, barley, and sorghum) databases. EST-SSRs are mostly anchored in more conserved transcribed regions across species compared to the untranscribed regions (Caudrado and Schwarzacher 1998). Therefore, they are more transferable to closely related genera (Cordeiro, Casu, McIntyre, *et al.* 2001; Decroocq, Fave, Hagen, *et al.* 2003; Hempel and Peakall 2003). There is higher probability that EST-SSRs are more functionally related to differences in gene expression than genomic SSRs (Gao, Jing, Huo, *et al.* 2004). For developing EST-SSRs, most of the multispecies, large-scale mining efforts have focused on monocotyledonous crops (Kantety, Rota, Matthews, *et al.* 2002; Thiel, Michalek, Varshney, *et al.* 2003; Varshney, Thiel, Stein, *et al.* 2002). However, ESTs of a few dicot species have been also explored for SSR mining (Eujayl, Sledge, Wang, *et al.* 2004; Morgante, Hanafey, and Powell 2002; Qureshi, Saha, and Kantety 2004; Saha, Karaca, Jenkins, *et al.* 2003; Scott, Eggler, Seaton, *et al.* 2000; Varshney, Graner, and Sorrells 2005). SNPs based on ESTs have been developed using two strategies. In the first technique, ESTs from the 3' end of cDNA clones (which mainly consist of 3'-UTRs) are used to maximize the likelihood of finding sequence variations. From the EST sequences, primer pairs can be derived, and SNPs may be revealed by amplifying the corresponding regions from several genotypes followed by sequence comparison. The second strategy involves identifying computationally potential SNPs by using clusters of ESTs that contain sequences from different cultivars.

5.9 SEQUENCE-CHARACTERIZED AMPLIFIED REGION

A sequence-characterized amplified region (SCAR) marker is a genomic DNA fragment recognized by PCR amplification with a pair of specific oligonucleotide primers (McDermott, Brandle, Dutly, *et al.* 1994; Paran and Michelmore 1993). SCAR markers can be derived by cloning and sequencing the two ends of RAPD markers that appeared to be diagnostic for specific purposes (e.g., a RAPD band absent in susceptible lines but seen in disease-resistant lines). SCAR markers are preferable to RAPD markers because they can detect only a single locus and can potentially be converted into co-dominant markers. Moreover, amplification of SCAR markers is less sensitive to reaction conditions.

Olsen, Hood, Cantor, *et al.* (1989) were the first to develop a sequence-tagged site (STS) as a DNA landmark in the physical mapping of human genome. Later it was adopted in plants. STS is a unique short sequence, and its exact sequence is not found anywhere else in the genome. When two or more clones have the same STS, they should overlap. The overlap should also contain the STS. If a clone contains a unique sequence and can be sequenced, it may be used as STS. In plants, a pair of PCR primers characterize STS, which are designed by sequencing either an RFLP probe that represents a mapped sequence with low copy number (Blake, Kadyrzhanova, Shepherd, *et al.* 1996) or AFLP fragments. Despite being a technical challenge and often frustrating in polyploids (such as hexaploid wheat), AFLP markers have been successfully converted into STS markers in several crops (Meksem, Leister, Peleman, *et al.* 1995, 2001). Primers designed on the basis of an RAPD are also sometimes referred to as STSs (Naik, Gill, Rao, *et al.* 1998); however, they should be more appropriately called SCAR markers. STS markers are technically simple in use, highly reproducible, co-dominant, and suitable for high throughput and automation (Reamon-Buttner and Jung 2000).

5.10   SINGLE NUCLEOTIDE POLYMORPHISM

Studies on sequence variations among individuals, cultivars, and subspecies have been made possible by the easy accessibility to the genome sequences of many organisms. These studies have revealed that SNPs and InDels are plenty and spread throughout the genome in various species, including plants. For example, Yu, Hu, Wang, *et al.* (2002) compared sequences from a japonica rice cultivar to those from an indica cultivar. On an average, they identified 1 SNP in every 170 bp and 1 InDel in every 540 bp. The SNP marker system is an attractive tool for marker-assisted breeding, mapping, and map-based cloning due to the abundance of these polymorphisms in plant genomes. In a DNA sequence, an SNP marker is only a single base change, with the usual alternative of two possible nucleotides at a given position. Thus, contrary to all earlier methods, allele discrimination does not depend on size differences on a gel. During the last few years, researchers have developed different SNP genotyping methods and chemistries based on various methods of allelic discrimination and detection platforms. Two elements are essential in all methods of SNP genotyping: (i) generation of an allele-specific product and (ii) analysis thereof. Various methods of SNP detection have been broadly categorized into two broad groups (Vignal, Milana, Sancristobala, *et al.* 2002):

1. Direct hybridization techniques
2. Techniques involving the generation and separation of an allele-specific product (restriction enzyme cutting, single-strand DNA conformation and hetero-duplexes, primer extension, oligonucleotide ligation assay, pyrosequencing, exonuclease detection or TaqMan, invasive cleavage of oligonucleotide probes, or invader assay)

Based on molecular mechanism, most SNP genotyping assays belong to one of the four groups: (i) allele-specific hybridization, (ii) primer extension, (iii) oligonucleotide ligation, and (iv) invasive cleavage. Also called ASO hybridization, allele-specific hybridization distinguishes between two DNA targets that differ at one nucleotide position by hybridization (Wallace, Shaffer, Murphy, *et al.* 1979). Generally with the polymorphic base in the central position in the probe sequence, two allele-specific probes are designed. If the assay conditions are optimal, the perfectly matched probe–target hybrids are stable and hybrids with one-base mismatch are unstable. Most hybridization techniques are based on the dot blot, in which the DNA to be tested is attached to a membrane and hybridized with a probe, usually an oligonucleotide. The oligonucleotide probes are immobilized in the reverse dot blot technique. Using ASOs, one can infer genotypes from hybridization signals. However, since hybridization techniques are prone to errors, they require meticulously designed probes and hybridization protocols. The most recent development in this family of techniques is the introduction of DNA chips (microscopic DNA spots fixed to a rigid surface such as glass, plastic, or silicon chips). On these chips, the probes are directly synthesized through a parallel procedure involving masks and photolithography. By using detection methods with high accuracy, sensitivity, and throughput, one can take full advantage of the new ASO probe formats for SNP typing. Primer extension depends on whether DNA polymerase can incorporate specific deoxyribonucleotides complementary to the template DNA sequence. Primer extension reactions can be divided into the following three main types:

1. The single-nucleotide primer extension or minisequencing reaction, where the polymorphic base is determined by adding dideoxynucleotide triphosphate (ddNTP) complementary to the base interrogated by a DNA polymerase.
2. The allele-specific extension, where the DNA polymerase amplifies only when the primers have a perfect match with the template.
3. Pyrosequencing is a method of DNA sequencing (determining the order of nucleotides in DNA) based on the 'sequencing by synthesis' principle. It differs from Sanger sequencing, in that it relies on the detection of pyrophosphate release on nucleotide incorporation, rather than chain termination with dideoxynucleotides.

The method of ASO ligation for SNP typing depends on the ability of ligase to bind two oligonucleotides covalently when they hybridize on a DNA template next to one another. The invader assay is based on the specificity of recognition and cleavage by a flap endonuclease of the three-dimensional structure that is formed when two overlapping oligonucleotides hybridize perfectly to a target DNA. Several detection methods can be used for analysing the products of each type of allelic discrimination reaction including fluorescence resonance energy transfer (FRET), fluorescence polarization, arrays or chips, gel electrophoresis, mass spectrophotometry (matrix-assisted laser desorption/ionization time-of-flight, or MALDI-TOF, mass spectrometry), luminescence, and chromatography. Two different categories relate to the assay format or detection: (i) homogenous reactions in solution and (ii) reactions on rigid surface such as a glass slide, chip, or bead. Allelic discrimination and PCR reactions occur in the same reaction in technologies such as Light Cycler, TaqMan, and Molecular Beacon, which are based on homogenous hybridization with FRET detection. Due to this advantage, further manipulation steps are not required and the automation and throughput of the process are favoured, more so when the high-throughput equipment for TaqMan assays are utilized. Since no separation or purification steps follow the allele discrimination reaction, homogenous reactions are generally more amenable to automation. However, the limited multiplex capability is a major disadvantage. In contrast, further manipulations are needed despite greater multiplex capability being witnessed in reactions on a solid support.

Researchers have access to a variety of different SNP-typing protocols, and no single protocol satisfies all the research needs. The aspects that should be considered while choosing the best technology, include reproducibility, accuracy, sensitivity, capability of multiplexing for high-throughput analysis, cost effectiveness in the initial investment for equipment and cost per data point, flexibility of using the technology for purposes other than SNP discovery, and time consumption for analysis. Predicting which technique will emerge as a standard in the future is difficult because academic laboratories performing medium-scale research and commercial companies or genome centres aiming at high throughput have different priorities. Also the technique has to be selected based on the type of project; performing genotypes with a limited number of SNPs on very large population samples is quite different from that with a large number of SNPs on a limited number of individuals. For studies that include large sets of samples, primer extension techniques analysed by MALDI-TOF mass spectrometry are highly promising with regard to automation accuracy, throughput, and price (Tost and Gut 2002). Methods for SNP genotyping based on mass spectrometry have continuously improved. One of the most efficient and automated detection platforms now is the matrix-assisted laser desorption/ionization mass spectrometry (MALDIMS). It is price competitive when used at high throughput and delivers highly accurate and reliable results. The robustness, accuracy, reproducibility, and success rates of mass spectrometry assays based on primer extension reactions are better than most other methods. One can apply the same platform to the analysis of DNA methylation, expression profiling, and proteomics. This makes the mass spectrometer a very versatile tool in the post-genome-sequencing era. The four allele discrimination methods—hybridization, ligation, cleavage, and primer extension—have been combined with MALDI analysis. Most of the commercially available mass spectrometers come with software to interpret the spectrometric data. Peak identification is carried out for DNA analysis, followed by labelling and assignment. Also a very promising technique, pyrosequencing is comparable to MALDI-TOF with regard to prices and throughput. However, pyrosequencing has very low multiplex capability and automation capacity because many steps are required before detection. Its main advantage is that the contribution of each allele can be quantified, which is very useful in analysing mixtures' profiles.

5.11   AMPLIFIED FRAGMENT LENGTH POLYMORPHISM

In the AFLP technique, the power of RFLP and the flexibility of PCR-based technology combine through the ligation of prime recognition sequences (adaptors) to the restricted DNA. The AFLP has the capacity to simultaneously screen the randomly distributed representative DNA regions throughout the genome. One can generate AFLP markers for the DNA of any organism without initially investing in sequence analysis and primer/probe development. For digestion, both good quality and partially degraded DNA can be utilized. However, the DNA should not have restriction enzyme and PCR inhibitors.

Various scientists have reviewed the AFLP methodology. In AFLP analysis, the first step is the restriction digestion of genomic DNA (about 500 µg) with a mixture of frequent cutter (*Mse*I or *Taq*I) and rare cutter (*Eco*RI or *Pst*I) restriction enzymes. Then the double-stranded oligonucleotide adaptors are designed keeping in mind that after ligation, the initial restriction site is not restored. For providing the known sequences for PCR amplification, such adaptors are ligated to both ends of the fragments. Vos, Hogers, Bleeker, *et al.* (1995) described that PCR amplification would occur only where the primers can anneal to fragments that have the adaptor sequence and the complementary base pairs to the additional nucleotides that are called selective nucleotides. Under stringent conditions, an aliquot undergoes two subsequent PCR amplifications, with primers that are complementary to the adaptors and that possess 3′ selective nucleotides of 1–3 bases. In the first PCR, also known as pre-amplification, the primer combinations contain a single base-pair extension, while in the final (selective) amplification, primer pairs with up to 3 base-pair extensions are used. Due to the high selectivity, primers that differ by only a single base in the AFLP extension amplify a different subset of fragments. A primer extension of one, two, or three bases decreases the number of amplified fragments by factors of 4, 16, and 64, respectively. Depending on the genome size of the species, ideal primer extension lengths vary and result in an optimal number of bands: not too many bands so that smears or high levels of band co-migration do not occur during electrophoresis, but sufficient enough to provide adequate polymorphism.

Researchers visualize AFLP fragments on either agarose gel or denaturing polyacrylamide gels through autoradiography, $AgNO_3$ staining (Figure 5.6), or automatic DNA sequencers. With PAGE, the maximum resolution of AFLP banding patterns to the level of single-nucleotide length differences can be obtained. However, on agarose gels, fragment length differences of less than 10 nucleotides are difficult to estimate.

For automatic separation of AFLP products through fluorescent detection systems on DNA sequencers (such as ABI Prism), one of the selective primers should be marked with dyes (fluorophore) of different colours at the 5′ end, such as 6-carboxyfluorescine (6-FAM), hexachloro-6-carboxy-fluorescine (HEX), tetrachloro-6-carboxy-fluorescine (TET) (Figure 5.6). The sequencers will detect only fragments that contain a priming site complementary to the fluorophore-labelled primer. A fluorescence detection system has the following four essential elements:

1. An excitation source
2. A fluorophore

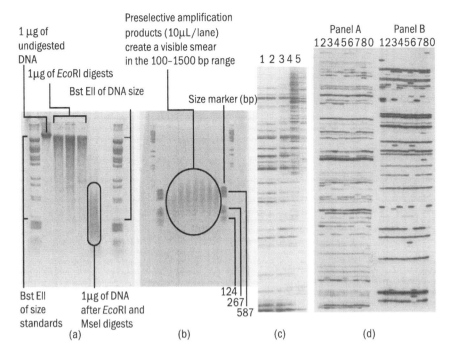

Fig. 5.6 Example of AFLP analysis in plants. Agarose gel showing (a) digested and undigested DNA; (b) pre-selective amplification products for AFLP analysis; (c) the effect of partial digestion on selective amplification (the same sample was digested with five different concentrations of *Eco*RI and *Mse*I enzymes, with lanes 1 and 5 representing amplification products after digestion with the highest and lowest enzyme concentrations, respectively. Banding pattern for lanes 1 and 2 are the correct ones but not that for lanes 4 and 5; and (d) selective amplification products for nine samples (1–9) and two primer pairs (E-AA/M-CAC for Panel A and E-AC/MCAC for Panel B).

3. Wavelength filters to isolate emission photons from excitation photons

4. A detector for registering emission photons and producing a recordable output, such as an electrical signal or a photographic image

For optimizing fluorescence detection, above-mentioned four elements should be compatible regardless of the application. In a high-throughput analysis, one can multiplex and load three to nine different reactions labelled with different dyes in a single lane (such as ABI Prism® 377 DNA Sequencer) or in a single injection (such as ABI Prism 3730 DNA Analyzer). To estimate the size of AFLP amplification fragments, an internal size standard labelled with a different colour is loaded and computer programs such as GeneScan and GeneMapper (from PE Applied Biosystems) are used. AFLP analysis is not easy compared to RAPDs but is more efficient than RFLPs. The advantages of AFLP include the following:

- Highly reliable and reproducible.
- Does not require any DNA sequence information from the organism under study.
- Rich in information because AFLP can simultaneously analyse a large number of polymorphic loci (effective multiplex ratio) with a single primer combination on a single gel as compared to RAPDs, RFLPs, and microsatellites.
- Mostly homologous and locus-specific co-migrating AFLP amplification products, with exceptions in polyploidy species.

  In contrast to RAPD, the limitations of AFLP are as follows:

- Requires more steps to produce results.
- The template DNA must be free of inhibitor compounds that interfere with the restriction enzyme.
- Requires the use of polyacrylamide gel in combination with $AgNO_3$ staining, radioactivity, or fluorescent methods of detection, which are more expensive and laborious than agarose gels.
- Involves additional cost since restriction and ligation enzymes as well as adapters have to be purchased.
- Like RAPD, most AFLP loci are dominant, which fail to differentiate dominant homozygotes from heterozygotes. Thus, the accuracy of AFLP markers in genetic mapping, marker-assisted selection, and population genetic analysis gets reduced.

5.12 QUANTITATIVE TRAIT LOCI

For more than a century, a major area of study in genetics has been quantitative characteristics. These are a common feature of natural variation in eukaryotes, including crop plants. Until 1980, quantitative traits were studied through statistical techniques such as means, variances, and covariances of relatives. The conceptual base provided by these studies enabled the partitioning of the total phenotypic variance into genetic and environmental variances as well as the further analysis of genetic variance with regard to additive, dominant, and epistatic effects. This information enabled researchers to estimate the heritability of the trait as well as predict the response of the trait to selection. Researchers could also estimate the minimum number of genes that controlled the trait of interest. However, the researchers could not identify these genes, their location, and how they controlled the traits. What they understood was the fact that for any given trait, several such genes segregated in a Mendelian fashion in any given population and that their effects were approximately additive in most cases.

Mather (1949) termed these genes 'polygenes'. Sax (1932) experimented with beans and found that the effect of an individual locus on a quantitative trait could be isolated through a series of crosses. This randomized the genetic background with respect to all genes not linked to the genetic markers under study. Although Sax used only morphological seed markers with complete dominance, he could show a considerable effect of some of his markers on seed weight. Unfortunately, this demonstration was followed by only a limited number of successful detections of linkages between markers and QTL in crop plants during 1930–80, and of these, even fewer were repeated. The major limitation was the absence of requisite polymorphic markers. The scenario changed after two major developments during the 1980s: (i) extensive, yet easily visualized, variability at the DNA level was discovered, which could be used as markers; and (ii) statistical packages were developed, which could analyse variation in a quantitative trait based on the molecular marker data generated in a segregating population.

The phenomenal improvements in molecular marker technology have led to identification and use of polymorphic DNA markers as a framework for locating the polygenes. Due to these improvements, a genetic map can be generated for almost any species, which is saturated with polymorphic co-dominant Mendelian markers. For many species of economic or scientific interest, nearly saturated genetic maps have already been generated. Polygenes are now referred to as 'QTL', a term coined by Gelderman (1975). A QTL is defined as 'a region of the genome that is associated with an effect on a quantitative trait'. Thus, a QTL may be a single gene or a cluster of linked genes that affect the trait. Studies on QTL mapping have been carried out on most crop plants for traits such as quality, yield, disease and insect resistance, abiotic stress tolerance, and environmental adaptation. This section provides a brief overview of the principle and salient requirements of QTL mapping, statistical tools and techniques employed in QTL analysis, and the strengths, constraints, and applications of QTL mapping for crop improvement.

5.12.1 Principle of QTL Mapping

In most crop plants, a considerable number of segregating markers (10–50) per chromosome can be easily identified and mapped. Being in the non-coding regions of the genome, most of these markers might not directly affect the trait of interest. However, a few of these markers might be linked to genomic regions (QTLs) that influence the trait of interest. The marker locus and the QTL will co-segregate where such linkages occur. Thus, to establish whether a QTL is linked to a marker, the mapping population should be partitioned into different genotypic classes on the basis of the genotypes at the marker locus. Then correlative statistics should be applied to determine whether the trait being measured differs significantly between the individuals of one genotype and the individuals of the other genotype. Situations in which genes fail to segregate independently are called 'linkage disequilibrium'. Hence, QTL analysis depends on linkage disequilibrium. In natural populations, association between QTL and marker genotype will not be generally consistent, except in a rarest situation in which the marker is entirely linked to the QTL. Therefore, QTL analysis is used in mapping populations such as F2-derived populations, backcross populations, recombinant inbred lines (RILs), near-isogenic lines (NILs), and doubled haploid lines (DHs).

5.12.2 Objectives of QTL Mapping

Molecular marker research on quantitative traits is mostly carried out to map QTL. These experiments have the following objectives:

- To identify the regions of the genome that affect the trait of interest.
- To analyse the effect of the QTL on the trait:
  - How much variation a specific region causes in the trait?
  - What is the gene action associated with the QTL?
  - Which allele is associated with the favourable effect?

5.12.3 Salient Requirements for QTL Mapping

- A suitable mapping population generated from phenotypically contrasting parents
- A saturated linkage map based on molecular markers

- Reliable phenotypic screening of mapping population
- Appropriate statistical packages to analyse the genotypic information along with phenotypic information for QTL detection

5.12.4 Types and Size of Mapping Population

Linkage disequilibrium is a key to detect QTLs with markers. This makes random-mating populations more difficult for QTL mapping. An appropriate experimental mapping population must be developed using parental lines that are phenotypically highly contrasting for the target trait (e.g., highly resistant and susceptible lines). These parental lines should also be genetically divergent so that the possibility of recognizing a large set of polymorphic markers well distributed across the genome is enhanced. To fulfil this criterion, a molecular polymorphism survey might be carried out across a set of potentially useful lines so that the most suitable ones can be identified for generating mapping population. The selection of a mapping population is governed by the goals of the experiment, the timeframe, and the resources available for QTL analysis. However, the ability to detect QTLs or the information in F2 or F2-derived populations and RILs is considerably greater than others.

F2:3 families have the advantage that they can measure the effects of additive and dominance gene actions at specific loci. Since RILs are essentially homozygous, only additive gene action can be measured. The RILs have the advantage of larger experiments being performed at several locations and multiple years. Many crops cannot generate enough seeds for a multi-location experiment with population of F2:3 families. For accommodating different types of populations, the genetic model has to be modified. Several factors determine the size of the mapping population for QTL analysis: the type of mapping population meant for analysis, genetic nature of the target trait, the experiment's objectives, and the resources available for managing a sizeable mapping population during phenotyping and genotyping. When a large number of individuals (~500 or more) are analysed, even QTLs having minor effects on the target trait can be detected. However, the basic purpose of QTL mapping is served when QTLs with major effects are identified. This requires a mapping population of 200–300 individuals (Figure 5.7).

5.12.5   Generating a Reasonably Saturated Linkage Map

The segregation patterns for each marker can be analysed by screening the mapping population with polymorphic molecular markers, also known as genotyping. The segregation patterns and the type of mapping population used are usually compatible. A statistical package such as MAPMAKER or JOINMAP is then used to analyse the genotypic data for creating a linkage map of the molecular markers analysed in the

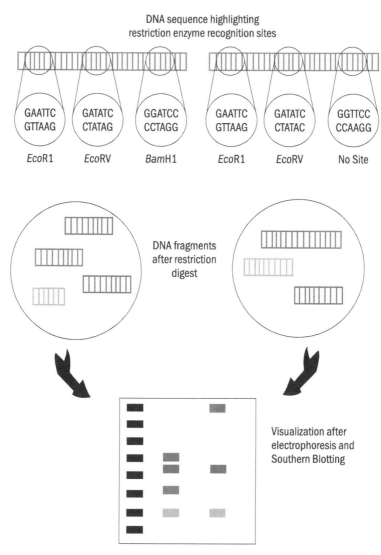

Fig. 5.7   *Major events in RFLP analysis*

study. In mapping, the markers are placed in order, which indicates the relative genetic distances between them, and are assigned to their linkage groups depending on the recombination values from all pairwise combinations between the markers. A saturated marker map is desirable to perform a whole-genome QTL scan, which has markers available for each chromosome, from one end to the other. Also in such a map, recombination events between adjacent markers occur only rarely since they are spaced sufficiently close. This is considered to be fewer than 10 recombinations per 100 meioses, or less than 10 cM in terms of map distance, for practical purposes. The model plant *A. thaliana* has a small genome and requires as few as 50 markers. Manifold more markers are required for plant genomes such as wheat and maize. A rule of the thumb for crops such as maize is covering each chromosomal (bin) location with a minimum of one or two polymorphic molecular markers.

5.13   MAPPING FUNCTIONS

Recombination fraction can compute the distance between two genes. Morgan is the unit used for mapping. The distance over which one crossover occurs per meiosis (on average) is equal to one Morgan. More than two points on a genetic map can be mapped conveniently if the distances on the map are additive. However, recombination fractions are not additive. For loci A, B, and C, the sum of the recombination fractions AB and BC is not equal to the recombination fraction between A and C. Let the distance A–B be r_1, B–C be r_2, and A–C be r_{12}, depending on interference. There is no interference if the recombination between A and B (with probability r_1) and the event of the recombination between B and C (with probability r_2) are independent of each other. In that case, the recombination between A and C is equal to $r_{12} = r_1 + r_2 - 2 \times r_1 \times r_2$. The last term reflects the double crossovers. In the case of complete interference, a crossover in one region completely suppresses recombination in adjacent regions. In that case, $r_{12} = r_1 + r_2$, that is, the recombination fractions are additive. Also the term '$2r_1r_2$' may be ignored for small distances, and recombination fractions are nearly additive.

More generally, one cannot ignore double recombinants, and recombination fractions are not additive. If distances are not additive, a genetic map has to be redone each time new loci are discovered. This problem is avoided by mapping the distances on the genetic map with a mapping function. Recombination frequencies between two loci are translated into a map distance in cM using a mapping function. It gives the relationship between the distance between two chromosomal locations on the genetic map (in cM) and their recombination frequency. A good mapping function has the following two properties:

1. Distances are additive, which means the distance AC should be equal to AB + BC if the order is ABC.
2. A distance of more than 50 cM should translate into a recombination fraction of 50%.

Generally, a mapping function is dependent on the assumed interference. In the case of complete interference and within small distances, a mapping function is simply distance (d) = recombination fraction (r). With no interference, the Haldane mapping function, $d = -1/2 \ln(1 - 2r)$, is appropriate.

Given the map distance (d), the recombination fraction can be calculated as follows:

$$r = 1/2\,(1 - e - 2d)$$

Kosambi's mapping function allows some interference:

$$d = 1/4\,\ln[(1 + 2r)/(1 - 2r)]$$

REFERENCES

Armour, J. A. L., S. A. Alegre, S. Miles, L. J. Williams, and R. M. Badge. 1999. Minisatellites and Mutation Processes in Tandemly Repetitive DNA. In: *Microsatellites: evolution and applications*, D. B. Goldstein and C. Schlotterer (eds), pp. 24–33. Oxford: Oxford University Press

Blake, T. K., D. Kadyrzhanova, K. W. Shepherd, A. K. M. R. Islam, P. L. Langridge, C. L. McDonald, J. Erpelding, S. Larson, N. K. Blake, and L. E. Talbert. 1996. STS-PCR markers appropriate for wheat-barley introgression. *Theor. Appl. Genet.* 82: 715–21

Bruford, M. W. , D. J. Cheesman, T. Coote, H. A. A. Green, S. A. Haines, C. O'Ryan, and T. R. William. 1996. Microsatellites and their application to conservation genetics. In: *Molecular Genetic Approaches in Conservation*, T. B. Smith, and R. K. Wayne (eds), pp. 278–97. New York: Oxford University Press

Caetano-Anolles, G. 1994. MAAP: a versatile and universal tool for genome analysis. *Plant Mol. Biol.* 25: 1011–26

Caudrado, A. and T. Schwarzacher. 1998. The chromosomal organization of simple sequence repeats in wheat and rye genomes. *Chromosome* 107: 587–94

Chambers, G. K, and E. S. MacAvoy. 2000. Microsatellites: consensus and controversy. *Comparative Biochem. Physiol.* (Part B) 126: 455–76

Cho, Y. G., T. Ishii, S. Temnykh, X. Chen, L. Lipovich, S. R. McCouch, W. D. Parl, N. Ayers, and S. Cartinhour. 2000. Diversity of microsatellites derived from genomic libraries and GenBank sequences in rice (Oryza sativa L.). *Theor. Appl. Genet.* 100: 713–22

Cooper, G. 1995. *Analysis of Genetic Variation and Sperm Competition in Dragon flies.* D. Phil. Oxford: Thesis, Oxford University

Cordeiro, G. M., R. Casu, C. L. McIntyre, J. M. Manners, and R. J. Henry. 2001. Microsatellite markers from sugarcane (Sacharum spp.) ESTs cross transferable to erianthus and sorghum. *Plant Sci.* 160: 115–23

Davis, G. L., M. D. Mcmullen, C. Baysdorfer, T. Musket, D. Grant, *et al.* 1999. A maize map standard with sequenced core markers, grass genome reference points, and 932 expressed sequence tagged sites (ESTs) in a 1736 locus map. *Genetics* 152: 1137–72

Decroocq, V., M. G. Fave, L. Hagen, L. Bordenave, and S. Decroocq. 2003. Development and transferability of apricot and grape EST microsatellite markers across taxa. *Theor. Appl. Genet.* 106: 912–22

Dellaporta, S. L., J. Wood, and J. B. Hicks. 1983. A plant DNA minipreparation: version II. *Plant Mol. Biol. Rep.* 1: 19–21

Devos, K. M. and M. D. Gale. 1992. The use of random amplified polymorphic DNA markers in wheat. *Theor. Appl. Genet.* 84: 567–72

Eisen, J. A. 1999. Mechanistic basis for microsatellilte instability. In: *Microsatellites: evolution and applications,* D. B. Goldstein and C. Schlotterer (eds), pp. 34–48. Oxford: Oxford University Press

Eujayl, I., M. K. Sledge, L. Wang, G. D. May, K. Chekhovskiy, J. C. Zwonitzer, and M. A. Mian. 2004. Medicago truncatula EST-SSRs reveal cross- species genetic markers for Medicago spp. *Theor. Appl. Genet.* 108: 414–22

Eujayl, I., M. Sorrells, M. Baum, P. Wolters, and W. Powell. 2001. Assessment of genotypic variation among cultivated durum wheat based on EST-SSRs and genomic SSRs. *Euphytica* 119: 39–43

Gao, L. F., L. R. Jing, N. X. Huo, Y. Li, X. P. Li, R. H. Zhou, X. P. Chang, J. F. Tang, Z. Y. Ma, and J. Z. Jia. 2004. One hundred and one new microsatellite loci derived from ESTs (EST-SSRs) in bread wheat. *Theor. Appl. Genet.* 108: 1392–1400

Gelderman, H. 1975. Investigations on inheritance of quantitative characters in animals by gene markers. *Theoritical and Applied Genetics* 46: 319–30

Goldstein, D. B. and D. D. Pollock.1997. Launching microsatellites: a review of mutation processes and methods of phylogenetic inference. *J. Hered.* 88: 335–42

Gupta, P. K, R. K. Varshney, P. C. Sharma, and B. Ramesh. 1999. Molecular markers and their applications in wheat breeding. *Plant Breed.* 118: 369–90

Gupta, P. K., S. Rustgi, S. Sharma, R. Singh, N. Kumar, and H. S. Balyan. 2003. Transferable EST-SSR markers for the study of polymorphism and genetic diversity in bread wheat. *Mol. Genet. Genomics* 270: 315–23

Hammond, R. L., I. J. Saccheri, C. Ciofi, T. Coote, S. M. Funk, W. O. McMillan, M. K. Bayes, E. Taylor, and M. W. Bruford. 1998. Isolation of microsatellite markers in animals. In *Molecular Tools for Screening Biodiversity*, A. Karp, P. G. Issac, and D. S. Ingram DS (eds), pp. 279–87. London: Chapman and Hall

Harushima, Y., M. Yano, A. Shomura, M. Sato, T. Shimano, Y. Kuboki, *et al.* 1998. A high-density rice genetic linkage map with 2275 markers using a single F2 population. *Genetics* 148: 479-494

Hempel, K. and R. Peakall. 2003. Cross-species amplification from crop soybean Glycine max provides informative microsatellite markers for the study of inbreeding wild relatives. *Genome* 46: 382–93

Huff, D. R., R. Peakall, and P. E. Smouse. 1993. RAPD variation within and among natural populations of outcrossing buffalograss. *Theor. Appl. Genet.* 86: 927–934

Jaccoud, D., K. Peng , D. Feinstein, and A. Kilian. 2001. Diversity arrays: a solid state technology for sequence information independent genotyping. *Nucleic Acids Res.* 29: e25

Jongeneel, C. V. 2000. Searching the expressed sequence tag (EST) databases: panning for genes. *Briefings in Bioinformatics* 1: 6–92

Kantety, R.V., M. L. Rota, D. E. Matthews, and M. E. Sorrells. 2002. Data mining for simple-sequence repeats in expressed sequence tags from barley, maize, rice, sorghum, and wheat. *Plant Mol. Biol.* 48: 501–10

King, R. C. and W. D. Stansfield. 1990. *A Dictionary of Genetics,* 4th edn, p. 188. New York: Oxford University Press

Litt, M. and J. M. Luty. 1989. A hypervariable microsatellite revealed by in vitro amplification of a dinucleotide repeat within the cardiac muscle actin gene. *Am. J. Hum. Genet.* 44: 397–401

Markert, C. L. and F. Moller. 1959. Multiple forms of enzymes: tissue, ontogenetic, and species specific patterns. *Proc. Natl. Acad. Sci.* USA 45: 753–73

Mather, K. 1949. *Biometrical Genetics.* London: Methuen & Co.

McDermott, J. M., U. Brandle, F. Dutly, U. A. Haemmerli, S. Keller, K. E. Muller, and M. S. Wolf. 1994. Genetic variation in powdery mildew of barley: Development of RAPD, SCAR and VNTR markers. *Phytopathology* 84: 1316–21

McDonald, D. B. and W. K. Potts. 1997. DNA Microsatellites as Genetic Markers for Several Scales. In: *Avian Molecular Evolution and Systematics,* D. P. Mindell (ed.), pp. 29–49. San Diego: Academic Press

McDonald, D. B. and W. K. Potts.1997. DNA microsatellites as genetic markers for several scales. In: *Avian Molecular Evolution and Systematics,* D. P. Mindell (ed.), pp. 29–49. San Diego: Academic Press

Meksem, K., D. Leister, J. Peleman, M. Zabeau, F. Salamini, C. Gebhardt. 1995. A high resolution mapping of the vicinity of the R1 locus on chromosome V of potato based on RFLP and AFLP markers. *Mol. Gen. Genet.* 249: 74–81

Meksem, K., E. Ruben, D. Hyten, K. Triwitayakorn, D. A. Lightfoot. 2001. Conversion of AFLP bands into high-throughput DNA markers. *Mol. Genet. Genomics* 265: 207–14

Morgante, M., M. Hanafey, and W. Powell. 2002. Microsatellites are preferentially associated with non-repetitive DNA in plant genomes. *Nat. Genet.* 30: 194–200

Murray, M. G. and W. F. Thompson.1980. Rapid isolation of high molecular weight plant DNA. *Nucleic Acids Res.* 8: 4321–25

Naik, S., K. S. Gill, V. S. Prakasa Rao, V. S. Gupta, S. A. Tamhankar, S. Pujar, B. S. Gill, and P. K. Ranjekar. 1998. Identification of a STS marker linked to the *Aegilops speltoides*-derived leaf rust resistance gene Lr17 in wheat. *Theor. Appl. Genet.* 86: 424–39

Nakamura, Y., M. Leppert, P. O'Connell, R. Wolff, T. Holm, M. Culver, C. Martin, E. Fujimoto, M. Hoff, E. Kumlin, and R. White. 1987. Variable number tandem repeat (VNTR) markers for human gene mapping. *Sci.* 235: 1616–22

Olsen, M., L. Hood, C. Cantor, and D. Botstein. 1989. A common language for physical mapping of the human genome. *Science* 245: 1434–35

Paran, I. and R. W. Michelmore. 1993. Development of reliable PCR-based markers linked to downy mildew resistance genes in lettuce. *Theor. Appl. Genet.* 85:985–93

Paterson, A. H., C. L. Brubaker, and J. F. Wendel. 1993. A rapid method for extraction of cotton (Gossypium spp.) genomic DNA suitable for RFLP or PCR analysis. *Plant Mol. Biol. Rep.* 11: 122–27

Potter, R. H. and M. G. K. Jones.1991. Molecular Analysis of Genetic Stability. In: In Vitro *Methods for Conservation of Plant Genetic Resources*, J. H. Dodds (ed.), pp. 71–91. London: Chapman and Hall

Queller, D. C., J. E. Strassman, and C. R. Hughes. 1993. Microsatellites and Kinship. *Trends Ecol. Evol.* 8: 285–88

Qureshi, S. N, S. Saha, R. V. Kantety, and J. N. Jenkins. 2004. EST-SSR: a new class of genetic markers in cotton. *J. Cotton Sci.* 8: 112–13

Reamon-Buttner, S. M. and C. Jung. 2000. AFLP-derived STS markers for the identification of sex in Asparagus officinalis L. *Theor. Appl. Genet.* 100: 432–38

Roder, M. S., V. Korzun, K. Wendehake, J. Plaschke, M. H. Tixier, P. Leroy, and M. W. Ganal. 1998. A microsatellite map of wheat. *Genetics* 149: 2007–23

Saghai-Maroof, M. A., K. M. Soliman, R. A. Jorgensen, and R. W. Allard. 1984. Ribosomal DNA sepacer-length polymorphism in barley: Mendelian inheritance, chromosomal location, and population dynamics. *Proc. Natl. Acad. Sci. USA* 81: 8014–19

Saha, S., M. Karaca, J. N. Jenkins, A. E. Zipf, U. K. Reddy, and R. V. Kantety. 2003. Simple sequence repeats as useful resources to study transcribed genes of cotton. *Euphytica* 130: 355–64

Sax, K. 1932. The association of size differences with seed-coat pattern and pigmentation in Phaseolus vulgaris. *Genetics* 8: 552–60

Schlotterer, C. 1998. Microsatellites. In: *Molecular Genetic Analysis of Populations: a practical approach*, A. R. Hoelzel (ed.), pp. 237–261. IRL Oxfor: Oxford Press

Schlotterer, C. 1998. Microsatellites. In: *Molecular Genetic Analysis of Populations: a practical approach*, A. R. Hoelzel (ed.), pp. 237–261. Oxford: IRL Press

Scott, K. D., P. Eggler, G. Seaton, M. Rosetto, E. M. Ablett, L. S. Lee, and R. J. Henry. 2000. Analysis of SSRs derived from grape ESTs. *Theor. Appl. Genet.* 100: 723–26

Squirrell J., P. M. Hollingsworth, M. Woodhead, J. Russell, A. J. Lowe, M. Gibby, and W. Powell. 2003. How much effort is required to isolate nuclear microsatellites from plants? *Mol. Ecol.* 12: 1339–48

Tautz, D., M. Trick, and G. A. Dover. 1986. Cryptic simplicity in DNA is a major source of genetic variation. *Nature* 322: 652–56

The Rice Genome Sequencing Project. 2005. The map based sequence of the rice genome. *Nature* 436: 793–800

Thiel, T., R. K. Michalek, R. K.Varshney, and A. Graner. 2003. Exploiting EST databases for the development and characterization of gene-derived SSR markers in barley (*Hordeum vulgare* L.). *Theor. Appl. Genet.* 106: 411–22

Tost, J. and I. G. Gut. 2002. Genotyping single nucleotide polymorphisms by mass spectrometry. *Mass Spectrom. Rev.* 21: 388–418

Varshney, R. K., T. Thiel, N. Stein, P. Langridge, and A. Graner. 2002. In silico analysis on frequency and distribution of microsatellites in ESTs of some cereal species. *Cell. Mol. Biol. Lett.* 7: 537–46

Varshney, R. K., A. Graner, and M. E. Sorrells. 2005. Genic microsatellite markers in plants: features and applications. *Trends Biotechnol.* 23: 48–55

Vignal, A., D. Milana, M. Sancristobala, and A. Eggenb. 2002. A review on SNP and other types of molecular markers and their use in animal genetics. *Genet. Sel. Evol.* 34: 275–305

Vos, P., R. Hogers, M. Bleeker, M. Reijans, T. Van De Lee, M. Hornes, A. Frijters, J. Pot, J. Peleman, M. Kuiper, M. Zabeau. 1995. AFLP: a new technique for DNA fingerprinting. *Nucleic Acids Res.* 23: 4407–14

Wallace, R. B, J. Shaffer, R. F. Murphy, J. Bonner, T. Hirose, and K. Itakura. 1979. Hybridization of synthetic oligodeoxyribonucleotides to phi 174 DNA: the effect of single base pair mismatch. *Nucleic Acids Res.* 6: 3543–57

Weising, K., H. Nybom, K. Wolff, and W. Meyer. 1995. *DNA Fingerprinting in Plants and Fungi*, p. 321. Boca Raton/Ann Arbor/ London/ Tokyo: CRC Press

Welsh, J. and M. McClelland. 1990. Fingerprinting genomes using PCR with arbitrary primers. *Nucleic Acids Res.* 18: 7213–18

Wexelsen, H. 1933. Linkage between quantitative and qualitative characters in barley. *Hereditas* 17: 323–41

Wu, K., R. Jones, L. Dannaeberger, and P. A. Scolnik. 1994. Detection of microsatellite polymorphisms without cloning. *Nucleic Acids Res.* 22: 3257–58

Yu, J., S. Hu, J. Wang, G. K. S. Wong, S. Li, B. Liu, *et al.* 2002. A draft sequence of the rice genome (Oryza sativa L. ssp. indica). *Science* 296: 79–92

Zane, L., L. Bargelloni, and T. Patarnello. 2002. Strategies for microsatellite isolation: a review. *Mol. Ecol.* 11: 1–16

Application of Plant Biotechnology

6.1 INTRODUCTION

The only difference between fats and oils is in their physical characteristics; at the room temperature, fats are solids and oils are liquid. In plants and animals, fats and oils occur as mixtures of triacylglycerols (also referred to as triglycerides), which are water-insoluble, non-polar, fatty acid triesters of glycerol. The composition of their fatty acids depends on the organism. Unlike animal fats, plant oils have unsaturated fatty acids in rich quantities. Fatty acids are carboxylic acids containing long-chain hydrocarbon side groups. There are one or more double bonds between carbon atoms in unsaturated fatty acids. However, saturated fatty acids have all carbon atoms bonded to the highest possible number of hydrogen atoms. Some saturated biological fatty acids include arachidonic acid, palmitic acid, lauric acid, and stearic acid, and some unsaturated fatty acids include arachidonic acid, oleic acid, palmitoic acid, linoleic acid, α-linolenic acid, and γ-linolenic acid (GLA).

The energy required for seedling growth in plant species is stored in different polymers, generally oil and starch. The oil might be found in seeds (such as groundnut, rapeseed, mustard, soya bean, and sunflower, which contain about 50% of the seed dry weight) or in fruits (such as avocado, olive, and oil palm fruits, which have high levels of oil). Vegetable oil is a major commodity for both food and industrial needs.

The composition of fatty acids of the oil used for food or in industrial applications varies. The quality (fatty acid composition) and yield of oil can be enhanced by genetic engineering to suit specific purposes. An example of such genetic modification is high lauric acid (a fatty acid with 12 carbon atoms). There is 60% oleic acid along with other fatty acids in the wild-type rapeseed. The genetically engineered rapeseed variety produces lauric acid, a 12-carbon fatty acid used in producing detergents and soaps. Scientists added only one gene coding for a thioesterase enzyme from the California bay tree, which contains high levels of lauric acid. Fatty acid synthesis should be preferably terminated after 12 carbons instead of allowing the acid to grow to 18 carbons (oleic acid), which is usual for the plant and has little effect on productivity. The fatty acids in the oil drastically change from 60% oleic acid to 60% lauric acid. The commercial production of the genetically engineered oil has already begun. Monounsaturated fatty acid level in plants enhances the net value of oils. Transgenic soya bean with enhanced oil composition (from about 25% oleic acid to more than 85%) has been produced, which has been used for important food and non-food purposes.

Rapeseed contains low amounts of stearate and stearic acid. Into the embryos of rapeseed, an antisense deoxyribonucleic acid (DNA) constructed for the critical enzyme was inserted. The transgenic rapeseed oil produced had a high level of stearate and was used for making margarine (a butter-like solid oil). The transgenic plant produced when a $\Delta 15$ desaturase gene was introduced into rapeseed was rich in α-linolenic acid. The cocoa gene involved in the biosynthesis of oil was transferred to soya by disarmed Ti plasmid. The transgenic soya bean produced cocoa oil that can be used for making chocolates. The first intermediate in the bioconversion of linolenic acid to arachidonic acid is GLA, which helps alleviate hypercholesterolaemia and coronary heart diseases. It is not produced in oilseed crops. This conversion is catalysed by $\Delta 6$ desaturase, and the gene encoding this enzyme can be cloned from the cyanobacterium *Synechocystis*. When this gene was expressed in transgenic tobacco, GLA was produced. In 1998, this desaturation gene was cloned from *Mortierella alpina* and introduced in canola. Ricinoleic acid, a hydroxy fatty acid, is used for making varnishes, nylon paints, lubricants, resins, and cosmetics. Transgenic *Arabidopsis*, which expresses the castor oleate hydroxylase gene, accumulates ricinoleic acid up to 17% of the total lipid content. Researchers at the Michigan State University have cloned the *Arabidopsis* gene, which codes for an enzyme that catalyses the polyunsaturated fatty acid synthesis. Polyunsaturated fatty acids found in food have a role in lowering blood cholesterol and are required for human growth. So far eight desaturase genes from *Arabidopsis* that regulate the polymerization of plant oils have been identified and introduced into plants such as canola, soya bean, and flax, which generally produce more saturated oils. In the crops, this has led to the production of nutritionally enhanced oils. This shows that plant oils can be modified to specific nutritional needs.

6.2   INDUSTRIAL ENZYMES

Using biological molecules (usually enzymes) to catalyse specific chemical reactions is referred to as biocatalysis. Enzymes are complex protein molecules produced by living organisms to catalyse biochemical reactions necessary for life. Enzymes can function *in vitro* even though they are produced within living cells. Since enzymes have the ability to perform specific chemical transformations, they are being increasingly used in industrial processes. Biocatalysis is often carried out in the house, often without realization. Using 'biological' detergents or contact lens cleaners or even cooking ham with pineapple are examples of employing the principles of biocatalysis. In each of these applications, a proteinase enzyme is used to hydrolyse a protein substrate. Even for making bread, there is application of biocatalysis. In the commercial production of yeast, a purified xylanase is employed, which partially breaks down the hemicellulose polymer xylan that can make bread heavy. The yeast itself, *Saccharomyces cerevisiae*, utilizes its own α- and β-amylases to hydrolyse the starch in the flour. The yeast then converts sugars into ethanol and CO_2, and the CO_2 makes the bread fluffy. Food production and processing industries most commonly use biocatalysis. Some familiar examples of such applications include the production of yoghurt, cheese, beer, and wine. However, many food additives and sugars are also prepared using enzymes.

Earlier, the chemical industry restrained itself from applying this technology, mostly because enzymes were considered too delicate to survive extreme conditions in reaction vessels. Enzymes function in specific pH and temperature, and often these conditions are quite different from those found in the industrial plant. However, attitudes are changing now. There has been new awareness about the diversity of microbial life. This knowledge has made microbiology to move out of academic research and find place in practical applications in industry. It is now known that microorganisms can produce certain enzymes that survive and function in extreme conditions. According to an estimate, less than 1% of all known microorganisms have been discovered so far, and possibly more than 50 million bacterial species are yet to be identified. New research and development (R&D) has now focused on isolating enzymes that can function in existing industrial processes. Finding a natural environment resembling the conditions in the reaction vessel can be advantageous as it is most likely to have many organisms that produce enzymes fitting the requirements. This chapter focuses on biocatalysis and analyses a single case study—application of enzymes in the pulp and paper industry. However, at the beginning, the chapter gives a broader view of the history and structure of biotechnology.

Chemical processes are being gradually replaced by enzyme catalysed processes in the industry. Enzymes have all the properties of true catalysts. An appropriate enzyme increases the rate of chemical reaction to a great extent without being consumed in the reaction. The capacity of enzymes to perform specific chemical transformations (biotransformations) has made them very popular in industries, where less specific chemical processes generate unwanted by-products. In food manufacturing industries, purity and predictability are of particular importance because by-products might harm or affect the flavour. Because of their specificity, pharmaceutical companies favour biotransformations for developing novel therapeutic agents. Furthermore, enzymes are chiral catalysts, which can be used to produce optically active homochiral compounds that are often difficult to generate through the traditional organic chemistry. The recent awareness on conservation issues has prompted industries to reduce their contribution to pollution and consider cleaner methods. This has resulted in significant growth of biotechnology outside the pharmaceutical and food industries.

Currently, on the industrial scale, only about 20 enzymes are being produced. These enzymes are produced by 35 producers, with four major companies holding 75% of the market. In 1992, these enzymes had a market value of approximately $800 million, with a predicted annual increase of around 10%–15%. The market suffers from overcapacity, is highly competitive, has small profit margins, and is technologically intensive. However, there is scope for growth as long as new markets emerge.

The greatest challenge in an R&D programme is matching an enzyme with a process. An industrial plant can be adapted to take care of the limitations of an enzyme, but this can be costly.

So a better approach is finding an enzyme that suits the existing process. New organisms living in unusual environments are being discovered, which are providing an excellent source of novel enzymes. Living organisms are generally categorized into three groups (or domains): Eukarya (often called eukaryotes), bacteria, and Archaea (or archaebacteria). Eukarya, which include all animals and plants, have limited ability to overcome hostile conditions such as extreme temperature or pH. Although some worms surviving at temperatures above 60°C have been discovered around deep-ocean volcanic vents, these are exceptional cases. However, bacteria and Archaea are not so constrained and can thrive in unbelievable conditions—from frozen ice to boiling water and from an acidic pH 2 to alkaline pH 12. *Pyrococcus furiosa* is an example of such an Archaea, which shows optimal growth at around 113°C and finds it chilly even at 100°C. These organisms hold the future of biotechnology. Many industrial processes are designed to operate at high temperatures in which viscosity is decreased and chemical reactions are faster. For enzymes that exhibit optimal activities at these temperatures, changes to the existing plants can be minimized, contamination issues can be reduced, and less cooling is required if it is an exothermic reaction. Earlier it was perceived that thermophiles survive by having a rapid turnover of cellular components as nobody really believed that proteins or other biomolecules could withstand such extremes of temperature. However, this theory was proved to be largely incorrect in the early 1980s.

Thermophilic organisms use macromolecules that are intrinsically heat stable and optimally active at the same temperature as the environment in which they grow. Many reactions that need a huge amount of energy to produce macromolecules are energetically favoured (with a negative ΔG) in the physical and chemical environment of a geothermal hot pool. Therefore, it appears that thermophiles gain energy while building the same molecules that cost us so much energy. The properties of the proteins of an 'extremophile' (such as thermal stability) directly result from the protein's sequence of amino acids. This implies that enzymes can be isolated and purified and can still function at high temperatures. The use of thermostable biocatalysis is still limited, and most thermostable enzymes are being applied in detergent or starch industries. However, their application is speedily expanding into other areas and ousting their mesophilic counterparts.

Beyond industry, thermostable enzymes find applications in medical diagnostics and genetic research. The best possible example is a DNA polymerase isolated from *Thermus aquaticus,* which is utilized in polymerase chain reaction (PCR). The half-life of this enzyme is 40 min at 95°C. Therefore, *Taq* DNA polymerase can be heated again and again during the DNA denaturation steps of the reaction. PCR has brought about such a revolution in molecular genetics that its inventor, Kary Mullis, shared the Nobel Prize for Chemistry in 1993. The future seems fairly straightforward; one has to identify an organism growing at a suitable temperature. Unfortunately, it is not so simple because many enzymes are produced in extremely small quantities or it is very difficult to grow the organisms that produce them in the laboratory. In fact, there is still no knowledge about growing most of the microorganisms. Although appropriate growth conditions can be provided, many organisms thrive in such difficult conditions as 110°C, 2000 atm pressure, and total absence of oxygen. Also, they are produced in pitiful cell densities, which make it almost impossible to extract as little as a few milligrams of enzyme, whereas industrial end users require kilograms or tonnes of enzymes. This difficulty can be overcome by recombinant DNA technology or genetic engineering.

6.3   GENETIC ENGINEERING

Genes are the fundamental basis of all life forms. The genetic code is almost similar to a computer program. The only difference is that instead of binary codes 1's and 0's, the basis of the genetic code is the chemical sequence within DNA. DNA is a heteropolymer; that is, a polymer containing more than one monomeric unit. DNA consists of four units or bases—adenine (A), cytosine (C), guanine (G), and thymine (T). DNA is a set of instructions that cannot be interpreted at a glance, and the DNA of one organism appears to be similar to that of another. With increasing knowledge about DNA, the ability to recognize and discriminate between genes and organisms increases. The sequence of the four bases along a DNA molecule is translated into the amino acid sequence of proteins. The rules of translation are the same in all organisms, except a few minor differences. As a result, a gene can be transferred from one organism to another and its translation will be accurate. The sequence of amino acids defines the characteristics of an enzyme, and the amino acid sequence is defined by the gene from which it is translated. Hence, if a gene can be located and isolated from the chromosome of an organism that is difficult to grow, the gene can be inserted into a new host which is much easier to handle. New methods have been devised to simplify the isolation process of genes. In fact, advanced methods do not even require the original organisms. Genes can be directly isolated from environmental samples through PCR technique. The mixture of organisms can be simply treated as a gene pool. The strategies for obtaining appropriate enzymes for industry are as follows.

- Find a pool that has pH and temperature similar to those within an existing process.
- Use PCR to directly extract genes from the pool water or sediment.
- Transfer the genes to a production host.
- Engineer the new host to generate the gene product in large quantities.

If a pool that is a natural counterpart to the process undertaken cannot be found, suitable organisms living within the existing industrial plant might be used. If reactions take place in less harsh conditions, microorganisms may grow on vessel walls or in outlet and waste pipes. Once the gene is cloned, it is relatively simple to increase the production levels of the 'recombinant' enzyme if the organism is well understood. A gene contains the translatable code and DNA sequences upstream and downstream, which determine the activation time of the gene and its quantity of production. These control sequences can be exchanged and the host can be tricked into believing that it is producing something beneficial for its survival. The gene can also be modified so that the organism releases the protein into the surrounding medium through the cell wall that greatly simplifies purification. It is now possible to generate more than 20 g of protein from every litre of culture with the help of new genetic systems. With this level of production, costs are dramatically reduced and the large amounts of enzymes required by the industry can be produced.

6.3.1 Applications in Food and Food Ingredients

6.3.1.1 *Sugar syrups from starch*

In the 19th century, sugar syrups were produced by boiling starch with strong acids, such as sulphuric acid. This difficult method became a predominant technique of producing a range of starch syrups. However, enzymes rapidly supplemented the use of strong acids in producing sugar syrups by the middle of the 20th century. The use of enzymes offers a number of advantages, such as energy efficiency, higher quality products, and a safer working environment. The processing equipment also lasts longer because corrosion is minimized due to the milder conditions of the processes.

A syrup was developed in the 1970s that closely resembled the sweetness of sucrose (table sugar). This syrup was termed high-fructose corn syrup (HFCS). The syrup can be prepared chemically with sodium hydroxide, but the extremely high alkalinity lowers the yield due to the production of large amounts of by-products. These limitations encouraged the use of enzymes having greater specificity and mild reaction conditions, which became the preferred production method. At present, the production of HFCS constitutes a major industry, which transforms large quantities of corn (maize) and other plant starches to HFCS and other edible sweeteners. These artificial sweetening agents are used in baking, candies, soft drinks, jams and jellies, and many other food products.

Environmental benefits. Environmental benefits include the reduced use of strong acids and bases, decreased energy consumption, less greenhouse gases, less corrosive waste, and safer environment for workers.

Consumer benefits. The discovery of a source of starch as an alternative to sugarcane and sugar beets has enhanced the availability of sweetener and stabilized prices. Also, consistent and high quality syrups are available.

6.3.1.2 Dairy applications

6.3.1.2.1 Cheese making

Rennet is a mixture of enzymes obtained from the stomach of calves and other ruminating mammals. It is a vital element in cheese making. For thousands of years, rennet has been the main ingredient facilitating the separation of curd (cheese) from whey. Considerable knowledge has been gathered about the functional properties of rennet and other similar enzymes since their first use. Chymosin, which is the purified form of the major enzyme in rennet, can be produced from genetically modified (GM) microorganisms containing the gene of calf chymosin. It is now commercially available, and thus the need for sacrificing young animals is not necessary. This chymosin has the same properties as that isolated from calves.

Environmental benefits. Cheese makers no longer have to rely on enzymes extracted from slaughtered calves and lambs for rennet, which is required for making cheese. For the current demand for chymosin, the commercial requirement of rennet cannot be met only from animal sources.

Consumer benefits. Consistently high-quality enzyme (chymosin) is available at an affordable price in abundance. This assures the availability of excellent cheese at a reasonable cost. Moreover, people following kosher and vegetarian practices can consume cheese since the enzyme is extracted from a microbe and not from an animal.

6.3.1.2.2 Cheese flavours

The different varieties of cheese available in the market today are partly due to the action of enzymes called lipases, which give a distinctive flavour to the product during the ripening stage. Lipases act on the butterfat in cheese and produce characteristic flavours for different types of cheese. Particular lipases are responsible for the particular flavours of various cheeses, such as the piquant flavour typical of Romano and Provolone cheeses and the distinct flavours of blue and Roquefort cheeses.

Customer benefits. A wide variety of flavourful, high-quality cheeses are obtained.

6.3.1.2.3 Lactose-free dairy products

Many adult people cannot consume normal dairy products because they cause gastrointestinal upset such as bloating, gas, diarrhoea, or a combination of all the three. Lactase occurs naturally in the intestinal tract of children and many adults. However, it is either absent or present in inadequate levels in adults intolerant to lactose. Milk sugar found in dairy products such as milk, ice cream, yogurt, and cheese is converted by lactase into two easily digestible sugars, namely, galactose and glucose. In the absence of sufficient lactase, lactose in the food ferments in the intestine resulting in unwanted side effects.

People who cannot consume dairy products can now enjoy them since the digestive enzyme lactase is commercially available. At present, many milk products are pre-treated with lactase and labelled as 'lactose free'. Moreover, lactase can be bought from retail stores for removing lactose from dairy products at home.

Consumer benefits.   In the United States, around 20%–30% of adults are intolerant to lactose. They can now have the taste and nutritional benefits of dairy products avoiding gastrointestinal side effects by buying lactose-free or low-lactose products or by using commercially available lactase at home.

6.3.1.3   *Baking applications*

6.3.1.3.1   Bromate replacers

Oxidative compounds that produce dough consistency is widely used in modern bread production techniques. Ascorbic acid, azodicarbonamide, and bromates are common chemical oxidants used to strengthen gluten during bread making. For many years, potassium bromate has also been used as it was the first inorganic compound known to be used for improving quality of flour. There has been a high consumer acceptance of good quality bread made using bromate. However, the use of bromate in bread has been recently questioned, and many countries have abandoned its use. As an alternative, enzymes such as glucose oxidase have been used to achieve the unique effect of bromate. These enzymes help in making bread that meets the quality standards expected by consumers.

Consumer benefit.   Bromate has been removed from the food supply without affecting the quality of the bread.

6.3.1.3.2   Softer bread products

Consumers prefer soft bread. Good quality bread can be produced by using enzymes that modify the starch, which in turn keep the bread softer for a long time. Bread becomes stale due to certain changes in the starch. The moisture in the starch gets unbound over time when starch granules convert from a soluble to an insoluble form. On losing the ability to hold water any longer, bread loses flexibility and becomes hard and brittle. This subsequently reduces the taste appeal of the bread, making it stale. A right enzyme can modify the starch during baking and retard staling. Thus, the bread remains flavourful and soft for a longer time (3–6 days).

Environmental benefits. These benefits include less waste and better use of raw materials.

Consumer benefits. Consumers enjoy better quality bread and there is less waste due to improved stability.

6.3.1.4 *Brewing and juicing applications*

6.3.1.4.1 Low-calorie beer

Special enzymes are used in the brewing industry to produce low-calorie beer. Barley and rice are the major ingredients used in the preparation of beer. During yeast fermentation, carbohydrates in these grains get converted to alcohol. Simple carbohydrates are converted first, followed by carbohydrates of increasing complexity, till the desired alcohol content is obtained. The leftover carbohydrate becomes a component of the end product. Enzymes can help in transforming complex carbohydrates to simpler sugars. Hence, a smaller amount of added grain can yield the desired alcohol content. This results in obtaining a beer with fewer carbohydrate calories, that is, lower calorie beer.

Environmental benefits. There has been a reduction in the agricultural demand for grains used in brewing.

Consumer benefits. Good-tasting, lower-calorie beers are available.

6.3.1.4.2 Clear fruit juice

A significant amount of pectin is present in juices extracted from ripe fruits. Pectin gives a cloudy appearance to the juice and results in characteristics not appealing to many consumers. The problem can be overcome by using pectinases, which are naturally occurring enzymes that act on pectin. With the use of pectinases, a crystal clear juice is obtained that has the appearance, stability, taste, and texture preferred by consumers. Although fruits used to make juice naturally contain pectinases, juice manufacturers often add more enzymes to make clear juice in less time.

Environmental benefits. Enzymes used in juice processing help in producing the maximum quantity of juice from a fruit, thereby reducing waste and controlling costs.

Consumer benefits. Aesthetically pleasing, clear, sediment-free juices from many varieties of fruit are widely available at retail stores.

6.3.1.5 *Other food applications*

6.3.1.5.1 Meat tenderizing

Some pieces of meat are tenderer than others. Meat mostly consists of a complex set of proteins with defined structure(s). Myofibrillar proteins and connective tissue proteins are the major proteins that make the meat tender. These proteins are modified by protease enzymes. For many years, people have used proteases such as papain and bromelain to tenderize tougher cuts of meat. Controlling this process is quite difficult because there is a fine line between tender and mushy. This process has been improved by the introduction of more specific proteases to make the tenderizing process robust.

Environmental benefits. Less waste is produced and raw materials are used in a better way.

Consumer benefits. Tougher cuts of meat can be tenderized, thereby making less expensive cuts more attractive menu items.

6.3.1.5.2 Confections

The shelf life of soft candies and other treats made of sugar, particularly soft-centred candies, is often short because the sugar sucrose in the product crystallizes soon after the confection is made. A similar change occurs in soft cookies and other specialty bakery items. Invertase is an enzyme that converts sucrose into two simple sugars, glucose and fructose. This inhibits the formation of sugar crystals that severely shorten the shelf life of the product or make some products costly.

Environmental benefits. Hydrochloric acid is replaced by enzymes in the manufacturing process, which eliminates the need for harsh chemical processing and reduces the risk to environment. The workplace also becomes safer if a strong acid is not used.

Consumer benefits. The use of invertase in confections and baked goods having excellent taste has made them easily available at affordable prices. This food enzyme keeps the soft centres of chocolates smooth and creamy and thus makes soft cookies available on grocer's shelves.

6.3.1.6 *Household and personal care applications*

6.3.1.6.1 Cloth washing at low temperature and with no phosphate

All over the world, there has been a trend towards reducing wash temperatures and banning phosphates. Detergent manufacturers have compensated for this reduced cleaning ability by introducing various types of enzymes into their products.

The energy required to wash a load of clothes can be significantly reduced by lowering the wash temperature. In northern Europe, wash temperatures have been decreased to 40–60°C (100–140°F) from around 90°C (195°F). The energy input is considerably reduced, but the same wash performance is retained by using enzymes. Moreover, the phosphate load to rivers and lakes is reduced, which minimizes the human-induced degradation of these systems.

Environmental benefits. Less phosphate is discarded in rivers and lakes. Also, low-temperature washing reduces energy consumption.

Consumer benefits. The advantages of gentler wash conditions include less colour fading, enhanced fabric life, and lower energy bills. There is a wider range of fabric types to choose from due to the lower temperature washing.

6.3.1.6.2 Milder dishwashing detergents

Dishwashing detergents used in some countries have high alkaline content, which is required to assure the complete cleansing of dishes and utensils. Now the same cleaning quality can be maintained by replacing the harsh chemicals with enzymes.

Environmental benefits. Chemical load has been reduced.

Consumer benefits. Products are more user-friendly and safer because of reduced alkalinity.

6.3.1.6.3 Contact lens cleaner

Many naturally occurring proteinaceous and lipid materials from the eye slowly gather on contact lens. The enzyme lysozyme is secreted by the eye ducts that keep the eye surface clean. This normal cleansing process gets interfered when a contact lens is put on the eye surface. If protease and lipase enzymes are incorporated in the lens cleaning system, they can significantly enhance the removal of dirt from the contact lens.

Environmental benefits. Contact lenses can be cleaned with biodegradable enzymes.

Consumer benefits. Better contact lens washing can be achieved.

6.3.1.7 *Digestive aids in food and feed*

6.3.1.7.1 Alpha-galactosidase for improved nutritional value of legume- and soy-based foods

The nutrition of food for humans and animals can be improved by using enzymes. The occurrence of non-digestible sugars such as raffinose and stachyose in legume- and soy-based foods prevents the complete utilization of their nutritional value. The chemical linkages in these sugars cannot be broken by the enzymes produced in the body. As a result, these sugars reach the large intestine undigested, where the intestinal microflora hydrolyses them. Gas is produced during metabolism of these sugars by the organisms, causing flatulence and discomfort.

The enzyme alpha-galactosidase breaks down stachyose and raffinose to simple sugars, which are readily adsorbed by the human digestive tract. This prevents flatulence often caused by legumes, such as beans and soy-based foods. This enzyme can hydrolyse raffinose and stachyose during soy processing or during the food preparation process, or it can be added to the food immediately before consumption.

Consumer benefits. The physical discomfort that results from consuming soy products and legumes containing poorly digested sugars can be avoided by adding enzymes. Wider acceptance of soy-based foods will make their nutritive value to be fully realized.

6.3.1.7.2 Reduced-phosphorous animal feed

Grains fed to poultry and pigs contain phosphorous which is bound to phytic acid. This phosphorous cannot be absorbed by animals and is excreted. As animals and humans require phosphorous for growth of bones and various biochemical processes, usually extra phosphorous is added to the diet. Phytase is an enzyme that releases the bound phosphorous, making it digestible to chicken or hog. Adding phytase to the feed removes the need for compensating levels of phosphorous, which decreases the phosphorous concentration in the animal waste.

Environmental benefits. Low-phosphorous animal waste decreases the harmful impact on rivers and lakes.

Consumer benefits. Reduced cost of feed helps in controlling consumer prices.

6.3.1.8 *Industrial applications: energy*

6.3.1.8.1 Ethanol fuel from renewable resources

Before petroleum was discovered, food, clothing, and energy were produced from natural carbohydrates. Ethanol fuels can be produced from renewable sources such as corn, sugar cane, and sugar beet or from by-products such as whey from cheese making and waste from potato processing. Ethanol can completely replace petroleum fuels or can be used as an extender. It can also be introduced in petroleum fuels as a replacement for methyl tert-butyl ether (MTBE), a toxic oxygenate.

Enzymes such as glucoamylase, alpha-amylase, lactase, and invertase hydrolyse sucrose, starch, and lactose into fermentable sugars. Ethanol is produced when these sugars are fermented with yeast. A substantial quantity of agricultural crop residues from the production of grain, oilseed, and textile fibres remain unutilized. Even though some of this cellulosic residue should be returned to the soil, much of this material can be used for producing ethanol. The best technology available today for ethanol production utilizes acid hydrolysis of the biomass. The use of strong acids can be avoided by employing enzymes such as cellulose and hemicellulose. These enzymes generate a cleaner stream of sugars for fermentation and produce fewer by-products.

Environmental benefits. Safer working conditions, greater use of natural and renewable resources, and decreased harmful emissions.

Consumer benefits. Safer alternative to MTBE boosts the existing supply of liquid fuel, that is, petroleum.

6.3.1.9 *Industrial applications: textiles*

6.3.1.9.1 Textiles processing

Use of enzymes in textile processing has proved to be greatly beneficial in terms of both the environment and the product quality. The warp yarns are coated with a sizing agent before weaving to lubricate and protect the yarn from abrasion. Previously, the main sizing agent used for cotton fabrics was starch because it has excellent film-forming capacity, is easily available, and is relatively cheap. Before the fabric is dyed, it has to be freed of the applied sizing agent and the natural non-cellulosic materials. Prior to the discovery of amylase enzymes, starch-based sizing was removed through treatment with caustic soda at a high temperature. This chemical treatment was not completely effective in removing the starch. It also degraded the cotton fibre and destroyed the natural, soft feel of the cotton. In the textile industry, the use of harsh chemicals has been decreased by the use of amylases. The use of amylases has resulted in reducing the discharge of waste chemicals to the environment, improving the safety of textile workers, and enhancing the quality of fabric. New enzymatic processes are being developed to completely replace the use of other chemicals in the textile preparation.

Environmental benefits. Discharge of chemicals and wastewater into the environment and handling of hazardous chemicals for textile workers have decreased.

Consumer benefits. Fabric quality has improved.

6.3.1.9.2 Stonewashed jeans without stones

In the jeans industry, pumice stones have been traditionally used to get the look and feel of stonewashed jeans. However, introduction of cellulase enzymes has reduced and even eliminated the use of stones. As the jeans industry is driven by fashion, enzymes provide manufacturers with a newer, easier set of tools to create new looks. Pumice stones used to stonewash denim clothes can also damage the garment. With the use of enzymes, jeans producers can give consumers the look they want without degrading the garment.

Environmental benefits. The benefits of using enzymes include reduced mining, less waste generation, decreased energy usage, reduced clogging of municipal pipes with stones and stone dust, and fewer worn-out machines and pipes.

Consumer benefits. People now have more fashion choices and longer garment life due to lower damage of original fabric.

6.3.1.9.3 Yarn treatment

When cotton yarn is prepared for dyeing and making garments, it is bleached with hydrogen peroxide. For proper dyeing, hydrogen peroxide should be removed. Generally, either a reducing agent is used to neutralize the hydrogen peroxide or water is used to rinse out the hydrogen peroxide bleach. The enzyme catalase can disintegrate hydrogen peroxide to water and oxygen. The use of catalase can eliminate the reducing agent or considerably reduce the amount of rinse water, which results in less polluted wastewater and water consumption.

Environmental benefits. Reduced chemical load, reduced water consumption, and lower energy consumption reduce the adverse impacts on the environment.

6.3.1.10 *Industrial applications: leather processing*

6.3.1.10.1 Leather tanning with enzymes: dehairing and bating

Before being used as leather, it is necessary to remove hairs attached to hides and skins. The conventional method of removing hair from hides is by using harsh chemicals such as lime and sodium sulphide, which completely dissolve the hair and open up the fibre structure. It is possible to reduce the requirement of chemicals for dehairing by the use of enzymes and obtain a cleaner product and a higher area yield with less chemical waste in water. Unlike chemicals, enzymes do not dissolve the hair. So it is possible to filter out the hair, thus reducing the chemical and biological oxygen demand of wastewater.

In addition, proteins and fats present between the collagen fibres must be removed before tanning the hides and skins. The leather can be made pliable by treating the hide to an enzyme before tanning to dissolve selective protein components. This is called bating. Conventional bating agents included dog or pigeon dung, which made the process difficult, unreliable, and stinky. Since the softening effect of dung bates was due to the action of a protease enzyme, the leather industry switched to using bacterial proteases and pancreatic trypsin during the 20th century.

Environmental benefits. Chemical load to waste system and odour during processing have decreased.

Consumer benefits. The quality of leather has improved, and odour during the processing has reduced.

6.3.1.10.2 Degreasing of leather

The traditional method of degreasing sheep skin is the solvent extraction process using paraffin–solvent systems. The leather industry has now adopted a new technique based on the breakdown of fats by a lipase enzyme. Since enzymes interfere less with the skin structure, products with improved quality, such as improved tear strength and more uniform colour, are produced.

Environmental benefits. The enzymatic process replaces the solvent-based system and lowers the volatile organic chemical load.

Consumer benefits. Higher quality leather is obtained; the improved tear strength should be very meaningful to anyone with leather furniture.

6.3.1.11 Industrial applications: paper

6.3.1.11.1 Helping papermakers reduce their load on the environment

A lot of chemical processing is required to make white paper from tree pulp. In the paper industry, chlorine oxidants are used to bleach pulp. This produces chlorine-containing organics, a class of toxic compounds, as by-products. Chlorine bleaching for whitening the paper generates a waste stream containing a range of chlorinated organic compounds, some of which are detrimental to our ecosystem. With enzymes, papermakers have the option to reduce the use of harsh chemicals. The bleaching efficacy can be enhanced by hemicellulase enzymes such as xylanase, which dramatically reduces the consumption of chlorine. The enzymes open up the pulp matrix and allow better penetration of the bleaching chemicals, leading to better extraction or washout of lignin and the associated dark brown compounds.

Environmental benefits. Less chlorine bleach is used; therefore, waste stream contains less chlorinated organics.

Consumer benefits. White paper with lower environmental impact is obtained. Rivers and streams remain cleaner.

6.3.1.11.2 Deinking of waste paper

The process of recycling waste paper is relatively straightforward. Cellulosic fibres can be easily separated by re-pulping and cleaning and converted to new paper. Most fillers and binders present in the original paper can be easily recovered during reprocessing. The most difficult components to remove are the leftover printing inks and adhesives. Earlier, ink was separated from cellulosic fibres by using caustic surfactants and large quantities of wash water. The amount of chemicals and wash water used can be significantly reduced if enzymes such as cellulase and hemicellulose are used. Some of the linkages that entrap the ink can be hydrolysed by these enzymes.

Environmental benefits. Owing to improved processes of deinking, there is more opportunity for recycled paper, less wash water use, less chemical discharge to waste streams, decreased load on landfills, and a better utilization of natural resources. These examples are a few of the different ways commercial enzymes can improve our lives. These tools of nature provide daily products in an environmentally friendly manner. Current commercial use of enzymes, as well as new applications, will help in enhancing and maintaining the quality of life, without degrading the environment for the future generations.

6.4 EDIBLE VACCINES

Edible vaccines provide a readily acceptable vaccine delivery system, which is cost effective and easy to administer and store, especially in developing countries. It involves introduction of selected genes into plants and then inducing these altered plants to produce the encoded proteins. It was introduced as a concept about a decade ago, but has become a reality now. A number of delivery systems have been developed. Initially it was thought to be useful only for inhibiting infectious diseases. However, now it has been applied in preventing autoimmune diseases, birth control, and cancer therapy. Edible vaccines are being developed for a variety of human and animal diseases. In both industrial and developing countries, there is growing acceptance of transgenic crops. Resistance to GM foods may harm the future of edible vaccines. They have overcome the major obstacles in the path of an evolving vaccine technology. All technical obstacles, regulatory and non-scientific challenges, have to be overcome.

Vaccines have provided a revolutionary method for the inhibition of infectious diseases. Although children have been immunized worldwide against six devastating diseases, 20% of infants are still unimmunized, and approximately 2 million deaths occur every year, particularly in the impoverished parts of the globe (Landridge 2000). This is the result of the constraints on vaccine production, distribution, and delivery. Complete coverage is imperative because unimmunized people might spread infections to the immunized areas. There is still no immunization for some infectious diseases or those available are unreliable or expensive. Even though immunization through DNA vaccines provides an alternative, it is a costly approach having unsatisfactory immune response. Therefore, there is a search for cost-effective, fail-safe, easy-to-administer, easy-to-store, and socioculturally acceptable vaccines and their delivery systems. Scientists are genetically engineering plants and plant viruses to produce vaccines against diseases such as dental caries and life-threatening infections such as diarrhoea and AIDS (Moffat 1995). The concept of edible vaccines entails this. The following sections will discuss issues relating to their commercial development, especially in preventing infectious diseases in developing countries.

6.4.1 Concept of Edible Vaccines

Edible vaccines can be created by introducing desired genes into plants and then inducing these modified plants to produce the encoded proteins. This process is known as 'transformation', and the altered plants are termed 'transgenic plants'. Similar to traditional subunit vaccines, edible vaccines consist of antigenic proteins and do not have pathogenic genes. Hence, they cannot establish infection, assuring its safety, particularly in immunocompromised patients. Conventional subunit vaccines are costly and technology dependent, require purification and refrigeration, and have poor mucosal response. On the other hand, edible vaccines can increase compliance, especially in children, and can reduce the need for doctors because they can be administered orally by trained people. Their production is very efficient and can be easily enhanced. For example, the hepatitis B antigen required to annually vaccinate the entire China can be grown on a 40 acre plot and that required for vaccinating babies of the entire world can be produced on just 200 acres of land. They are cheaper, do not require purification, can be produced locally by standard techniques, and do not require capital intensive manufacturing facilities. Mass production will reduce dependence on imports. They show genetic stability, are heat stable, do not need cold chain maintenance, and can be stored near the site of use. As syringes and needles are not required, there is less risk of infection. Apprehensions of contamination with animal viruses, such as the mad cow disease, which is a threat in vaccines produced from mammalian cells, are eliminated because plant viruses do not infect humans.

On coming in contact with the digestive tract lining, edible vaccines activate both mucosal and systemic immunity. This dual effect acts as the first line of defence against pathogens infecting through mucosa, such as *Mycobacterium tuberculosis*, and microorganisms causing diarrhoea, pneumonia, STDs, and HIV. Researchers give high priority to fighting diarrhoeal agents, such as Norwalk virus, rotavirus, *Vibrio cholerae*, and enterotoxigenic *Escherichia coli*, responsible for about 3 million infant deaths per year, mostly in developing countries. When mothers are administered with edible vaccines, the foetus might be successfully immunized by transplacental transfer of maternal antibodies or the infant through breast milk. Even in the presence of maternal antibodies, edible vaccines can seroconvert and protect infants against diseases such as respiratory syncytial virus (RSV) and Group B *Streptococcus*, which are under study. Neglected and rare diseases can also be treated with edible vaccines.

Other vaccine approaches might be integrated with them, and multiple antigens might also be delivered. Currently, the plants under research include rice, potato, tomato, banana, and lettuce. The human and animal diseases for which edible vaccines are being developed include cholera, measles, foot-and-mouth disease, and hepatitis B, C, and E (Giddings, Allison, Brooks, *et al.* 2000).

6.4.2 Mechanism of Action

Antigens in transgenic plants can be introduced through bioencapsulation, that is, through the tough exterior wall of plant cells which protect them from gastric secretions. On being released, the antigens are absorbed by the M cells in the intestinal lining overlying Peyer's patches and gut-associated lymphoid tissue (GALT) and passed on to macrophages, other antigen presenting cells, and local lymphocyte populations, generating serum immunoglobulin G (IgG), immunoglobulin E (IgE) responses, and local immunoglobulin A (IgA) response, which would promptly nullify the attack of the original infectious agent (de Aizpura and Russel-Jones 1988).

6.4.3 Preparation of Edible Vaccines

Foreign DNA can be introduced into plant genome either by bombarding embryonic suspension cell cultures by a gene gun or through *Agrobacterium tumefaciens*, a soil bacterium having the ability to get into plants through a wound such as scratch. It has a circular tumour-inducing (Ti) plasmid that helps in infecting plant cells, integrating into their genome, and producing a hollow tumour (crown gall tumour), where it can reside. This ability can be utilized to insert foreign DNA into plant genome. However, prior to this, the plasmid should be disarmed by deleting the genes for auxin and cytokinin synthesis so that no tumour is produced. Genes for antibiotic resistance are utilized to select transformed cells and whole plants containing the foreign gene and to express the desired product, which can then be regenerated from them. The DNA randomly integrates into the plant genome and results in a separate antigen expression level for each independent line, and so 50–100 plants can be transformed together at a time. The plant expressing the highest levels of antigen and the least number of adverse effects can be selected from them. The generation of transgenic plants is dependent on species and takes three to nine months. Studies are under way to decrease this time to six to eight weeks. Some antigens such as viral capsid proteins have to be self-assembled into virus-like particles (VLPs). VLPs imitate the virus without carrying DNA or RNA and are, therefore, not infectious. Every antigen expressed in plants should be tested for its proper assembly, verified by animal studies and Western blot, and quantified by enzyme-linked immunosorbent assay (ELISA) (Haq, Mason, Clements, *et al.* 1995).

6.4.4   Second-generation Edible Vaccines

The possibility of producing plants expressing more than one antigenic protein has been expanded by the successful expression of foreign genes in plant cells and edible portions. Multicomponent vaccines can be produced by crossing two plant lines having different antigens. Adjuvants might also be co-expressed along with the antigen in the same plant. The B subunit of *V. cholerae* toxin (VC-B) has the tendency to associate with copies of itself and form a doughnut-shaped, five-member ring with a hole at the centre. This property can bring various antigens to M cells at a time; for example, a significant immune response can be successfully elicited against cholera, ETEC, and rotavirus by a trivalent edible vaccine. The Global Alliance for Vaccines and Immunization (GAVI) gives high importance to such combination vaccines for developing countries.

6.4.5 Various Strategies

The approaches to formulate mucosal vaccines include the following: (1) genetically inactivating antigens by deleting an essential gene; (2) gene fusion technology, which creates non-toxic derivatives of mucosal adjuvants; (3) co-expression of antigen and a cytokine, which modulates and controls mucosal immune response; and (4) genetic material itself, which allows uptake of DNA/RNA and its endogenous expression in the host cell. Different mucosal delivery systems include liposomes, mucosal adjuvants, live bacterial/viral vectors, and biodegradable microparticles and nanoparticles. The prime-boost strategy of vaccination integrates different routes of administration and vaccine types, especially where more than one antigen or dose is required. For instance, a single parenteral dose of MV-H DNA (measles virus haemagglutinin), followed by multiple oral MV-H boosters, can induce larger amounts of MV neutralizing antibodies than with either vaccine alone (Webster, Cooney, Huang, *et al.* 2002).

6.4.6 Chimeric Viruses

It is possible to redesign certain viruses to express fragments of antigenic proteins on their surface, for example, alfalfa mosaic virus, cowpea mosaic virus (CPMV), cauliflower mosaic virus (CaMV), tobacco mosaic virus (TMV), tomato bushy stunt virus, and potato virus X. For this, overcoat and epicoat technologies are employed. The overcoat technology allows the plant to generate the entire protein, whereas the epicoat technology involves the expression of only foreign proteins. It has been shown that a plant-derived mink enteritis virus (MEV) injectable vaccine protects minks. When fragments of gp41 surface protein of HIV virus are put into CPMV, it could induce a powerful neutralizing antibody response in mice. A cocktail of specific HIV epitopes can also be presented on the surface of the plant virus. CPMV is thermally and genetically stable. It can endure acidity (pH 1) for 1 h. A wide variety of epitopes have been expressed in CPMV, including HIV-1 (gp41 and gp120), human rhinoviruses, canine parvovirus, foot-and-mouth disease virus, mammalian epitopes from hormones or colon cancer cells, and fungal epitopes and protozoan epitopes from *Plasmodium falciparum*. A wide range of routes are available, for example, oral (formulated leaf extracts), parenteral and nasal (purified particles), and whole and homogenized leaves, fruits, or vegetable tissues. Plant viruses are engineered in all these cases to carry the desired genes. They are used to infect their natural hosts, such as the edible plants, in which cloned genes are expressed to different degrees in different portions of the plant, including edible parts.

6.4.7 Challenges

For developing a plant-based vaccine, many questions have to be answered (Table 6.1). Successful clinical trials in humans have shown that suitable doses of antigen can be obtained with plant-based vaccines. For determining the right dosage, a person's weight and age and the plant's size and protein content have to be considered. The amount to be consumed is crucial, especially for infants, who might spit after eating a part or the whole. A very low dose will fail to induce antibodies, and a high dose will result in tolerance. Concentrating the vaccine into a teaspoon of baby food might be more effective than administering it in an entire fruit. The transformed plants can also be processed into pills, puddings, and chips. The regulatory concerns include consistency from lot to lot, uniformity of dosage, and purity.

Foreign proteins gather in low amounts (0.01%–2% of total protein) in plants and are less immunogenic. Hence, the oral dose is far larger than the intranasal/parenteral dose. For instance, oral hepatitis B dose is 10–100 times the parenteral dose. Also, to be immunogenic, 100 g potato in three different doses is required for expressing B subunit of labile toxin of ETEC (LT-B). When attempts are made to boost the amount of antigens, it leads to stunted growth of plants and reduced formation of tuber and fruit. This is because too much mRNA from the transgene results in gene silencing in plant genome. The following techniques can be used to overcome these drawbacks: (1) optimization of coding sequence of bacterial/viral genes for expression as plant nuclear genes; (2) plant viruses expressing foreign genes; (3) expression in plastids; (4) viral-assisted expression in transgenic plants; (5) coat-protein fusions; and (6) promoter elements of bean yellow dwarf virus with reporter genes GUS (β-glucuronidase) and GFP (green fluorescent protein), substituted later with target antigen genes. Antigen genes might be linked with regulatory elements that more readily switch on the genes or do so only at selected times

TABLE 6.1 Considerations in developing a plant-based vaccine

Antigen selection
- Is the antigen safe and non-pathogenic in all circumstances?
- Can the antigen induce a protective immune response?
- Is the antigen suitable for expression in plants?

Efficacy in model systems
- Does the antigen accumulate in plants in sufficient quantities?
- Is the plant-derived antigen immunogenic?
- Do trial animals develop protective immune responses?
- Is plant cell interference with antigen presentation possible?
- Is induction of immune tolerance possible?

Choice of plant species for vaccine delivery
- Is the food plant the best one?
- Can it be eaten raw and unprocessed?
- Is it suitable for infants?
- Can it be widely and easily grown?
- Can it be easily stored? Is it resistant to spoiling?
- Is it amenable to transformation and regeneration?
- What is the possible cost to produce plants of multiple transgenes?

Delivery and dosing issues
- What are the requirements of mucosal adjuvants for protective response?
- Can a large enough dose be delivered by simply eating the plant?
- What is the number of doses required?

Safety issues
- What is the allergenic and toxic (such as glycans and nicotine) potential of plant components?
- What is the potential for interference?
- It is possible to produce oral tolerance?
- What are the risks of atypical measles (in plants with cloned measles virus genes)?
- What are the health and environmental risks of genetically modified organisms?
- How to prevent misuse/overuse?

Public perceptions and attitudes to genetic modification
- Will negative attitudes to genetically modified organisms influence vaccine acceptability?
- What are the legal and ethical considerations regarding products from plants with status like tobacco?

Quality control and licensing
- Can antigen expression be consistent in crops?
- Who will control vaccine availability and production?

(after the plant is nearly fully grown) or only in its edible regions. Exposure to some outside activator molecule can also be carried out.

Immunogenicity can be enhanced by using mucosal adjuvants, which are better targeted to the immune system, like molecules that bind to M cells in the intestine lining and transfer them to immune cells. These include LT-B (ETEC), CT-B (cholera toxin B subunit), mammalian/viral immunomodulators, and plant-derived secondary metabolites. Toxicity and allergic potential can be decreased by using mutant forms of *E. coli*-labile toxins, such as LT-K63, LT-R72, and hinge cleavage mutant LT-G192. Another challenge would be in dealing with diseases caused by multiple serotypes (dengue) or by complex parts from different life cycles of parasites (malaria) or by quickly mutating organisms (influenza, trypanosomiasis, and HIV). Each plant has its advantages and disadvantages (Table 6.2).

Currently, small companies are carrying out most of the research since edible vaccines are targeted to markets of developing nations. Large companies prefer livestock market to human application. Only a few global aid organizations and some national governments are giving support, but the effort remains largely underfunded. The lack of investors' confidence in returns on investments in GM foods has discouraged some companies carrying out research on edible vaccines. There is also a shortage of researchers in pharmaceutical companies. Moreover, recombinant (injectable) vaccines against diphtheria and tetanus are so inexpensive today that companies find little interest in developing edible vaccines for them.

There is still no clarity on whether edible vaccines should be regulated under food, drugs, or agricultural products and what part of vaccine should be licensed—antigen itself, genetically engineered fruit, or transgenic seeds. Regulatory bodies would subject them to a very close scrutiny to ensure that they never enter the food supply. This would require segregation of medicinal plants from food crops in a greenhouse to prevent out-crossing and would need separate storage and processing facilities. Even though edible vaccines come under GM plants, it is hoped that they would avoid controversy because they are intended to protect lives.

6.4.8 Clinical Trials

In the past, expression of antigens has been successfully shown in plants, such as rabies virus-G protein in tomato, LT-B (ETEC), HBsAg, and Norwalk virus in tobacco and potato, and CT-B (*V. cholerae*) in potato. Except in a few cases, ethical considerations prohibit clinical

TABLE 6.2 Advantages and disadvantages of different plants

Plant/fruit	Advantage	Disadvantage
Tobacco	• Good model for evaluating recombinant proteins • Low-cost preserving system (numerous seeds can be stored for a long time) • Easy purification of antibodies stored in the seeds, at any location • Large harvests, number of times per year	• Produces toxic compounds
Potato	• Dominated clinical trials • Easily manipulated/transformed • Easily propagated from its 'eyes' • Stored for long periods without refrigeration	• Needs cooking, which can denature the antigens and decrease immunogenicity
Banana	• Cooking not required • Proteins not destroyed even if cooked • Inexpensive • Grown widely in developing countries	
Tomato	• Quick growth • Cultivated broadly • High content of vitamin A boosting immune response • Spoilage problem overcome by freeze-drying technology • Heat-stable, antigen-containing powders made into capsules • Different batches blended to give uniform doses of antigen	• Trees take two to three years to mature • Transformed trees take about 12 months to bear fruit • Spoils rapidly after ripening • Contains very little protein, so unlikely to produce large amounts of recombinant proteins • Spoils readily
Rice	• Commonly used in baby food because of low allergenic potential • High expression of proteins/antigens • Easy storage/transportation • Heat-stable expressed protein	
Lettuce	• Fast growing • Direct consumption	
Soya bean and alfalfa	• Large harvests, number of times/year	
Musk melon (cantaloupe)	• Fast growing • Easily propagated by seed • Easily transformed	• Grows slowly • Requires specialized glasshouse conditions

trials from directly assessing protection. However, veterinary scientists can assess immune protection more directly.

6.4.8.1 Cholera

It has been found that transgenic potato with CT-B gene of *V. cholerae* is effective in mice. Eating one potato every week for a month, with periodic boosters, has been found to provide immunity. Co-expression of mutant cholera toxin subunit A (mCT-A) and LT-B in crop seed is effective by nasal administration and is extremely practical.

6.4.8.2 Measles

When mice are fed with tobacco expressing MV-H (from Edmonston strain), they could attain antibody titres five times the level considered protective for humans. Secretory IgA is also seen in their faeces. Titres 20 times the human protective levels could be induced through prime-boost strategy by combining parenteral and subsequent oral MV-H boosters. These titres were considerably larger than those attained with either vaccine administered alone. MV-H edible vaccine does not cause atypical measles, which may occasionally be seen with the current vaccine. Hence, it might prove more effective in achieving its eradication. The success in mice has encouraged similar experiments in primates. Transgenic rice, lettuce, and baby food against measles are also developed. When administered with CT-B (adjuvant), 35–50 g of MV-H lettuce is enough; however, an enhanced dose might be required if given alone.

6.4.8.3 Hepatitis B

Parenteral VLPs can generate specific antibodies in mice for hepatitis B. First human trials of a potato-based vaccine against hepatitis B have shown encouraging results. The quantity of HBsAg required for one dose can be achieved with a single potato. Levels of specific antibodies in humans considerably exceeded the protective level of 10 mIU/mL. When cloned into CaMV, plasmid HBsAg subtype ayw exhibited higher expression in roots compared to leaf tissue in transgenic potato. Further research is needed to increase the production of antigen by using different promoters, such as patatin. It has been proved that the resulting plant material is superior to the yeast-derived antigen in inducing and boosting immunity in mice. The prime-boost strategy in mice with a single sub-immunogenic parenteral dose of yeast-derived recombinant HBsAg and the subsequent oral transgenic potatoes led to antibody development that immediately peaked above 1000 mIU/mL and was maintained above 200 mIU/mL for five months. This can be a beneficial immunization strategy for developing countries. Tomatoes expressing hepatitis B are being produced in guarded greenhouses. Sufficient antigens for 4000 vaccines can be obtained from only 30 tomato plants. Transgenic lettuce is also being developed.

6.4.8.4   Rabies

Antibodies can be induced in mice by tomato plants that express rabies antigens (Prakash 1996). Instead, TMV might also be used. It has been shown that transformed tomato plants using CaMV with the glycoprotein (G-protein) gene of rabies virus (ERA strain) were immunogenic in animals.

6.4.8.5   HIV

Initial attempts at splicing HIV protein into CPMV have been successful. Two HIV protein genes and CaMV as promoter were successfully injected into tomatoes. The expressed protein was demonstrated by PCR in different parts of the plant, including the ripe fruit, and in the second-generation plant. Of late, spinach has been successfully inoculated for Tat protein expression cloned into TMV. Up to 300–500 µg of Tat antigen was present in each gram of leaf tissue of spinach. When mice were fed with this spinach followed by DNA vaccinations, they produced higher antibody titres than the controls, with the levels peaking in four weeks after vaccination.

6.4.8.6   STDs

Human papillomavirus type 11 (HPV-11) recombinant VLPs produced in insect cells is immunogenic when administered orally to BALB/c mice. The response is dependent on dose and restricted by genotype. Hence, VLPs might be potent oral immunogens for preventing anogenital HPV disease.

6.4.8.7   Anthrax

A protein structurally similar to the main protein present in the existing vaccine can be expressed by tobacco leaves bombarded with *pag* gene (anthrax protective antigen—PA) with a gene gun. Billions of units of anthrax antigen can be produced. Moreover, this vaccine did not have the oedema factor and lethal factor responsible for toxic side effects. The same anthrax antigen is now being introduced into tomato plants. Researchers are also attempting to transform spinach by inoculating it with TMV-expressing PA, as spinach might be a safer vaccine.

6.4.8.8 Others

Transgenic *Arabidopsis thaliana* plants can generate a fusion protein that consists of LT-B and early secretory antigen ESAT-6 of *M. tuberculosis*, demonstrating both the antigens by ELISA. Children below 2 years die in large numbers every year due to RSV pneumonia. Mice have shown encouraging results for RSV expressed in tomato and potato plants, and further clinical trials are planned. An RSV vaccine based on apple juice is also being developed. Programmes for the production of new vaccines against rotavirus and *Streptococcus pneumoniae* have been instituted. In case of rotavirus, transgenic potatoes expressing VP7 can induce high titres of IgG and mucosal IgA in mice. In case of parasites such as those of malaria, progress in vaccine generation has been inhibited by the complicated multistage life cycle of the parasite, its inaccessibility to study, and its large genome. However, chimeric coat proteins of CPMV expressing malarial and foot-and-mouth disease epitopes have been reported.

6.4.9 Veterinary Sciences

The first patented edible vaccine to demonstrate efficacy in animal trials was against the transmissible gastroenteritis virus (TGEV) in pigs. Plans are under way to produce it commercially. Researchers are studying vaccines against porcine reproductive and respiratory syndrome (PRRS) and other diseases such as parvovirus. Many transgenic animal feeds are currently under clinical trials in pigs. Other candidates include diseases of pets, animals in swine and poultry industries, and draft and wild animals. Different plant viral system patents exist (Table 6.3).

6.4.10 Passive Immunization

Antibody molecules can also be expressed in plants like any antigenic protein. Protective antibodies produced in plants are also called 'plantibodies'. To date, only four plant antibodies have been found to have the potential to be used as therapeutic agents, and only IgG–IgA against *Streptococcus mutans* has been experimented on humans. There

TABLE 6.3 Plant viral system patents

Virus	Target host	Plant	Mode of administration
ETEC	Humans	Tobacco	Oral
ETEC	Humans	Potato	Oral
ETEC	Humans	Maize	Oral
Vibrio cholerae	Humans	Potato	Oral
Hepatitis B virus	Humans	Tobacco	Injection (extracted protein)
Hepatitis B virus	Humans	Potato	Oral
Hepatitis B virus	Humans	Lettuce	Oral
Norwalk virus	Humans	Tobacco	Oral
Norwalk virus	Humans	Potato	Oral (VLPs)
Rabies virus	Humans	Tomato	Intact glycoprotein
CMV cytomegalo virus protein	Humans	Tobacco	Immunologically related
Rabbit haemorrhagic disease virus	Humans	Potato	Injection
Foot-and-mouth disease	Rabbits	Arabidopsis	Injection
Foot-and-mouth disease	Agricultural domestic animals	Alfalfa	Injection or oral
TGEV	Agricultural domestic animal	Arabidopsis	Injection
TGEV	Pigs	Tobacco	Injection
TGEV	Pigs	Maize	Oral

are efforts to introduce these secretory antibodies into toothpaste to prevent tooth decay and human trials are also planned. Other antibodies have also been tried on animals, and they have shown considerable promise.

It was quite difficult to produce complete secretory antibodies because they contain two extra proteins apart from four protein chains (two heavy and two light chains). One of the extra proteins was added *in vivo* during secretion. This was finally done by crossing four different lines of transgenic tobacco plants. It is possible to utilize this process occurring in plants to a great advantage. Complex antigens having high molecular weight can be expressed in different lines of the same plant, optimally as component parts. Cross-pollination between these lines of plants might yield a progeny producing the complex antigen in its native configuration, thus eliminating the need for any *in vitro* manipulation. This would be of huge economic advantage. The major proportion (57%) of IgA plantibodies occur in the dimerized form, unlike the IgA produced in insect cells using Baculovirus, which is generally monomeric. Antibodies expressed in seeds are preferred to antibodies expressed in leaves because the latter have the risk of early decay and require immediate extraction. On the other hand, antibodies in seeds can be stored at room temperature for a long time till they are extracted or consumed. This is possible by linking antibody genes to a genetic 'switch'. Injectable materials are now being studied, and they are posed with challenges of purification. Researchers are investigating HBV antibodies in tobacco plants. A variety of other plantibodies such as anti-HIV antibodies are also under investigation. It has been proposed to develop a cream containing anti-HIV antibodies, which could possibly reduce the risk of HIV infection during intercourse. Expression of monoclonal antibodies (Mabs) against STDs such as genital herpes has been possible in soya bean. On comparing with anti-HSV-2 Mab expressed in mammalian cell culture, researchers found that they were similar in stability in human semen and cervical mucus for more than 24 h, with similar ability in avoiding vaginal HSV-2 infection in mice.

6.4.11   Other Potential Domains

6.4.11.1   *Birth control*

A protein found in mouse zona pellucida (ZB3 protein) has been expressed in TMV. When TMV is injected, the antibodies that result can prevent fertilization of eggs in mice.

6.4.11.2   *Cancer therapy*

Sufficient quantities of monoclonal antibodies can be produced from plants for cancer therapy. Scientists have genetically engineered soya bean to create monoclonal antibody (BR-96) as a vehicle for administering doxorubicin for lung, colon, ovarian, and breast tumours. Non-Hodgkin's lymphoma is also under investigation.

6.4.11.3   *Chloroplast transformation*

The chloroplast genome is inherited from the mother. Hence, after transformation of chloroplast, the protein would be absent in pollen, thereby decreasing the possibility of transmission of the transgene to neighbouring plants or weed species by cross-pollination. It may also result in the accumulation of considerably larger quantities of transgenic protein.

6.4.11.4 *Role in autoimmune diseases*

For unidentified reasons, oral delivery of autoantigens can inhibit immunity by activating suppressor cells of the immune system, resulting in immunological tolerance. For example, arthritis patients found relief after taking collagen. This result is frequently observed in test animals and is time and dose dependent. Tolerance is induced when the antigen is repeatedly administered in doses greater than the immunogenic dose. Tolerance can be unknowingly induced by misidentification and misadministration of plants and fruits. Therefore, it is necessary to recognize the transgenic plant or fruit and determine effective, safe doses and feeding schedules, which would decide whether immunity will be stimulated or depressed by an antigen. Type 1 diabetes, rheumatoid arthritis, multiple sclerosis, and transplant rejection are some autoimmune disorders that might be prevented or relieved. Potatoes producing insulin and a protein glutamic acid decarboxylase (GAD), linked to CT-B subunit, can suppress immune attack in a mouse strain prone to become diabetic and can postpone the onset of high blood sugar. Scientists are also working on another diabetes-related protein. Research on lupus, rheumatoid arthritis, multiple sclerosis, and transplant rejection is under progress. However, it is still required to produce increased amounts of self-antigen in plants for autoimmune diseases.

6.4.11.5 *Recombinant drugs/proteins*

In addition to antibodies and vaccines, plants are also being inoculated with engineered virus to generate drugs and enzymes such as albumin, interferon, and serum protease, which are otherwise difficult or expensive to produce, for example, production of glucocerebrosidase (hGC) in tobacco plants for the treatment of Gaucher's disease. This advancement would considerably bring down its cost. Tests are under way for transgenic tobacco plants producing interleukin-10 to treat Crohn's disease. Industrial methods have been developed for the large-scale generation of recombinant therapeutic proteins in plants. Some other examples of novel compounds are an antiviral protein that suppresses the HIV virus *in vitro*, trichosanthin (a ribosome inactivator), and angiotensin I (antihypersensitive drug). There is now commercial production of transgenic plant-derived hirudin (an antithrombin). Pharmaceuticals and drugs are being increasingly produced from food crops.

6.4.12 Future of Edible Vaccines

The resistance to GM foods is an impediment in the future development of edible vaccines. Despite facing the threat of famine, Zambia refused to take GM maize in food aid from the United States, which reflects the enormity of the problem. It is not possible to avoid transgenic contamination because, besides pollen, transgenes may spread by feeding insects and transferred to soil microbes and pollute surface and groundwater. In a recent case, a GM corn meant only for animal consumption was found in human food. Seeds left behind from previous plantings caused the contamination. Quarantine done to avoid further spread resulted in financial loss, penalties, and criminal charges for alleged violation of permit to grow gene-altered crops. The United States has lost approximately $12 billion because of transgenic contamination. Such vaccines should be endorsed for human use only after addressing the concerns of the World Health Organization regarding quality assurance, efficacy, and environmental impact. Moreover, GM crops in greenhouse facilities must be properly controlled and surrounded by protective buffer crops.

Genetically modified crops have not been able to enhance production or significantly decrease the use of herbicides and pesticides as proposed. Moreover, major crop failures have resulted from the instability of transgenic lines and transgene inactivation in a few cases. Harmful gene products or potent immunogens or allergens might also be incorporated. Some of them even have side effects. For example, cytokines cause sickness and CNS toxicity, and α-interferon induces neurotoxicity, dementia, and mood/cognitive changes. Terminator crops containing 'suicide' genes for male sterility can prevent the spread of transgenes. However, they actually spread both male sterile suicide genes and herbicide-tolerant genes through pollen. Emerging biotech-resistant pests and multiple herbicide-tolerant weeds (superweeds) require the application of toxic herbicides such as atrazine, glufosinate ammonium, and glyphosate.

The safety of GM foods has not been established by credible studies. An investigation showed growth factor-like effects of GM foods on the stomach and small intestine of young rats. However, follow-up studies were not carried out to find whether CaMV 35S promoter was responsible for this. Most commercial GM crops grown today possess this promoter, which is unstable and prone to horizontal gene transfer, recombination, or mutations by random insertion. The genomes of plant and animal hosts can get destabilized by random insertion of genes, and the effects can be transferred to the neighbouring ecosystem.

By assisting horizontal gene transfer/recombination, genetic engineering might contribute to the emergence and re-emergence of drug-resistant and infectious diseases, the rise of autoimmune diseases and cancers, and the revival of dormant viruses. Bacteria might assimilate the transgenic DNA in food in the human gut. Infections might become very difficult to treat if antibiotic resistance marker genes transfer to pathogenic bacteria from transgenic food. Minor genetic alterations in pathogens can cause drastic changes in host spectrum and disease-causing potential, which might make plants their unintentional reservoirs. Moreover, new strains of infectious agents, such as super-viruses, might emerge. Geneticists can create millions of recombinant viruses that have never existed in billions of years of evolution by DNA shuffling in a very short time. This might be misused for the production of bioweapons. The inherent hazard of genetic engineering is that it creates vectors/carriers specially designed to cross wide species barriers, such as between plant and animal kingdoms, and transfer genes by overcoming defence mechanisms that are operative against genetic assaults. Perhaps the biggest risk comes from naked DNA vaccines because these short pieces of DNA are easily absorbed by cells of all species and get integrated into the cell's genome material. These short DNA fragments can mutate and multiply for an indefinite period. Animals feeding on GM crops such as maize carry the risk of transferring mutations to human beings consuming these animals or their products. The environmental and ecological risks that edible vaccines pose should be considered. The science of edible vaccines is still at a nascent stage and has to be thoroughly tested before being applied to people for combating infectious diseases and inducing autoimmunity.

The acceptance of transgenic crops has increased in both industrial and developing countries. This is evident from the increase in the global area under cultivation of transgenic crops from 1.7 to 44.2 million ha from 1996 to 2000 and rise in the number of countries producing these crops from 6 to 13. In the United States and Canada, research is being conducted on at least 350 genetically engineered pharmaceutical products. In a case of urgency, edible vaccines offer great economic and technical benefits because they can be easily produced in large quantities within a limited period of time.

The process of immunization might become safer and more effective with the help of edible plant-derived vaccines. They would eliminate some of the drawbacks associated with traditional vaccines in production, distribution, and delivery, and they can be integrated into the immunization of plants. They have overcome the main hurdles in the way of an evolving vaccine technology. However, the technical obstacles have to be overcome before making them a reality. With limited presence of essential health-care facilities in most parts of the world and with scientists still struggling with complex diseases, edible vaccines will provide a safe, efficacious, and cost-effective delivery system for prevention of diseases.

6.5 PLANTIBODIES

Antibodies are bioactive molecules having individual and specific binding properties and so have a large diversity of potential applications. They are applicable in medical diagnosis and therapy, control of pathogens, sensitive detection and elimination of environmental pollutants, and industrial purification processes (Stoger, Sack, Fischer, *et al.* 2002). Antibodies have the ability to interfere with the metabolic processes within an organism, which makes them an excellent tool in fundamental research. For more than a decade, plants have been used as heterologous expression hosts for recombinant antibodies (plantibodies). The integration of antibody and plant engineering has led to the generation of a variety of molecular forms in different plant species.

For specific applications of recombinant antibodies, recent studies have shifted focus towards strategies for formulating well-defined objectives (commercial or otherwise) involving specific types of recombinant antibodies. Important research areas include the advancement and comparison of various expression systems in terms of feasibility and efficacy. These include assessing the product quality and using antibody molecules with improved properties (e.g., fusion proteins with enhanced or novel functions). Immunomodulation of physiological systems and engineering of antibody-mediated resistance to infection by pathogen are some contemporary applications in agronomic research. However, the use of plants as bioreactors to produce antibodies required for medical or industrial processes is the most advanced application.

6.5.1   Advances in Transformation and Expression Technology

Antibody genes have been introduced into plants using both *Agrobacterium*-mediated transformation and particle bombardment. In particle bombardment, multiple constructs are introduced simultaneously, which expedites the recovery of transgenic lines expressing multimeric antibodies such as secretory IgA (sIgA). The recombinant protein can get accumulated in the entire plant or in specific organs. It has been demonstrated that antibody molecules get deposited and stored in the seeds of various crop plants. High accumulation of a single-chain Fv fragment (scFv) antibody in pea seeds was recently found using the seed-specific USP promoter. Within a period of one and a half years, the high stability of scFv antibody was established in tobacco seeds (Ramirez, Oramas, Ayala, *et al*. 2001).

Although the cost involved in plant cell or organ culture in bioreactors is more than that in agricultural production, it has advantages since recombinant antibodies can be produced in controlled conditions. Recently, a murine Ig G1 was expressed in hairy root cultures, which led to the secretion of recombinant antibodies in the medium. Transformation of organelles provides an alternative to nuclear gene transfer. The plastidial expression of a multimeric vaccine was possible because of the recent advances in the methodology of chloroplast transformation. Likewise, a recombinant antibody has also been expressed in chloroplasts. There are some advantages of expression of recombinant proteins in chloroplast genome compared to nuclear gene transfer (e.g., high levels of expression and containment). Moderate quantities of recombinant products can be obtained within a very short period (a few weeks) by transient expression systems involving viral vectors or agro-infiltration. Compared to regular small-scale bacterial expression systems, such systems might prove to be advantageous for producing correctly folded, soluble proteins. For the commercial application of plant-based expression systems, the stability of transgene expression over multiple generations is necessary. Therefore, attempts to understand and remove factors responsible for transgene silencing are imperative. Dosage-dependent, homology-based, post-transcriptional gene silencing has been recently found in *A. thaliana* lines expressing a Fab fragment. However, such problems seem sporadic since a large volume of literature refers to stable antibody production. Specific targeting signals are often useful to direct expression and deposition of a plantibody. The accumulation of most scFv antibodies is very high when expression is targeted to the apoplast or the endoplasmic reticulum (ER), rather than to the cytosol. The capacity of some scFv antibodies to accumulate in the cytosol seems to be dependent on their intrinsic characteristics. An in vivo assay has been developed based on the yeast two-hybrid system to study scFv fragments in the cytosol having a reducing intracellular environment. There is great expectation from efforts to identify and engineer stable scFv scaffolds. Accumulation of recombinant antibodies in the cytosol is promising for many intracellular antibody applications, such as engineering viral resistance. Assembly and post-translational alterations in the ER are essential for the production of full-size antibodies and Fab fragments, and these molecular forms should be directed to the secretory pathway. Peeters, de Wilde, and Depicker (2001) showed that a Fab fragment was effectively secreted to the apoplast of leaves and roots of *Arabidopsis*. However, results obtained with a hybrid immunoglobulin (IgA/G) show that complex molecules get sorted within the secretory pathway based not only on particular signal sequences, but also on the protein itself. Tissue-specific variations in deposition of protein have been observed with an scFv having an ER retention signal. This scFv was found in protein fragments derived from ER as well as in the protein storage vacuoles of the endosperm cells of transgenic rice. The attachment of a transmembrane anchor to the heavy chain led to the association of an assembled IgG to the plasma membrane.

6.5.2 Disease resistance and Metabolic Traits in Plants

Applications that depend on the *in vivo* modulation of antigen levels are based on the accurate expression and accumulation of antibodies in particular subcellular compartments and specific tissues. It has been found that passive immunization of plants decreases infection and symptoms caused by viruses and mollicutes. There has been significant progress towards engineering resistance against insects. Antibody-mediated fungal resistance in plants is yet to be demonstrated, but monoclonal antibodies that inhibit fungal growth have been identified. A great tool for analysing or modifying the function of an antigen *in vivo* is immunomodulation. For this, an antigen, say a metabolite or an enzyme, can be either blocked or stabilized in its action. On creating an artificial abscisic acid (ABA) sink by the generation of an ABA-specific scFv in the ER of potato and tobacco plants, morphological and physiological changes were noticed in plants.

6.5.3 Plants as Bioreactors

At present, plant-based expression systems produce antibodies that are mostly high-value products for pharmaceutical use. Certainly, plants provide economical systems for the mass production of pharmaceuticals and have additional safety levels compared to animal production systems. Moreover, plants offer the only commercially feasible system for large-scale manufacturing of many complex molecular forms, including SIgA. CaroRX is the most advanced product produced in tobacco, which is an SIgA that prevents dental caries. Clinical studies of CaroRX in humans have shown favourable results. A humanized antibody against HSV glycoprotein B is another plantibody that might result in a product for medical applications in human beings. This antibody, expressed in soya bean, proved to be fruitful in a model study on mice. In a more recent study, a diabody against carcinoembryonic antigen (CEA) was produced by agro-infiltration of tobacco. Attachment of the purified antibody to human colon carcinoma cells was demonstrated *in vitro*. Examples of plant-derived pharmaceutical antibodies, which have been extensively studied previously, are listed in Table 6.4. In addition to uses in human health care, plantibodies might prove useful as additives in feed or for phytoremediation.

TABLE 6.4 Examples of pharmaceutical antibodies produced in transgenic plants

Antigen	Plant	Antibody form	Application	Company
Streptococcus surface antigen SAI/II	Tobacco	SIgA/G	Therapeutic	Plant biotechnology
Herpes simplex virus	Soya bean, rice	IgG	Therapeutic	
Non-Hodgkin's lymphoma idiotypes	Tobacco, virus vector	scFv	Vaccine	Large-scale biology corporation
Human IgG	Alfalfa	IgG	Diagnostic	
Carcinoembryonic antigen	Tobacco, rice, wheat, pea, tomato	scFv	Therapeutic/ diagnostic	

6.5.4 Remaining Challenges in Biopharming

6.5.4.1 *Product quantity and quality*

The level of product accumulated per unit biomass and the scope for increasing production are crucial factors that determine the economic feasibility of any production system. It is equally important to fully assess the quality of the end product based on functionality and homogeneity.

Accumulation of a recombinant antibody in an expression system can be increased by using suitable regulatory elements in the expression construct, by optimizing codon usage, and by enhancing the stability of the antibody. During the last few years, scientists have developed a range of different plant systems for use as bioreactors to produce recombinant proteins as well as recombinant antibodies. For large-scale production, the choice of the system depends on the effectiveness of the expression system and its aptness for storage, scale-up, and downstream processing. Issues such as use and value of the product, expected production scale, geographical production area, closeness to processing facilities, safety considerations (containment and self-pollination levels), intellectual property, and economic considerations are of significance in this context.

For commercially producing pharmaceutical antibodies, many organizations have chosen seed crops, particularly corn (Table 6.4).

Antibodies need to be homogeneous, functional, non-immunogenic, stable, and free from contaminants to satisfy the quality requirements for a pharmaceutical product. The levels of functional product in plant tissues get reduced by protein degradation even if the molecules are appropriately synthesized and assembled initially. In addition to decreasing the efficiency of the system, the quality of the product is compromised by the presence of inactive protein fragments, which might not be separated during purification. Degradation during extraction can be lessened by adding protein stabilizers and proteinase inhibitors, whereas proteolysis in planta poses a greater challenge. Fragmentation of IgG occurs mostly in the apoplast of root and shoot tissues and also within the cell. It has been recently proposed that proteolytic degradation in leaves is partially related to the process of senescence. Therefore, the physiological state of the plant host might have a bearing on the integrity of antibody, and different tissues, plants, and conditions might be more or less suitable for producing high-quality antibodies.

6.5.4.2 *Purification*

For biopharmaceutical production, the purification cost is high and it directly affects the choice of expression system. The preliminary processing steps of plant material in many cases benefit from equipment and technology commonly employed for food processing, whereas the final steps typically consist of standard chromatography procedures. Significant progress has been reported for recombinant antibodies expressed in tobacco and corn. Expression in seeds guarantees tremendous storage properties, and thus provides extra flexibility in processing management and batch production. As endogenous proteins are limited in seeds, it is an advantage in separation and helps in formulation of strategies to reduce native extraction of protein. Other methods, such as fusions of oleosin or polymer, to assist in the purification of recombinant proteins have been recently reported and might also be applicable to antibody molecules.

6.5.4.3 *Glycosylation*

Glycosylation of antibodies differs depending on the production system. Variations in the glycosylation patterns of proteins created in humans and plants are reasons for concern over the possible immunogenicity of plant-specific complex N-glycans, which are found in the heavy chain of plant-derived antibodies. Although a significant serum immune response was not obtained when a plantibody was injected into mice, differential glycosylation poses a major limitation for application in humans. One of the strategies is the elimination of appropriate peptide recognition sequences for N-glycosylation. This strategy has been widely adopted in cases where effector functions dependent on glycosylation are not required. It is possible to avoid Golgi-mediated modifications by ER retention of a recombinant antibody through adding a C-terminal Lys-Asp-Glu-Leu (KDEL; in single-letter code for amino acids) sequence. In a well-designed strategy towards the humanization of plant glycans, human β-1,4-galactosyltransferase was expressed in tobacco plants in a stable manner. A plantibody with partially galactosylated N-glycans was produced when these plants were crossed with those expressing a murine antibody. The glycosylation profiles of endogenous proteins and of a recombinant Ig in tobacco leaves were also influenced by senescence, which demonstrates that physiological factors may cause subtle qualitative variations.

6.5.4.4 *Timelines and regulatory issues*

Plants can be compared to animal systems (such as goat and chicken) with regard to timelines for protein production. It takes about 20 months for the clinical production of proteins maintaining good manufacturing practice (GMP). It is projected for corn that the first commercial lot (~1 kg of antibody) will be available 36 months after the initiation of gene transfer. From then, each generation provides a 100× scale-up. Species with an exceptionally high number of seeds per plant, such as tobacco, may exhibit an even faster scale-up. However, other problems might be posed by constraints created by the use of a system rich in toxic secondary metabolites.

The regulatory framework for plant-derived recombinant pharmaceuticals has to be completely established. GMP regulations can be applied, but these have to be refined to make them relevant to plant-based expression systems. There is no longer an issue of contamination risk with mammalian viral pathogens and prions; however, regulatory guidelines specific to plants have to take care of the presence of herbicide and pesticide residues. An issue of concern is environmental impact. There is a need to formulate standard procedures for labelling and containment to prevent the inadvertent entry of transgenic crops expressing recombinant pharmaceutical proteins into the food chain. Some of the containment strategies include use of self-pollinating crops, counter-selectable markers, and male sterility. From a technical point of view, it is now possible to produce a wide range of antibodies in plants. Researchers have demonstrated the engineering of enhanced pathogen resistance and modification of phenotypes by immunomodulation. Moreover, the significance of antibodies as a tool for *in vitro* research has been extended to *in vivo* applications in studies of proteins and other compounds. Many approaches have been suggested to exploit plants as bioreactors for producing pharmaceutical antibodies. In recent developments, focus has been on the thorough characterization of recombinant products, including the effects of glycosylation. New insights and production strategies might be evolved from recent discoveries that tissue-specific and physiological factors might have an influence on the quality and glycosylation pattern of a plantibody.

Purification remains the major cost factor in biopharmaceutical production although significant progress has been made. It is probably not mere coincidence that the two products nearest to market at this point, CaroRX and genital herpes antivirus formulation, have been designed for ectopic application. Even though there have been successful clinical trials of some plant-derived antibody products, several issues are yet to be resolved, including public acceptance and regulatory guidelines. At present, over 200 novel antibody-based products worldwide are undergoing clinical trials, and market demand will definitely strain the capacities of the current production systems. Plants are extremely competitive regarding cost, timelines, productivity, and safety. Hence, they represent a viable substitute for prokaryotic and mammalian expression systems. Transgenic plant systems have the advantage of easy scale-up of production; hence, with regard to cost effectiveness, the full potential of plants might be realized at higher production requirements. Therefore, long-term targets for plant bioreactors may include low-cost, high-volume antibodies, which do not need extensive purification.

6.6 PHENYLPROPANOID PATHWAY

Phenylpropanoids are natural products that result from the deamination of amino acid L-phenylalanine by phenylalanine ammonia-lyase (PAL). Simplest examples of phenylpropanoids, having only the C6 phenylpropane skeleton, are hydroxycinnamic acids (e.g., sinapic acid) and monolignols (e.g., coniferyl alcohol). More complex phenylpropanoids are produced by the condensation of a phenylpropane unit with a unit derived from acetate via malonyl coenzyme A. These phenylpropanoids include stilbenes, flavonoids, and isoflavonoids. Benzoic acids are believed to have biosynthetic origin via side-chain shortening of hydroxycinnamic acids. However, this might not be the only method for synthesizing them in plants. For the important biosynthetic routes to the different kinds of phenylpropanoid compounds, which also exhibit the primary metabolic pathways that provide precursors for phenylpropanoid biosynthesis, note the organization of the pathways into a 'core' phenylpropanoid pathway, from phenylalanine to an activated (hydroxy)cinnamic acid derivative via the actions of PAL, cinnamate 4-hydroxylase (C4H), and 4-coumarate:coenzyme A ligase (4CL), and the specific branch pathways for the formation of monolignols/lignin, coumarins, benzoic acids, stilbenes, and flavonoids/isoflavonoids. All plant species do not contain all classes of phenylpropanoid compounds. Even though the hydroxycinnamic acid and flavonoid classes are very common in higher plants, members of these classes having certain substitution patterns might be specific to certain genera or species. Other phenylpropanoid classes, such as isoflavonoids and stilbenes, are limited to individual plant families. Isoflavonoids are mostly limited to the subfamily Papilionoideae of the Leguminosae. They have large structural variations, involving the complexity and number of substituents on the 3-phenylchroman framework, different oxidation levels of the heterocycle, and the presence of extra heterocyclic rings. The stilbenes occur sporadically in widely divergent species, including peanut (Leguminosae), grapevine (Vitaceae), and pine (Pinaceae).

Owing to the availability of extensive information about its regulatory and structural genes, the phenylpropanoid pathway is an ideal system to understand how to genetically alter complex natural product pathways in plants. However, important information relating to flux control at and within the various branch pathways and the potential cross-talk between pathways is not available. Moreover, the extent to which sets of reactions are organized in metabolic channels (or metabolons), which result in the sequestration of intermediates from diffusible cytosolic pools, is important. All these factors might strongly affect the consequence of efforts to increase or decrease the amount of a particular compound by transgenic paths. These questions can be addressed through interdisciplinary approaches involving molecular, cellular, and structural biology. The understanding of cross-talk and flux control in phenylpropanoid biosynthesis has become possible mainly from studies in which certain enzymes in the pathway were overexpressed or downregulated in transgenic plants. Such an approach shows that the entry point enzyme PAL is directly rate limiting for the production of chlorogenic acid (CGA, caffeoylquinic acid) in leaves of tobacco but that factors other than PAL control flux into flavonoids and lignin. CGA offers resistance to both microbes and insects, even though PAL overexpressing plants with elevated CGA appear to show decreased resistance to insect herbivory because of cross-talk between salicylate and jasmonate signal pathways.

In tubers of potato, the creation of an artificial sink for tryptophan through the transgenic expression of a tryptophan decarboxylase gene gave rise to lowered phenylalanine pools and decreased levels of wound-induced CGA and lignin, with a consequent increase in susceptibility to *Phytophthora infestans*. CGA levels are also decreased in tobacco by downregulating C4H, the second enzyme in the phenylpropanoid pathway. This is accompanied by a feedback inhibition of PAL activity, probably because of the feedback inhibition of PAL expression by cinnamate or some derivative. On the contrary, overexpressed C4H did not consistently enhance the levels of CGA, which confirmed that PAL, rather than C4H, is the flux control point into the phenylpropanoid pathway in tobacco leaves.

Chalcone isomerase (CHI) acts as catalyst to a near-diffusion-controlled reaction, which can also occur spontaneously at the cellular pH. Therefore, it is not generally viewed as a potential rate-limiting enzyme for flavonoid biosynthesis. However, overexpression of CHI in tomato fruit peel results in an 80-fold increase, and the expression of alfalfa CHI in *Arabidopsis* results in a threefold increase, both in the levels of flavonols. Therefore, CHI would seem to be a constituent of flux control into the flavonoid branch of phenylpropanoid biosynthesis. The phenylpropanoid pathway gives some of the best examples of metabolic channelling in plant metabolism. In metabolic channelling, successive enzymes in a metabolic pathway are physically organized into complexes through which pathway intermediates are channelled without diffusion into the bulk of the cytosol (Srere 1987). However, these complexes are loose, and many of the enzymes involved might be operationally soluble. It is possible with these complexes to efficiently control metabolic flux and protect unstable intermediates.

6.7   SHIKIMATE PATHWAY

The shikimate pathway is often regarded as the common aromatic biosynthetic pathway, although all aromatic compounds are not synthesized in nature by this route. This metabolic sequence converts the primary metabolites phosphoenolpyruvate and erythrose-4-P into chorismate, the last common precursor for the three aromatic amino acids: Phe, Tyr, and Trp and for *p*-amino and *p*-hydroxybenzoate (Figure 6.1). The shikimate pathway is present in bacteria, fungi, and plants. In monogastric animals, Phe and Trp are essential amino acids that have to come with the diet and Tyr is directly derived from Phe. Bacteria use more than 90% of their metabolic energy for protein biosynthesis; therefore, for most prokaryotes, the three aromatic amino acids constitute nearly the entire yield of aromatic biosynthesis, and regulatory mechanisms for shikimate pathway activity are induced by the intracellular levels of Phe, Tyr, and Trp. Higher plants do not exhibit this characteristic because in them the aromatic amino acids are the precursors for a wide variety of secondary metabolites with aromatic ring structures that often constitute a considerable amount of the total dry weight of the plant. Some of the many aromatic secondary metabolites include flavonoids, many phytoalexins, indole acetate, alkaloids such as morphine, UV light protectants, and lignin.

Even though the enzyme catalysed reactions of the shikimate pathway appear to be similar for prokaryotes and eukaryotes, plants have branches off the main pathway that have so far not been demonstrated in bacteria or fungi. Prokaryotes and eukaryotes have rather striking differences in this regard. The main trunk of the shikimate pathway involves reactions catalysed by seven enzymes. These include the penultimate enzyme 5-enol-pyruvoyl shikimate-3-P synthase, the primary target site for the herbicide glyphosate, and the first enzyme DAHP synthase that controls carbon flow into the shikimate pathway. The condensation of PEP (phosphoenolpyruvate) and erythrose-4-P is catalysed by DAHP (3-deoxy-D-arabino-heptulosonate-7-phosphate) synthase to yield DAHP and Pi. Although the enzyme was found in *E. coli* more than three decades ago and had been purified to electrophoretic homogeneity from a variety of sources, the fine structure of DAHP, the output of the enzyme catalysed reaction, was not defined until many years later.

6.8 TRANSGENICS

The development of methods for the production of transgenic crops started around 1983 as part of a wider technological movement for modifying organisms for medical, economic, military, and other human ends. There are significant and complex ethical issues associated with the modification of life. Production of transgenic crops has resulted in a considerable controversy among government agencies, researchers, business organizations, and non-profit organizations. Civil society groups are particularly more vocal. The measureable facts regarding GM foods appear less in dispute than the increasing number of implications. These often occur in ethical arguments, which some proponents of transgenic crops disregard as a defence of cultural artefacts. However, the new capacities created by transgenic crops reveal an absence of research into these implications. No uniformly accepted definition of biotechnology exists. The broadest definition of biotechnology includes the application of biological sciences and conventional plant and animal breeding techniques carried out since the dawn of civilization. Biotechnology refers to newly evolved scientific techniques to alter the genetic make-up of organisms and produce unique individuals or traits that are not readily obtained through traditional breeding techniques. These products are often termed transgenic, bioengineered, or GM because they contain the foreign genetic material. This new technology has radically affected agriculture on both a fundamental production level and a legal level. The products of transgenic engineering are often termed GM organisms (or GMOs).

All these terms describe methods through which biologists insert genes from one or more species into the DNA of crop plants to transfer selective genetic traits. The method is referred to as the recombinant DNA technology. Genes are fragments of DNA that contain information that partially determines the end function of a living organism. Genetic engineers alter DNA, usually by selecting genes from one species—an animal, plant, bacterium, or virus—and inserting them into another species, such as an agricultural crop plant. An intermediate organism or virus can be used to 'infect' the host DNA with the desired genetic material. Microparticle bombardment technology is also commonly applied to insert exogenous nucleic acids (DNA from another species) into plant cells. The desired genetic material is precipitated onto micron-sized metal particles and placed within one of a number of devices designed to accelerate these 'microcarriers' to velocities necessary to penetrate the plant cell wall. Transgenes can be inserted into the cell's genome in this manner. Electroporation can also be used to insert new DNA into a host cell. In this process, a jolt of electricity is applied to cells to create openings in the plasma membrane that surrounds a cell. A (typically antibiotic-resistant) marker gene is incorporated in the package to confirm the degree of effectiveness in introducing the foreign DNA.

Gene stacking is becoming increasingly popular, which adds an entire array of characteristics at once into the host (Stierle 2006). The whole technique can be explained by its application in the generation of transgenic rice using electroporation. When genetic engineering of plants began around 1983, it seemed that transgenic alteration might benefit and even revolutionize agriculture. The transfer of desired genetic characters across species barriers gave rise to promises to ease difficulties in the management of agricultural crops, provide new possibilities to improve human and animal health, and offer a new revenue stream to farmers through contract manufacturing of industrial and pharmaceutical crops. The prospective environmental

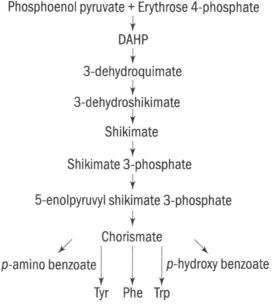

Fig. 6.1 *The shikimate pathway*

benefits included the decreased toxic pesticide use, enhanced weed control resulting in less tillage and soil erosion, and water conservation. In addition, the new technology promised enhanced yields. Transgenic crops were also patentable. Technology agreements would ensure that seeds could not be saved for planting the next year. The developer's intellectual property rights were thereby protected, which offered the potential to enhance profits and theoretically garner a monopoly over the transgenic seed supply.

REFERENCES

Brent, L. G. and P. A. Srere. 1987. The interaction of yeast citrate synthase with yeast mitochondrial inner membranes. *Journal of Biological Chemistry* 262: 319–25

de Aizpura, H. J. and G. J. Russel-Jones. 1988. Oral vaccination identification of classes of proteins that proke on immune response upon oral feeding. *The Journal of Experimental Medicine* 167: 440–451

Giddings, G., G. Allison, D. Brooks, and A. Carter. 2000. Transgenic plants as factories for biopharmaceauticals. *Nature Biotechnology* 18: 1151–55

Haq, T. A., H. S. Mason, J. D. Clements, and C. J. Arntzen. 1995. Oral immunization with a recombinant bacterial antigen produced in transgenic plants. *Science* 268: 714–16

Landridge, W. 2000. Edible vaccine. *Scientific American* 283: 66–71

Moffat, A. S. 1995. Exploring transgenic plants as a new vaccine source. *Science* 268: 658–60

Peeters, K., C. de Wilde, and A. Depicker. 2001. Highly efficient targeting and accumulation of a Fab fragment within the secretory pathway and apoplast of Arabidopsis thaliana. *European Journal of Biochemistry* 268: 4251–60

Prakash, C. S. 1996. Edible vaccines and antibody producing plants. *Biotechnology and Development Monitor* 27: 10–13

Ramirez, N., P. Oramas, M. Ayala, M. Rodriguez, M. Perez, and J. Gavilondo. 2001. Expression and long-term stability of a recombinant single-chain antibody fragment in transgenic *Nicotiana tobacum* seeds. *Biotechnology Letters* 23: 47–49

Stierle, A. 2006. Bioprospecting in the Berkeley Pit: the search for valuable natural products from a most unnatural world. Montana, *The Magazine of Western History* 56: 71

Stoger, E., M. Sack, R. Fischer, and P. Christou. 2002. Plantibodies: applications advantages and bottlenecks. *Current Opinion in Biotechnology* 13: 161–66

Webster, D. E., M. L. Cooney, Z. Huang, D. R. Drew, I. A. Ramshaw, and I. B. Dry. 2002. Successful boosting of a DNA measles immunization with an oral-plant derived measles virus vaccine. *Journal of Virology* 76: 7910–12

Biotechnological Applications

7.1 MASS MULTIPLICATION OF ECONOMICALLY IMPORTANT PLANTS

There are many medicinal plants that yield precious active ingredients in small quantities but are required in large quantities. Owing to this clonal mass multiplication becomes crucial. For example, only 1 g of alkaloid could be extracted using 2 tonnes of leaves of *Catharanthus roseus* (L.) G. Don (Madagascar periwinkle, Sada-bahar), for treating a leukaemia patient for 6 weeks. Similarly, for treating one patient of ovarian cancer, the bark of one fully mature tree of *Taxus brevifolia* Nutt. (Pacific yew), approximately 200 years old, is required. Again rapid micropropagation is necessary for endangered plant species, such as *Dioscorea deltoidea* Wall. (medicinal yam), an indigenous species with the highest diosgenin content (of approximately 10%) among all *Dioscorea* species; its regeneration cycle is approximately 10 years and is very slow propagating. Medicinal plants are more important, because of their increased requirement in the production of medicines.

7.1.1 *Rauwolfia serpentina* (L.) Benth. (serpentine, Sarpagandha)

Rauwolfia serpentina is an endangered medicinal plant found in India. Reserpine, one of the 50 indole alkaloids obtained from this plant, is responsible for the plant's medicinal property. Reserpine is used as a sedative and for treating hypertension and mental diseases, such as paranoia and schizophrenia. Ajmaline, another alkaloid extracted from the plant, possesses similar potent therapeutic qualities. *Rauwolfia serpentina* has poor viability of seeds, poor seed germination percentage and inefficient propagation by root cutting. This makes micropropagation the only favourable option to meet the demand for a huge amount of raw materials by the pharmaceutical industry. From somatic proliferating tissue, plants have been regenerated, where four morphogenic patterns of differentiation are somatic embryogenesis, caulogenesis, direct plantlet formation, and regeneration of shoot buds from roots differentiated from somatic tissue. Normal plants are regenerated from each organized structure. The rate of propagation has been moderate. In an alternative approach, shoot-to-shoot proliferation employing nodal stem segments has been used that resulted in a much faster multiplication rate of clonal plants. Plants raised by both procedures showed normal growth in field conditions.

7.1.2 *Withania somnifera* (L.) Dunal (Indian ginseng, winter cherry, Ashwagandha)

Withania somnifera is an endangered Indian ginseng plant. It is extensively used in Ayurveda for its anti-inflammatory, anti-arthritic, anti-tumour, immunomodulatory, and curative characteristics. In chemotherapy and radiotherapy, it is used as an adjuvant. The presence of alkaloids such as somniferine, withasomnine, and withanolide (withaferine A) is responsible for its medicinal properties, including antibacterial and anti-tumour compounds. The plant has poor seed germination ability and is excessively exploited. In India, approximately 1500 tonnes of the *W. somnifera's* root is produced annually, whereas the requirement is of more than 7000 tonnes. *W. somnifera* has been micropropagated using nodal and internodal explants from seedlings, by regeneration from somatic callus as well as from leaf segments taken from field-grown plants. In the last case, multiple shoots regenerated from leaf explants and the plantlets exhibit 87% survival under field conditions after 100% rooting.

7.1.3 *Aloe vera* L. (aloe, Gwarpatha)

In recent times, *Aloe vera* has gained prominence owing to its anti-ageing property such as anti-wrinkle effect on skin and as a safe laxative. It also helps in controlling bronchial asthma because of its anti-allergic and anti-histamenic properties. The gel from its fresh leaves soothes sunburns, mouth ulcers; acts as a topical pain killer due to its anti-inflammatory property; and can be used as an immunomodulator due to the presence of certain compounds in addition to bradykinase. *Aloe vera* leaves contain anthraquinones such as aloin and emodin; glycosides (barbaloin); and glucomannans, steroids such as lupeol, campesterol, and sitosterol as active ingredients. In one year, a few thousand plants of *Aloe vera* can be produced by clonal multiplication through the culture of shoot apices. Rapid micropropagation of elite aloe plants has been also carried out.

7.1.4 *Curcuma domestica* Val. (turmeric, Haldi)

Since ancient ages, Ayurveda and Unani systems of medicine have been using *Curcuma domestica*. However, it became important globally for its medicinal value only in recent times. *Curcuma domestica* has both antibacterial and antiprotozoal properties, acts as a blood purifier and stomachic, and is a constituent of antiseptic skin creams. Moreover, it works as an antioxidant and is useful in decreasing serum cholesterol level and treating Alzheimer's disease. Owing to its active component, curcumine, turmeric has been found to be effective against colon cancer. Its micropropagation is done by using newly sprouted buds and stem segments of aseptically established plantlets. Two varieties, Takurpeta and Duggirala, exhibited a high multiplication rate of around 88,000 and 800,000 plants, respectively, in 1 year from 1 rhizome. Large-scale micropropagation and field evaluation of the plants have been performed. The micropropagated plants exhibited genetic uniformity and were also superior to those multiplied through traditional methods due to having vigorous growth, greater rhizome fresh weight, and more number of fingers.

7.1.5 *Allium sativum* L. (garlic, Lahsan)

In Europe, *Allium sativum* is among the seven top-selling modern drugs with herbal formulations. Garlic has anti-atheroma and anti-tumour agents and helps in lowering serum cholesterol and triglycerides. It acts as a good antioxidant due to the presence of flavonoids and selenium.

Garlic is vegetatively propagated as it is sterile in nature, but the main concern is its viral infection. Many reports discuss the micropropagation of *A. sativum*, including shoot tip culture to mainly eliminate viruses with low multiplication rate. However, stem-disc culture has been found to be very effective for rapid propagation and production of virus-free plants. About 30 shoots regenerate in 1 month from 1 disc, and above 90% of these form bulblets *in vitro*. Under field conditions, the *in vitro* raised plants have been grown normally. Later, multiple shoots were directly regenerated from shoot apices of a late maturing cultivar, Howaito-roppen. In this, *in vitro* bulblets were produced, conditions were standardized for breaking their dormancy, and the resulting plantlets were cultivated under field conditions in large numbers.

7.1.6 *Zingiber officinale* Rosc. (ginger, Adrak)

Z. officinale is more valued for its medicinal properties than as a spice. Ginger's main medicinal properties include its ability to decrease intestinal contraction, function as an antiemetic, block prostaglandins, provide relief in migraine, and work as a blood thinner and a vasodilator. Gingerol and shogaol cause these remedies. Ginger is effective in curing rheumatoid arthritis and osteoarthritis and in keeping the prostate gland healthy. Several reports discuss micropropagation of ginger through excised shoot buds and somatic embryogenesis via callus. India is the main producer of ginger and will be benefitted by clonal multiplication through meristem culture. This method leads to production of plants free from fusarium yellow disease (caused by *Fusarium oxysporum* f. *zingiberi*), soft rot (caused by *Pythium* sp.), and wilt disease (caused by *Pseudomonas solanacearum*), as they have adverse effect on the production. A single shoot meristem explant has reported to generate about 300,500 plants in 30 weeks.

7.1.7 *D. deltoidea* Wall., *D. floribunda* Mart. & Gal., and *D. composita* Hemsl. (medicinal yams)

The main raw materials for the commercial production of steroidal drugs are *D. deltoidea* and *D. floribunda* Mart. & Gal., and *D. composita*. In India, steroidal drug is produced completely from diosgenin, which is provided by these plants. This makes these plants the most significant medicinal plants. *D. deltoidea*, the richest source of diosgenin, is an endangered plant species. Also called wonder drugs, steroidal drugs have wide application in modern medicine and provide relief from serious ailments such as rheumatoid arthritis, allergic conditions (such as asthma), rhinitis, inflammatory conditions, colitis, ulcers, hormonal deficiency resulting in gynaecological disorders and sexual inefficiency, pituitary dwarfism, muscular dystrophy, cancer, and many other diseases. *D. deltoidea* is also clonally multiplied through shoot tip culture and nodal stem cuttings, but at a much lower rate compared to *D. floribunda*. To produce *D. composita*, nodal stem segments are used. *D. deltoidea* have been developed under glasshouse conditions, while *D. composita* plants were grown under field conditions.

7.1.8 *Costus speciosus* (Koen.) Sm. (spiral ginger, Keu)

An alternative vegetable source of commercial diosgenin is C. *speciosus*. It can be easily generated under different agroclimates. C. *speciosus* can be cloned very quickly by proliferating the shoot buds excised from rhizome. Despite establishment of aseptic culture of shoot buds being problematic, once established in the culture the resulting shoots proliferated very fast. The excised shoots rooted easily *in vitro*, and the micropropagated plants flourished with 100% transplant success under field conditions, generating rhizomes having the same quantity of diosgenin content as that of the mother plant. Recently, a higher rate of multiplication has been recorded from thin sections of its rhizome.

7.2 BIOPESTICIDES (INSECTICIDES)

Synthetic chemical insecticides are hazardous despite proving beneficial for food production. Alternative methods of insect management are available that offer adequate levels of pest control and pose fewer hazards. Microbial insecticides are one such alternative. They are safer in comparison to chemical insecticides as these exhibit low toxicity to non-target animals and humans and are harmless to the pesticide user as well as the consumers of treated crops. Microbial insecticides are also known as biological control agents and biological pathogens. Microbial insecticides are composed of living microorganisms (bacteria, fungi, viruses, protozoa, or nematodes) or the toxins these organisms produce. They are used in the form of traditional insecticidal sprays, liquid drenches, dusts, liquid concentrates, wettable powders, or granules. Specific properties of each product determine the manner in which the product can be applied most effectively.

The following are the advantages of microbial insecticides:

- Microbial insecticides are safe as the microorganisms present in them do not exhibit toxic and pathogenic effect on species that are not related to the target pest, such as wildlife, humans, and other organisms. These insecticides target a single insect group or species and do not directly affect beneficial insects (including predators or parasites of pests) that are present in the treated areas.
- Both synthetic chemicals and microbial insecticides can be used together as per necessity. The residues of conventional insecticides

do not deactivate or damage the microbial product. The harmless nature of microbial insecticides allows them to be applied to crops ready for harvesting as they cause no damage to humans or other animals.

- Sometimes the pathogenic microorganisms might establish in a pest species or its habitat and provide control during subsequent pest generations or seasons.

The following are the disadvantages of microbial insecticides:

- A microbial insecticide targets a specific species or group of insects. It, thus, eliminates only that particular pest present in the field. If other types of pests are present, the microbial insecticide will not be able to target them. Conventional insecticides also have similar limitations as they are not equally effective against all pests. However, the negative aspect of selectivity is pronounced more often in microbial insecticide.

- Proper timing and procedure of application of microbial insecticides are especially important as they lose effectiveness against heat, desiccation (drying out), or exposure to ultraviolet radiation.

- Some microbial insecticides require special formulation and storage procedures. These procedures do not seriously limit the handling of widely available products but may complicate the generation and distribution of certain products.

- Pest specificity of microbial insecticides limits their potential market.

7.2.1 Bacteria

Spore-forming, rod-shaped bacteria in the genus *Bacillus* are bacterial bioagents that are used for controlling insects. They are commonly found in soils, and most insecticidal strains have been generated from soil samples. As bacterial insecticides are not contact poisons, they are effective only when ingested by the target insect. Moreover, an insecticide comprising a single *Bacillus* species may target an entire order of insects or only one or a few species. For instance, products having *Bacillus thuringiensis* var. *kurstaki* (Bt) eliminate the caterpillar stage of a large array of butterflies and moths whereas *Bacillus popillae* (milky spore disease) kills Japanese beetle larvae but does not act against the closely related yearly white grubs (masked chafers in the genus *Cyclocephala*) that commonly infest lawns.

Since the 1960s, microbial insecticides comprising bacterium Bt are widely used in the United States. Commercially, Bt products are made in large industrial fermentation tanks. When the bacteria live and multiply in the right conditions, each cell produces (internally) a spore and a crystalline protein toxin called endotoxin. Most of the commercial Bt products produce both the protein toxin and spores; however, some are cultured in a manner that only the toxin component gets generated.

When a susceptible insect ingests Bt, the alkaline condition inside the insect's gut activates the protein toxin and enzymatic activity. The presence of specific receptor sites on the insect's gut wall determines the toxicity of the activated toxin. The extent of complementarity between the active site of the receptor and toxin determines the strength of binding of the substrate and the enzyme, thereby determining the range of insect species eliminated by each Bt subspecies and isolate. If the activated toxin binds to the receptor sites, it destroys the gut wall cells of the insect, allowing the cell contents of the gut to enter the insect's body cavity and bloodstream. Poisoned insects may die immediately or in a few days after ingesting the endotoxin. Quick death may be due to septicaemia (blood poisoning) resulting from the activity of the toxin. After ingesting Bt, the insect might become lethargic and stop feeding (and therefore stop damaging crops) due to gut cell damage, and finally starve to death.

In the environment Bt does not colonize or cycle (reproduce and persist to infect the next generations of the pest) in the magnitude necessary to provide continuing eradication of target pests. The bacteria might multiply in the infected host (insect), but not result in the generation of sufficient spores or crystalline toxins. As a result, infective units released into the environment after the death of the poisoned insect are few or absent. Consequently, Bt products are used much like synthetic insecticides. In many outdoor situations, Bt treatments become inactive fairly rapidly (within one to a few days), and repeated applications may be essential for some crops and pests.

Commercial Bt products were effective only against caterpillars until the early 1980s. In recent times, additional isolates have been identified that kill other types of pests and have been developed for insecticidal use. The characteristics of the crystalline protein endotoxin vary among Bt subspecies and isolates, and the nature of the specific endotoxins determines the type of insects to be poisoned by each Bt product. Bt formulations are categorized into the following:

***Bt formulations that kill caterpillars*:** The popular and commonly used Bt insecticides are pathogenic only to the larvae of butterflies and moths. The common trade names of these are Dipel®, Javelin®, Thuricide®, Worm Attack®, Caterpillar Killer®, and SOK-Bt®. Smaller companies use different trade names to market similar products. Bt insecticides are mostly in the form of liquid concentrates, wettable powders, and ready-to-use dusts and granules. They target leaf-feeding caterpillars such as those found on vegetables (especially the 'worms' that attack cabbage, broccoli, cauliflower, and Brussels sprouts); larvae of the gypsy moth and other forest caterpillars; bagworms and tent caterpillars on shrubs and trees; European corn borer larvae; and Indian meal moth.

Bt effectively controls the following common caterpillar pests:
- European corn borer in corn
- Indian meal moth in stored grain
- Cabbage looper
- Imported cabbageworm
- Diamondback moth
- Tomato/tobacco hornworm
- Gypsy moth
- Corn earworm (on corn)
- Codling moth
- Squash vine borer

***Bt formulations that kill mosquito, black fly, and fungus gnat larvae*:** In the early 1980s, a second group of Bt insecticides was produced. In these products, the subspecies *B. thuringiensis* var. *israelensis* (Bti) is used, which eliminates the larvae of certain Diptera (the insect order containing flies and mosquitoes). Bt mainly targets mosquitoes, black flies, and fungus gnats in their larval stages of development. However, it cannot control the larval stages of 'higher' flies such as the housefly, stable fly, or blowflies. The mosquito genera most susceptible to Bt are *Aedes* and *Psorophora*. *Anopheles* and *Culex* species can be controlled by using higher-than-normal doses of Bti. Some commercially available Bti products are Vectobac®, Teknar®, Bactimos®, Skeetal®, and Mosquito Attack®.

A community-wide use of Bti is most effective in controlling mosquito or black fly. Bti products are developed for granular or spray applications. Bti products developed on corncob granules act effectively against mosquito larvae that grow in tires and other artificial containers (the 'Asian tiger mosquito', *Aedes albopictus*, develops in these containers). They also provide better residual control. For good penetration and uniform treatment, the corncob granules can be blown into tire piles. However, Bti is not able to control mosquito larvae in turbid water containing high levels of organic pollutants. Bti is also used to reduce fungus gnat larvae in greenhouses and mushroom culture beds, as a drench to potting soils or culture media. But not all Bti products can control fungus gnat.

Bacillus sphaericus is pathogenic against larvae of mosquitoes in the genera such as *Culex*, *Psorophora*, and *Culiseta*. It is not a Bt, but another bacterium species. Its effectiveness against larvae of *Aedes* varies across various species of this genus. *A. aegypti* (the yellow fever mosquito) and *A. albopictus* are not susceptible, while *Aedes vexans* is very susceptible. Laboratory tests showed pathogenic effect of *B. sphaericus* on *Anopheles* larvae, but field results were not promising. To be effective the bacterium has to remain on the water surface for an extended period because *Anopheles* mosquitoes are surface feeders. The *B. sphaericus* formulations tested so far fail to remain on the surface long enough to be fruitful. Along with infecting mosquito species against which Bti is ineffective, *B. sphaericus* is a potentially effective insecticide as it remains active in stagnant or turbid water. *B. sphaericus* occurs naturally; it cycles and maintains itself in the environment. However, currently developed insecticidal formulations do not cycle in water to infect subsequent generations of mosquito larvae.

Bt formulations that kill beetles: Certain beetles are susceptible to the toxicity of Bt isolates from *B. thuringiensis* var. *san diego* and *B. thuringiensis* var. *tenebrionis*. However, species within the order Coleoptera (beetles) exhibit vast differences in susceptibility to these isolates. This might be because of variations in the receptor sites of the insects' gut wall where the bacterial toxins should attach. Hence, the beetle-targeted Bt formulations do not eliminate all beetles or even all the species within a beetle family or subfamily.

Using Bt insecticides: In a variety of situations insecticides containing Bt can be used to control insects. Each Bt insecticide is effective against some types of insects; therefore, it is important that a specific Bt insecticide is identified that controls the target pest. Susceptibility to Bt also differs among separate stages of insects; isolates that are effective against larval stages of butterflies, moths, or mosquitoes will not be pathogenic for adults. Insects that are susceptible will be poisoned by Bt only after they consume it; hence, treatments must be oriented to the plant parts or other material that is food for the target pest. If pests bore into plant tissue without feeding on the surface foliage or fruits, then Bt becomes ineffective. Target insects are not killed by Bt immediately. Poisoned insects stay on plants for one day or two days after treatment, but they discontinue feeding and die soon. Sprays should be applied thoroughly to cover all plant surfaces, including the undersides of leaves as it will maximize the effectiveness of Bt treatments. As Bt gets deactivated by direct ultraviolet radiation when exposed to sunlight, it should be applied in the late afternoon or evening because the insecticide remains active on the foliage overnight. Application on cloudy (but not rainy) days yields a similar result. Bt spores or toxins encapsulated in a granular matrix (such as starch) or killed cells of other bacteria during production processes are protected against ultraviolet radiation.

Some Bt isolates (which are not yet applied in the currently produced insecticides) create significant quantities of an additional exotoxin called thuringiensin, which is released outside the bacterial cell wall. The development of commercial insecticides containing this toxin is under research stage. Thuringiensin is toxic to a wide range of animal species and humans though it might be lauded as 'natural' because it is produced by living organisms. However, thuringiensin insecticides are more toxic than other Bt products and should be handled cautiously.

When applying microbial insecticides, users should not breathe dusts or mists of microbial insecticides. They must wear gloves, long-sleeved shirts, and long trousers and should thoroughly wash their hands and feet after the application. These precautions can prevent unexpected reactions from unknown toxicity. If inhaled or rubbed into the skin, bacterial and mould spores and virus particles become 'foreign proteins', which can result in serious allergic reactions. Dusts or liquids used to dilute and deliver these microorganisms can also function as irritants or allergens.

Recent developments in biotechnology have made it possible to place Bt toxins within crop plants in a number of ways. For instance, genes regulating the production of Bt toxins can be put into certain bacteria dwelling in plants. Bt toxin is generated when these altered bacteria develop and multiply within an inoculated host plant. This technology may seem ideal, but high-level, season-long control would pose the risk of developing resistance to Bt toxin in insects. Developers should devise methods for stopping or managing insecticide resistance in target pests because genes are added to crop plants for the production of insecticidal compounds.

Other bacterial insecticides: The bacteria *B. popillae* and *B. lentimorbus* are found in insecticides sold under the trade names Doom®, Japidemic®, Grub Attack®, and the generic name 'milky spore disease'. These bacteria are 'cultivated' in laboratory-reared insect larvae as they cannot be cultured in fermentation tanks. Microbial insecticides having *B. popillae* and *B. lentimorbus* can be utilized on turf and 'watered in' to the soil below to regulate the larval (grub) stage of the Japanese beetle and some other beetle grubs. When a grub ingests spores of these bacteria, its internal organs are liquefied by the proliferating bacteria and turn milky white. These symptoms manifest gradually over a period of 3–4 weeks after the initial infection.

B. popillae and *B. lentimorbus* can cycle in the environment if a sizable population of grub is present during the application. A new batch of spores is released into the soil when grubs killed by these bacteria break apart. These spores can survive beneath undisturbed sod for 15–20 years. Hence, applications of milky spore disease bacteria might not be required every year.

7.2.2 Viruses

Many viral diseases affect the larvae of insect species at catastrophic level. These viruses specifically attack a single insect genus or single species. The viruses are mostly nuclear polyhedrosis viruses (NPVs) and target caterpillars and the larval stages of sawflies. In the NPVs, a number of virus particles are 'packaged' together in a crystalline envelope in the insect cell nuclei, or granulosis viruses (GVs) in which one or two virus particles are enveloped by a granular or capsule-like protein crystal present in the host cell nucleus.

The insect hosts get affected only when they ingest the viruses. The symptoms in the larvae of the sawfly differ from the larvae of the caterpillars. Virus infections in sawfly larvae are restricted to the gut. On the other hand, virus particles infect the body tissues of the caterpillars after passing through the gut wall and other tissues. The caterpillar's internal organs are liquefied with the progress of the infection, followed by discolouration and ruptures of the cuticle (body covering). Caterpillars killed by the viral infection appear limp and soggy. They release virus particles by remaining attached to foliage or twigs for days, which may be consumed by other larvae. In this way, the pathogen can be spread throughout an insect population (especially when raindrops help to splash the virus particles to adjacent foliage) and by infected adult females depositing virus-contaminated eggs. Exposure to direct ultraviolet radiation destroys virus particles, thus dissemination of viral pathogens is deterred by direct sunlight. Naturally occurring epidemics keep a check on insect population but rarely occur before pest populations have reached peak levels. Production and use of insect viruses are limited due to certain economic factors. Development of insect viruses requires live host insects, unlike Bt, which makes its production expensive and time consuming. Moreover, viruses are genus or species specific.

The effectiveness of some virus insecticides is less than that of available synthetic chemical insecticides. Nonetheless, several insect viruses have been developed and registered for use as insecticides though they are not well known or widely available. Most of them are specific to only one species or a small group of related pests found in forests, for instance gypsy moth, Douglas-fir tussock moth, spruce budworm, and pine sawfly. Although not commercially available, they are developed and used by the US Forest Service. Since the permanence of the forest environment contributes to the cycling of the pathogen (transmission through subsequent generation), forest pests are good targets for viral pathogens. The forest canopy also protects viral particles from ultraviolet radiation. Other insect viruses being examined for insecticides include those that infect alfalfa looper, soya bean looper, cabbage looper, armyworms, and imported cabbageworm. Though some of these viruses have been field tested, none of them has been registered or commercially sold.

Both the codling moth GV (Decyde®) and the *Heliothis* NPV (Elcar®) were once registered by the USEPA and commercially produced, but these are not registered or available any more. If new viral insecticides are registered or if the currently registered products become widely available, their effective use will be dependent on the following factors:

- Most of the viruses are host specific and active only against immature stages of the target species.
- Users should correctly match the pathogen and the target pest.
- Ultraviolet radiation kills virus particles, so applying them in the evening or on cloudy days would be more effective.

7.2.3 Fungi

Similar to viruses, fungi control insect populations. Fungi species use their asexual spores, also called conidia, to spread diseases in insects. The survival ability of conidia of different fungi differs. However, desiccation and exposure to ultraviolet radiation kill conidia. The fungi can cause infection in an insect without being ingested. The conidia can germinate on the insect cuticle and enter its body. The insects die because of the fungal toxins and not due to chronic effects of parasitism. Germination of viable conidia requires free water or very high humidity when it reaches a susceptible host.

Fungal pathogens have a wide range of life stages and species that they can infect. Various significant fungal pathogens infect a number of insect species at their egg, immature, and adult stages of life. Others are more effective in the immature stages or on a small number of insect species. In the United States, many factors have hindered the growth of fungal insecticides. Some of the fungal pathogens can be developed on artificial media, but their large-scale production has not yet been accomplished. Establishment and maintenance of production and storage conditions are essential for ensuring the production of infective spores and their storage without loss of viability before application. After application, pathogenic fungi can act only when environmental conditions are favourable, for example high humidity or rainfall. The negative effects of ultraviolet radiation and desiccation are minimized when fungal pathogens are applied to soil for killing belowground pests. However, other microorganisms might act as competitors or antagonists and often change the effectiveness of pathogen. In the United States, no fungal pathogens are currently registered or commercially available. One or more fungi are applied in Great Britain, China, Russia and former soviet countries, Eastern Europe, and South America. Some fungi that are applied as insecticides are as follows:

Beauveria bassiana: This soil fungus has a broad range of hosts, including beetles and fire ants. It affects both the larvae and adult stages of a number of species. Recently, a few companies are developing preparations of this fungus for EPA registration. For the successful use of this fungus, understanding the interactions between *B. bassiana* and other microorganisms of soil may be the key.

Nomuraea rileyi: In the south-eastern United States, naturally occurring epidemic infections of *N. rileyi* in soya bean result in a dramatic reduction in the populations of foliage-feeding caterpillars. Research on predicting disease outbreaks due to this fungus might help in determining when to apply insecticides.

Vericilliuim lecanii: To control aphids and whiteflies, this fungus (once commercially available under the trade name Vertelec®) was utilized in greenhouses in Great Britain.

Lagenidium giganteum: The larvae of several mosquito genera face the threat of infection from this aquatic fungus. Even when mosquito density is low, it effectively cycles in the aquatic environment (spores developed in the infected larvae persist and infect the larvae of coming generations). Its effectiveness is restricted by high temperatures.

Hirsutella thompsonii: This pathogen was once registered by the US EPA and sold under the trade name Mycar®; however, it is not commercially available any more. *H. thompsonii* is a pathogen of the citrus rust mite.

7.2.4 Protozoa

Protozoan pathogens infect a wide range of insects thereby playing an important role in limiting insect populations naturally. These pathogens can kill their insect hosts by producing chronic and debilitating effects, but only a few appear to be suitable for development as insecticides. A significant and common result of protozoan infection is a decrease in the number of offspring produced by infected insects.

For use as insecticides, species in the genera *Nosema* and *Vairimorpha* appear to offer the highest potential. Pathogens in these genera infect lepidopteran larvae and insects in the order Orthoptera (grasshoppers and related insects). *Nosema locustae* is the only protozoan pathogen of grasshoppers which is currently available in a registered insecticidal formulation. It is available in the market by the names NOLO Bait®

and Grasshopper Attack®. It is most effective against grasshoppers in their early nymphal stage. Infections progress slowly; all infected hoppers are not killed, but death takes place 3–6 weeks after the initial infection by the pathogen. *N. locustae* has been used to reduce the population of grasshoppers in rangeland areas. The application of this pathogen to large areas with a population of young hoppers has shown to achieve adequate control. Infected hoppers consume less forage, and infected females lay fewer viable eggs compared to uninfected females. *N. locustae* that persists on egg pods give varying degrees of infection in the subsequent season. The effectiveness and application of *N. locustae* for controlling rangeland grasshoppers will likely increase with progress in research. Unfortunately, negligible or much less utility is offered by small packages of *N. locustae* preparations weighing one pound, developed for gardeners and homeowners. Considering the mobility of grasshoppers and all the aforementioned facts, it can be concluded that use of baits with *N. locustae* in small lawns or gardens will not substantially reduce or damage grasshopper population.

Precisely, although insecticidal products comprising nematodes are nearly microscopic in size, nematodes are not microbial agents but multicellular roundworms. They are used much like the real microbial products discussed previously. Nematodes, called entomogenous nematodes used for insect control, infect only insects or related arthropods.

The species that are most commonly used in insecticidal preparations are *Steinernema feltiae* (sometimes identified as *Neoaplectana carpocapsae*), *S. scapteriscae*, *S. riobravis*, *S. carpocapsae*, and *Heterorhabditis heliothidis*. The ability and specificity to infect and kill insects vary among different strains of these species. In general, these nematodes kill a huge number of insects. Worldwide, laboratory and field applications have been found to be effective against more than 400 pest species, including beetles, fly larvae, and caterpillars.

The third juvenile stage is the infectious stage of these nematodes, which is often called the J3 stage or the 'dauer' larvae. In this stage, nematodes live in moist soil and similar habitats without feeding, sometimes for long periods. Host insects are infected by *Steinernema* species when they enter the host body through body openings such as mouth, spiracles (breathing pores), and anus. *Heterorhabditis juveniles* can also penetrate the cuticle of the host insects in order to enter their body. These nematodes can complete their lifecycle in the infected insect if the environment is warm and moist. Infective juveniles molt to form adults, which give birth to a new generation within the same host. The offspring mature to the J3 stage and leave the dead insect to seek a new host. The symbiotic bacteria linked with entomogenous nematodes actually kill the infected insects. After entering an insect host, nematodes empty their gut bacteria into the insect's blood. They then consume these bacteria as they multiply. Bacterial septicaemia kills the insect. After extensive tests on these entomogenous nematodes and their associated bacteria (*Xenorhabdus luminescens*) for their toxicity to non-target organisms, they have been found to be non-toxic and non-pathogenic to plants and mammals. Production of these nematodes in large scale has been started, and a few insecticides have been marketed. These products are formulated for application as sprays or drenches and have infective juveniles (J3 stage). Laboratory experiments have shown that *Steinernema* and *Heterorhabditis* species are effective against a large number of insect species, but temperature and moisture conditions in the target insect's habitat act as limitations in field applications in their range of host. Desiccation in the field usually eliminates entomogenous nematodes although controlled desiccation has successfully formulated some insecticidal products containing these nematodes. Therefore, applying nematodes to control insects in moist soil, plant tissues, or other protected habitats is most effective. Depending on soil moisture, temperature, and other organisms present, their viability and persistence in soil differ greatly. The following cases describe situations in which nematodes might provide effective insect control.

To control the annual white grubs in grass sod, entomogenous nematodes should be effective when applied at the appropriate time to sufficiently watered turf. If these are applied after the appearance of grubs (early to late August in Illinois; correct timing depends on the latitude and annual variations in weather), nematodes (and synthetic insecticides) need to remain effective only for several days to decrease grub damage. Most homeowners and turf managers can irrigate; therefore, they can effectively 'water in' nematode applications and keep soils moist enough to make the conditions favourable for the survival of nematodes and grub infection. Homeowners should expect nematodes to exist for only 2–4 weeks after application, although nematodes might survive for several months in cool, moist soils. Trials have found nematodes to be effective against root weevil larvae and similar soil pests that attack nursery plants, ornamental plantings, garden crops, and potted plants. In the success of these treatments, the capability to maintain soil moisture is an important factor.

Nematodes have also been effectively used to control flea larvae by eliminating them from lawns and outdoor areas. These do not exhibit toxicity to pets and humans. Entomogenous nematodes are not appropriate for termite control as these are infectious to termites but are not likely to provide the long-term persistence required in a termiticide. The low moisture levels and the possible long-term absence of host insects in soils around and under buildings would, in most cases, prevent nematode persistence. Therefore, for long-term control, repeated applications of nematodes would be required at intervals not yet determined. Nematodes are not recommended for termite control since repeated applications are impractical and undesirable. Treatments using nematodes can be effective if placed in moist, protected environments where insects are confined, such as tunnels bored into tree trunks or other plant tissues. The effectiveness of treatments is likely to decrease if nematodes are applied to exposed foliage or other locations where desiccation occurs rapidly. Though commercial availability will possibly increase, nematodes are now sold through mail order from gardening supply houses (or wholesale to large-volume purchasers) under the trade names Vector®, Scanmask®, and BioSafe®; some products simply have the scientific name of the nematode. The USEPA does not regard nematodes as pathogens, and nematode producers do not require completing the same EPA registration as required by the makers of other microbial pesticides. This exemption enables decreased start-up costs for making products containing entomogenous nematodes, but it also results in less standardization and outside regulation of quality. Now major producers are gaining a reputation for manufacturing high-quality products, so consumers can expect that these companies would practise proper quality control. Nonetheless, inferior products are likely to get accompanied by exaggerated claims in the market in the absence of extensive regulation. Care should be taken by the buyers as there exist differences among nematode products and lack of regulations. These products have great potential, but they do not possess unlimited efficacy as some promotions suggest.

7.3   BIOFERTILIZERS

Generally, after long-term cultivation, agricultural land gets impoverished if not supplemented properly. High doses of agrochemicals are applied in order to supplement the nutrient content of the soil under the conventional farming system. This adversely affects the ecosystem by polluting it. Consequently, for making agriculture sustainable, a balanced and responsible application of organic agriculture is vital. The principles of organic farming also follow the concept of developing sustainable plant growth in longer term by developing the soil health and biodiversity. Biofertilizers play a significant role in enhancing the supplies of nutrients and their availability in crop husbandry. *Rhizobium* is the maximum researched among all biofertilizers. The effectiveness of nutrients fixed by soil microbes is more than that of nutrients applied from outside. Nearly 80,000 tonnes of nitrogen are available on 1 ha of land (since atmospheric air contains 78% nitrogen). Biofertilizers trap some of this nitrogen and fix it in the soil. Application of biofertilizers improves the nutrient availability and content of soil as these contain many beneficial microbes. Use of biofertilizers in crop production also helps in the development of the biological properties of the soil under organic farming. Advantages of biofertilizers are as follows:

- Biofertilizers contain selective organisms such as bacteria, fungi, and algae.
- These can fix atmospheric nitrogen and solubilize native and added nutrients (for example, phosphorus) to the soil so that they are acceptable to plants.
- They are an ecofriendly, economic, and renewable source of plant nutrients.
- They can play a vital role in maintaining long-term soil fertility and sustainability, thus important for a healthy future for future generations.

7.3.1   What is a Biofertilizer?

A biofertilizer is a ready-to-use live formulation of beneficial microorganisms that mobilize the availability of nutrients by their biological activity, when applied to seeds, roots, or soil. It helps build up the soil microflora and thereby the soil health. Biofertilizers are recommended for enhancing soil fertility in organic farming.

A simple classification of biofertilizers on the basis of nutrients is as follows:

1. For nitrogen:
   - *Rhizobium* for legume crops
   - *Azotobacter/Azospirillum* for non-legume crops
   - *Acetobacter* for sugarcane only
   - Blue–green algae (BGA) and *Azolla* for lowland paddy
2. For phosphorous:
   - Phosphatika for all crops to be applied with *Rhizobium*, *Azotobacter*, vesicular–arbuscular mycorrhiza (VAM)
3. For enriched compost:
   - Cellulolytic fungal culture
   - Phosphotika and *Azotobacter* culture

Besides these common sources of biofertilizers, some recently identified microorganisms are also used successfully for nitrogen fixation, such as *Azorhizobium caulinodans* in rice and maize, *Acetobacter* in sugarcane, and so on. Microbes such as *Sinorhizobium* can be utilized for nodulating the soya bean crop. *Thiobacillus thiooxidans* are known to oxidize sulphur and iron. Plants have forged various relationships with bacteria, algae, and fungi, and the most common of them are *Rhizobium*, *Azotobacter*, *Azospirillum*, *Azolla* (BGA association), mycorrhiza, and phosphate-solubilizing bacteria (some examples are illustrated in Table 7.1).

Nitrogen-fixing bacteria function under two conditions: symbiotically (in beneficial association with some organism such as crop plant) and non-symbiotically (as free-living bacteria). The symbiotic association occurs through the formation of root nodules. The free-living bacteria are not involved in any association but fix atmospheric nitrogen living freely.

7.3.1.1 **Rhizobium**

Rhizobium is the commonest biofertilizer. By forming nodules, it thrives in the root hairs of legumes. In 1888, Beijirinck from Holland first isolated this bacterium from the nodules of a legume. Later, the bacterium was reported under the genus *Rhizobium* in Bergey's *Manual of Determinative Bacteriology*. In 1889, Frank established the name *Rhizobium*. Based on 'cross-inoculation group concept', this genus has

Table 7.1   Nitrogen-fixing biofertilizers

Contributing plant nutrients	Microbes	Suitable crops
Nitrogen	Symbiotic	
	Rhizobium (with legume) and its other groups	Pulse legume: gram, pea, lentil, arhar, green gram, black gram; Oil legume: groundnut, soya bean; Fodder legume: berseem and lucerne
	Azolla (fern–*Anabaena azollae* symbiosis)	Rice
	Associative symbiosis (*Azospirillum*)	Rice, sugarcane, finger millet, maize
	Non-symbiotic	
	Heterotrophs (e.g., *Azotobacter*)	Vegetables crops, wheat, rice, and other commercial crops
	Photo-autotrophs (e.g., BGA)	Rice
Phosphorous	Phosphate-solubilizing and mineralizes	For all crops
	Fungi: *Asergillus, Penicillium*	
	Bacteria: *Bacillus, Pseudomonas*	
	Phosphate absorber (root fungus symbiosis) VAM	For all crops
	Ectomycorrhizae: *Pisolithus, Rhizopogon*	
	Endomycorrhizae: *Glomus, Gigaspora*	

seven distinct species. So far, more than 20 cross-inoculation groups have been identified. Of this, only seven are prominent, as listed in Table 7.2.

Rhizobium has been categorized into two groups: slow-growing rhizobia known as *Bradyrhizobium* and fast-growing rhizobia called Rhizobium. This categorization is, however, still not discretely distinguishable because bacteria from one group may infect another group. This is called 'the principle of cross-inoculation', which depends on the assumption that legumes belonging to a particular infection group might be nodulated by another species of nodule-forming bacteria.

Table 7.2 Rhizobium cross-inoculations groups

Rhizobium spp.	Cross-inoculation grouping	Legume types
R. leguminosarium	Pea group	Pisum, Visia, Lens
R. phaseoli	Bean group	Phaseolus
R. trifolii	Clover group	Trifolium
R. meliloti	Alfalfa group	Melilotus, Medicago, Trigonella
R. lupine	Lupine group	Lupinus, Ornothopus
R. japonicum	Soya bean group	Glycine
Rhizobium spp.	Cowpea group	Vigna, Arachis

7.3.1.2 Rhizobium–*legume symbiosis*

Rhizobia are soil bacteria that can trap atmospheric nitrogen in the soil. They associate symbiotically with legumes and some non-legumes such as *Parasponia*. After entering the roots through root hairs, *Rhizobium* bacteria release some stimulatory root exudates and develop nodules. Rhizobia attack the expanded cells of cortex inside the root and then differentiate into nitrogen-forming 'bacteroids'. Living separately, neither the plant nor the bacterium can trap nitrogen. The effective nodules are filled with pink sap (leghaemoglobin pigment). This pigment preserves the rhythm of oxygen supply to the bacteria and aids the activity of nitrogenase enzyme. In the process of nitrogen fixation, nitrogenase helps reduce nitrogen to ammonia. This bacterium is classified into two genera: *Rhizobium* and *Bradyrhizobium*. Table 7.3 provides the details of their species.

Depending on the strains of rhizobium, plant species, and environmental factors, the quantity of atmospheric nitrogen fixed varies.

Recently, *Sinorhizobium* and *Azorhizobium* are the two new genera included in the family Rhizobiaceae, which form nodules in soya bean and dhaincha (*Sesbania*), respectively. From the stem nodules of *Sesbania rostrata*, *A. caulinodans* have been isolated. These can also colonize and form nodules in rice roots. It has been found that maize is also responsive to *A. caulinoidans*. In the free-living state, it can also fix nitrogen in large quantities.

7.3.2 Methods of Application of *Rhizobium* Inoculants

A suitable method for the inoculation of *Rhizobium* is seed treatment. An appropriate contact between seeds and inoculants (bacteria) is

Table 7.3 Classification of *Rhizobium* biofertilizer

Rhizobium species	Principal plant inoculated
R. leguminosarium	
Biovar phaseoli	Phaseolus (bean)
B. viceae	Vicea (vetch)
B. trifolii	Trifolium (berseem)
R. meliloti	Melilotus (senji)
	Trigonella (fenugreek)
	Medicago (lucene)
R. loti	Lotus (trefoils)
Bradyrhizobium japonicum	Glycine (soya bean)
Bradyrhizobium species	Lupinus (lupin)
	Vigna (cowpea)
	Cicer (gram)

made using an adhesive. In the case of legumes, about 900 g of soil base culture is enough to inoculate the seeds for 1 ha of area. A 10% jaggery (gur) solution is utilized as the sticker for *Rhizobium* cells to seed. Initially, the solution is spread over the seeds and mixed thoroughly to form a thin coat. Then the inoculant is sprinkled on the seeds, and the content is again vigorously mixed. By spreading thinly on a polythene sheet, the content is dried in shade for overnight at least.

7.3.3 Blue–Green Algae

Another important class of biofertilizers, blue–green algae (BGA) are small organisms visible as a single cell, colonies (large accumulation of cells), or trichomes (strings of cells) under the microscope. BGA are also called cyanophytes, cyanobacteria, and most recently cyanoprokaryotes.

They have an external appearance similar to that of algae, as well as requirements for nutrients, light, and carbon dioxide. Tiny gas vesicles are present in the cells of certain types of BGA, which allow BGA to rise to the water surface or sink to the bottom depending on the changes in the availability of light and nutrients.

The BGA (*A. azollae*) and *Azolla* (aquatic fern) share a symbiotic relationship and fix atmospheric nitrogen. BGA are linked with the *Azolla* present in the ventral pore in the dorsal lobe of each vegetative leaf. The endophyte resides inside the water fern tissue and fixes atmospheric nitrogen. In paddy fields, BGA and *Azolla* can also be utilized because in flooded rice ecosystem, these can perform photosynthesis as well as fix atmospheric nitrogen. *Azolla* is a water fern rich in organic manure. It is fast growing and can double its weight in a week. It rapidly mineralizes the soil nitrogen and makes it available to the crop within a very short time. *Azolla* acts as a protein-rich feed to fish and poultry because it releases nitrogen steadily, without any loss due to leaching. It is a free-living autotroph and uses the energy derived from photosynthesis to trap nitrogen. Apart from nitrogen fixation, BGA also synthesize some growth-enhancing substances such as auxin and amino compounds, which induce the growth of rice plants. In paddy fields, algae can be multiplied by broadcasting the inoculant at the rate of about 10 kg algal cultures after 1 week of transplanting.

7.3.3.1 Azospirillum

Azospirillum is a free-living or non-symbiotic bacterium that does not form nodules but forms association by living in the rhizosphere. *Azospirillum* species associates with many plants, particularly maize, sorghum, and sugarcane. As the most common organism, it is capable of forming associative symbiosis on a wide range of plants. *Azospirillum* (nitrogen-fixing *Spirillum*) was previously named *Spirillum lipoferum*, as reported by Beijerinck, in 1925. *Azospirillum* has been identified as a dominant soil microbe, which can also form a close associative symbiosis with higher plants. The bacteria do not form any visible nodules or outgrowth on the root surface but live on it and may penetrate the root tissues. They fix 10–40 kg of nitrogen per hectare. *Azospirillum* inoculation promotes better growth of plants and saves nitrogenous fertilizers by 25%–30%. Only four species of *Azospirillum* have been isolated so far: *A. lipoferum*, *A. brasilense*, *A. amazoniense*, and *A. iraquense*. In Indian soils, the common species are *A. brasilense* and *A. lipoferum*. *Azospirillum* inoculation has increased the yields of many vegetable crops.

7.3.3.2 Azotobacter

Azotobacter, found in alkaline and neutral soils, is a heterotrophic, free-living, nitrogen-fixing bacterium. *Azotobacter chrococcum* is the most commonly occurring species in arable soils of India. It can synthesize growth-promoting substances such as auxins and gibberellins, and vitamins to some extent, besides fixing atmospheric nitrogen in soils. Fungicidal properties are also exhibited by many strains of *Azotobacter*. *Azotobacter* has been found effective in maize, rice, sugarcane, cotton, pearl millet, vegetables, and a few plantation crops. Its population is very insignificant in uncultivated lands. Its multiplication and nitrogen-fixing capacity is promoted by the presence of organic matter in the soil.

From field experiments on *Azotobacter*, it was found that this is effective when inoculated with seeds or seedlings of crop plants such as onion, brinjal, tomato, and cabbage under various agroclimatic conditions. Under normal field conditions, inoculation with *Azotobacter* reduces the requirement of nitrogenous fertilizers by 10%–20%.

7.3.3.2.1 Features of *Azotobacter*

Moderate benefits are contributed by *Azotobacter*. As the heaviest breathing organism, it needs a large quantity of organic carbon for its development. It is not a good competitor for soil nutrients.

It can help crops by fixing nitrogen and releasing growth-promoting and fungicidal substances. *Azotobacter* has poor organic content, and thus it is less effective in soils. It stimulates seed germination and plant growth, and proliferates even in alkali soils.

7.3.3.3 Acetobacter

Acetobacter diazotrophicus is a recently isolated nitrogen-fixing bacterium associated with the sugarcane crop. *Acetobacter* is found in the apoplastic fluid of sugarcane stem and, to some extent, in xylem vessels. It was isolated from leaf, root, bud, and stem samples of sugarcane. This bacterium is a member of the alpha group of proteobacteria. It is a bacterium with high acid and salt tolerance and affinity for sucrose. It can fix up to 200 kg of nitrogen per hectare. The yield of sugarcane increased after inoculation under field conditions. After its application, generation of auxins and antibiotic-type substances has also been observed.

7.3.3.4 Frankia

Frankia is an actinomycete that forms a symbiotic association with some non-leguminous trees such as *Casuarina* and *Alnus* and fixes atmospheric nitrogen by forming root nodules. Soil fertility in wasteland can be enhanced by cultivating *Casuarina*. In nitrogen-deficient soils, *Frankia* can render *Casuarina* trees suitable for agroforestry systems. At the start of nodulation, *Frankia* appears as small lateral swelling on roots, which then morphs into new lobes at their apices and creates a cluster coralloid structure. In *Casuarina* and *Alnus* plants, its inoculation improves growth, nodulation, nitrogenase activity of nodules, and nodule dry weight.

7.3.4 Phosphorus-solubilizing Microorganisms

Phosphorus is another significant primary nutrient for plants after nitrogen. Crops can recover only 15%–20% of the applied phosphorus, and the rest gets fixed in the soil. The fixed form is not part of the available phosphorous content in the soil. These are made available to crops by a group of heterotrophic microorganisms that solubilize fixed phosphorous by releasing organic acids and enzymes. This class of microorganisms is termed phosphorous-solubilizing microorganisms (PSM).

Various species of *Bacillus*, *Aspergillus*, *Penicillium*, and *Trichoderma* are part of this group. These organisms solubilize the fixed soil phosphorus and make available the citrate and water-soluble phosphorus when applied to rock phosphate. Organic phosphate compounds present in the organic wastes are mineralized by these microorganisms. They can be applied to fasten the composting process after the completion of the thermophilic phase. Application of bacteria in neutral and alkaline soils and of fungus in acidic soils improves the efficacy of the applied soil phosphorus and prevents phosphorous fixation. We know that the fungus vesicular–arbuscular mycorrhiza (VAM) benefits crop plants in diverse ways under natural conditions.

7.3.5 Vesicular–Arbuscular Mycorrhiza

Vesicular–arbuscular mycorrhiza is the most fascinating group of fungi that are beneficial to plants. The term mycorrhiza was coined by Albert Bernhard Frank in 1885 and is derived from Greek, meaning 'fungus root'. Mycorrhiza is a mutualistic association between fungal mycelia and plant roots. VAM is formed by aseptated phycomycetous fungi, which is an endotrophic (live inside) mycorrhiza (Peters 2002).

With a majority of crops growing under broad ecological range, it is associated as an obligate symbiont. Various leguminous plants and plants belonging to the Graminae family are highly vulnerable to VAM colonization. Nutrients such as phosphorus, zinc, and sulphur are transferred with the help of VAM. Different nutrients such as Cu, K, Al, Mn, Fe, and Mg are mobilized by it from the soil to the plant roots. VAM enters the root cortex and forms intracellular obligate fungal endo-symbiont. For the storage and funnelling of nutrients into the root system, it possesses vesicles (sac-like structure). Although the hyphae of VAM fungi fail to solubilize the insoluble phosphorus, they absorb phosphorus and other nutrients from soil for their own need and transfer them to the host roots in different forms. VAM also enhances water absorption by roots.

The two main recognized groups of mycorrhiza are (i) ectomycorrhiza and (ii) endomycorrhiza. In ectomycorrhiza, the hyphae develop a cover both outside and inside the root in the intercellular spaces of epidermis and cortex. Trees are commonly infected with ectomycorrhiza. VAM is the most common endomycorrhiza among its three subgroups. It creates an internal network of hyphae between cortical cells, which penetrates the soil and absorbs water and nutrients. VAM develops an association with various crop plants, whether monocot, dicot, annual, or perennial.

Along with the use of FYM and cereal–legume crop rotations, VAM improves the performance of low-fertile soil, whereas chemicals, particularly fungicides, depress its survival.

The VAM culture is generated by utilizing conventional pot culture techniques containing all VAM fungal structures of highly infective nature. The selection of host plant and ambient conditions determines the mass multiplication of particular VAM fungus.

7.3.6   Mechanism of Action

Forming an association with plant roots, VAM enters the root cortex and spreads around the roots. As the name suggests, it has sac-like structure called vesicules, which contain phosphorus as phospholipids. Arbuscule, the other structure, helps in bringing the distant nutrients to the vesicules and roots.

7.3.6.1   *Actions of mycorrhiza*

- Improves the feeding areas of the plant root as the hyphae spread around the roots.

- Mobilizes nutrients from a distance to the roots.
- Stores nutrients (especially phosphorus).
- Removes toxic chemicals (such as phenolics), which hinder growth.
- Protects against other fungi and nematodes.

7.3.7 Methods of Biofertilizer Inoculation (Application)

Under ideal conditions, biofertilizers can be inoculated on seeds and roots of various crop plants. They can also be applied directly to the soil.

7.3.7.1 Seed inoculation

Seed inoculation is the most common practice for the application of biofertilizers. In this technique, a 10% solution of jaggery is mixed with biofertilizers. The seeds are mixed properly with the slurry on a cemented floor so that a thin layer is formed around the seeds. The treated seeds must be dried in shade overnight before application. About 750 g biofertilizer is needed to treat legume seeds for an area of 1 ha.

7.3.7.2 Root and seedling treatment

Before transplanting in the field, the seedling roots of transplanted crops are treated in a solution of biofertilizers for half an hour. In this technique, seedlings needed for 1 acre are inoculated with 2–2.5 kg of biofertilizers. Adequate quantity of water is taken in a bucket and the biofertilizer is mixed properly. To enable the roots to get inoculum, roots of the seedlings are then dipped in this mixture. These seedlings are then transplanted. This technique has been proved suitable for tomato, rice, onion, cole crops, and flowers.

7.3.7.3 Soil application

This method is commonly used for crops where localized application is required, such as fruit crops and sugarcane. About 20 g of biofertilizers mixed with compost is added to the ring of one sapling while planting fruit trees. The same amount of biofertilizer may be applied to the ring soil of the seedling after it attains maturity. In some cases, the biofertilizers are also broadcasted in the soil; however, this requires 4–10 times more biofertilizers. The inoculants must be incubated with the desired quantity of well-decomposed, granulated FYM for 24 h before broadcasting.

7.4 VERMICULTURE

Very little information is found in literature about the direct use of compost worms in agriculture since the technique is very new. Some of the following possibilities are noted.

Seeding mulch with compost worms: This has been carried out in the orchards of both the United States and Australia. Carbon-rich organic materials are seeded with worms and are placed in rows under the canopies of the fruit trees. Periodic addition of various feedstock, such as precomposted manures, legumes, and fruit-processing wastes, is continued, and then more mulch is added to it. The worms grow in the mulch, eat the feed, and drop their castings near or in the soil. Rain helps the transfer of nutrients and beneficial microbes to the root zone.

Wintering worms in raised beds: A bedding/feed mixture is filled in a trough dug in the centre of raised vegetable beds in autumn and inoculated with worms. Straw or leaves are then used to cover the bed to the depth of half a metre or more. The covering is removed in spring, and the garden is planted. As a result, a deep vein of rich vermicompost runs through the middle of each bed. These worms then move into whatever is applied to mulch the garden (they will survive under plastic as well) and provide fertilization in all seasons.

Seeding pasture with cocoons: Since these worms do not burrow, they cannot live indefinitely in soil and require a loose, porous, and moist surrounding. However, cocoons are extremely durable. The cocoons will remain viable for long periods when vermicompost rich in worm cocoons is spread on a pasture, waiting for an animal to defecate at that spot. When the cocoons hatch, the animal manure turns into vermicompost on the spot. The worms then die but not without leaving cocoons to hatch during the next cycle. Therefore, the capacity of ecosystem to process droppings into high-quality fertilizer rapidly increases with spreading vermicompost on pasturelands.

7.5 PHYTOREMEDIATION

Phytoremediation is an emerging technology for cleaning up of contaminated soils, groundwater, and wastewater, which is both economic and low tech. Removal of environmental contaminants such as heavy metals, trace elements, organic compounds, and radioactive compounds from soil or water by engineered use of green plants, including grasses and woody species, can be defined as phytoremediation. This definition includes all plant-influenced physical, chemical, and biological processes that help in the sequestration, degradation, uptake, and metabolism of contaminants, by either plants or free-living organisms that constitute the rhizosphere of the plant. The process of phytoremediation uses the unique uptake capacity of plant root systems, along with the abilities of translocation, bioaccumulation, and contaminant storage/degradation of the entire plant body. Plant-based soil remediation systems are solar-driven, pump-and-treat biological systems having a self-extending, extensive uptake network (the root system) that boosts the belowground ecosystem for subsequent productive use.

Examples of simpler phytoremediation systems to treat acid mine drainage or municipal sewage are often with the use of cattails on constructed or engineered wetlands. 'Farming' the soil with selected plants and 'biomine' the inorganic contaminants to convert them into plant biomass constitute the process of phytoremediation of a site contaminated with heavy metals and/or radionuclides. The approach taken for soils contaminated with toxic organics is similar, but the organic matter may be taken up or degraded by the plant. Concentrations of toxic inorganics or organics in the soil could also possibly be reduced to an extent by farming several sequential crops of hyperaccumulating plants such that residual concentrations would be acceptable environmentally and no longer regarded as hazardous. The hazardous organic component of mixed contamination in water can also be degraded using potential techniques, thus decreasing the waste (which may be sequestered in plant biomass) to a more manageable radioactive one. Hydroponic or aeroponic techniques are used to grow phytoremediation plants in a bed of inert granular substrate, such as sand or gravel for contaminated wastewater treatment. Wastewater can be supplemented with nutrients if necessary, which trickles through this bed and is ramified with plant roots that act as a biological filter and a system for contaminant uptake. Another advantage of phytoremediation of wastewater is the significant reduction in volume obtained through evapotranspiration. Most of the technologies currently utilized have shown a low degree of feasibility or success in soils with low permeability even in combination with more conventional clean-up technologies such as electromigration and foam migration, where the application of phytoremediation is well suited. In appropriate conditions, phytoremediation can provide an alternative to harsher technologies of thermal vaporization, incineration, solvent washing, or other soil-washing methods, which essentially devastate the biological component of the soil and largely alter its chemical and physical properties. It can create a relatively non-viable solid waste. Phytoremediation benefits the soil by creating an improved, functional soil ecosystem at nearly one-tenth of the cost of the currently adopted technologies.

The concept of phytoremediation is based on the capacity of plants and their associated rhizospheres to concentrate and/or degrade very dilute contaminants. Root exudates are a part of the critical components present in the rhizosphere, along with a variety of free-living microorganisms. These complex root secretions, which are the source of nutrition for many microorganisms also possess natural chelating agents such as citric, acetic, and other organic acids, which render the ions of both nutrients and contaminants more mobile in the soil. Enzymes such as nitroreductase, dehalogenases, and laccases may also be the part of root exudates, which may degrade organic contaminants that contain nitro groups such as trinitrotoluene (TNT) and other explosives containing nitro-group or halogenated compounds such as chlorinated hydrocarbons and many pesticides naturally. These enzymes also have many other important natural functions. There exists complex feedback mechanisms between plant roots and rhizosphere microorganisms, which permit them to adapt to changing conditions of their immediate soil environment as they grow. Some plants growing in phosphorus-deficient soil have root exudates containing large amounts of citric acid, which mobilize and make possible the uptake of any present phosphorus compounds. Some microorganisms in the rhizosphere secrete plant hormones that enhance growth of root, thereby secreting root exudates containing metabolites they use as energy source.

Through transpiration, large green plants absorb a large amount of water from soil via roots, move it through the entire plant body for its own utilization, and evaporate it through the leaves as pure water vapour. Transpiration helps plants to perform photosynthesis, transport water and minerals, and regulate temperature by keeping the plant cool. During this process, soil contaminants are also taken up and sequestered, metabolized, or vaporized out of the leaves together with the transpired water. In most economically important plants, the feature of low water use is considered desirable. However, plants such as hybrid poplars, willows, bulrush, and marsh grasses, which naturally grow in moist environments, are poor at water conservation. As they take up and 'process' large volumes of soil water, such species are good candidates for phytoremediation. For instance, data show that on a hot summer day, a single willow tree can transpire above 19 m of water (19,000 L or 5000 gal) and herbaceous plants (such as saltwater cordgrass) grown over a 1-ha area evapotranspire nearly 80 m of water (80,000 L or 21,000 gal) daily.

When adapted plants are grown selectively in contaminated substrates, the root system of the plant functions as a highly dispersed, fibrous uptake system. A large range of concentrations of contaminants are taken up along with the water and degraded, metabolized, and/or sequestered in the plant body. Evapotranspiration maximizes the movement of soil solution or wastewater through the plant. Concentration of the contaminants can be increased more than thousand times in the plant than in soil or wastewater body through the process of bioaccumulation. The volume of contaminated plant biomass can be reduced by up to 95% by digesting or converting it into ashes, resulting in a small volume of residual material to be processed as an 'ore' to recover the contaminant (such as valuable heavy metals, radionuclides). If economic recycling of the metal is not possible, the relatively small quantity of ash (compared to the original biomass or the extremely large amount of contaminated soil) can be appropriately disposed of.

7.6 NUTRACEUTICALS

A nutraceutical is a food or dietary supplement having a medical benefit, which includes prevention and treatment of a disease. Stephen DeFelice, founder and chairman of the Foundation for Innovation in Medicine, coined the term in the late 1980s. Such foods are commonly regarded as functional foods, which signify that the components they possess may provide a health benefit beyond basic nutrition. Fruits and vegetables as well as fortified or enhanced foods are some examples. Nutraceuticals contain health-promoting ingredients or natural components with potential health benefits, unlike all foods that provide only nutrients. Researchers are studying the 'functional' attributes of many traditional foods and developing new food products with beneficial components. The concept of nutraceuticals has evolved considerably over the years but is not entirely new. In the early 1900s, US food manufacturers began adding iodine to salt for preventing goitre (an enlargement of the thyroid gland), which can be regarded as one of the first attempts at producing a functional component through fortification. Today, researchers have found hundreds of compounds with functional qualities and are still discovering phytochemicals (non-nutritive plant chemicals with protective or disease-preventive characteristics) with complex benefits in foods.

Nutraceuticals are very popular among consumers in the United States and other regions of the world. In 2003, the estimated sales in the United States were about $31 billion, and the figure will likely grow substantially in the following years. Nutraceuticals constitute one of the fastest growing sectors of the food industry, especially among the affluent baby boomers.

Nutraceuticals have already occupied a significant part of the dietary landscape in Japan, England, and other countries. The increased demand for information on nutraceuticals has resulted from consumer interest in the relationship between diet and health. Among the factors fuelling the interest in nutraceuticals in the United States are rapid advances in science and technology, mounting health-care costs, amendments to food laws that affect label and product claims, an ageing population, and increasing interest in achieving wellness through diet. Credible scientific research indicates that food components possess many potential health benefits. These benefits have the capacity to expand the health claims now permitted by the US Food and Drug Administration (FDA).

7.6.1   Traditional versus Non-traditional

Nutraceuticals found in the market today have both traditional and non-traditional foods. Traditional nutraceuticals are natural, whole foods that have new information regarding their potential health qualities. No change to the actual foods has been made, other than the way the consumer perceives them (Table 7.4).

In some studies, tea and chocolate have been observed to have health-benefiting attributes. Non-traditional nutraceuticals, on the other hand, are foods that have added nutrients and/or ingredients or that result from agricultural breeding. Agricultural scientists have boosted the nutritional content of some crops by utilizing the same breeding techniques that develop other useful traits in plants and animals—everything from beta-carotene-enriched rice to vitamin-enhanced broccoli and soya beans.

To improve the nutritional quality of various other crops, research is currently being conducted.

Some examples of foods specially formulated with nutrients or other ingredients include orange juice fortified with calcium, cereals containing added vitamins or minerals, and flour containing added folic acid. As researchers continue to uncover more evidence about the role of added nutrients in reducing health and disease risk, more foods are being fortified with nutrients and physiologically active components (such as plant stanols and sterols).

7.7 COSMECEUTICALS

A new hot topic in the cosmetic industry is 'cosmeceuticals', the fastest growing segment of the natural personal-care industry. According to the Federal Food, Drug, and Cosmetic Act enforced by the FDA, cosmetics are defined as their intended use as 'articles intended to be rubbed, poured, sprinkled, or sprayed on, introduced into, or otherwise applied to the human body for cleansing, beautifying, promoting attractiveness, or altering the appearance'. Cosmeceuticals are topical cosmetic–pharmaceutical hybrids aimed to enhance the beauty through ingredients that provide additional health-related functional benefit. They are applied topically as cosmetics but contain ingredients that influence the skin's biological function (Grace 2002).

Table 7.4 Neutraceutical components with their sources and potential benefits

Class/Components	Source	Potential benefit
Beta-carotene	Carrots, various fruits	Neutralizes free radicals, which may damage cells; bolsters cellular antioxidant defence
Lycopene	Tomatoes and processed tomato products	May contribute to maintenance of prostate health

The products that are considered cosmetics as per the definition given by the FDA include skin moisturizers, lipsticks, perfumes, eye and facial make-up preparations, fingernail polishes, shampoos, hair colours, permanent waves, toothpastes, and deodorants, as well as any material intended for application as a component of a cosmetic product. Serving as a bridge between personal-care products and pharmaceuticals, these cosmeceuticals have been developed specifically for their medicinal and cosmetic benefits. The origin and recorded use of cosmetics can be traced back to 4000 BC and is attributed to Egyptians. The ancient Babylonians, Sumerians, and Hebrews also used cosmetics. European cosmetic known as Ceruse was used from the 2nd century to the 19th century. Big and small corporations engaged in pharmaceuticals, biotechnology, natural products, and cosmetics are constantly developing cosmetically active ingredients. Advances in the field of skin biology and pharmacology have helped the cosmetic industry in rapidly developing novel active compounds. The novel features of cosmeceutical agents include safety, efficacy, formulation stability, patent protection, novelty, metabolism within skin, and inexpensive manufacture (Dooley 1997). An effort has been made to review the different kinds of cosmeceuticals and their regulatory aspects.

7.7.1 Skin Cosmeceuticals

Cosmetics and skincare products form the part of everyday grooming. For good health, protecting and preserving the skin is essential. Our skin is the largest organ in the body that separates the external environment from the internal environment and provides protection. A cumulative damage is caused to the molecular and structural composition of skin—DNA, collagen, and cell membranes—due to environmental elements, air pollution, exposure to solar radiation, and normal ageing process. Use of cosmetics or beauty products does not change the skin or heal the damage, but these products are just meant to cover and beautify. Cosmeceuticals can affect the biological functioning of skin on the basis of their composition of functional ingredients. Certain skincare products go beyond colouring and adorning the skin by improving skin function and texture. These products encourage the growth of collagen and combat harmful effects of free radicals using antioxidants, thus maintaining keratin structure and making the skin healthier.

A well-known skincare product is the Olay, which includes vitamins A, C, D, E, selenium, lycopene, pycnogenol, and zinc and copper. Treating an ageing skin with a cream having a hormone such as oestrogen gives a fresh appearance with a rejuvenating effect. Kuno and Matsumoto had patented an external agent for skin. They have prepared an extract from olive plants, which has a skin-beautifying agent, particularly an anti-ageing component for the skin and/or a whitening component. For cosmeceutical purpose, a dry emollient preparation having monounsaturated jojoba esters has been used. For stimulating skin and/or hair pigmentation, a plant extract of genus *Chrysanthemum* was utilized in a cosmetic composition. To improve deteriorated or aged skin, novel cosmetic creams or gels containing active ingredients and water-soluble barrier disruption agents such as vitamin A palmitate have been prepared.

7.7.2 Sunscreens

For maintaining a healthy and youthful skin, regular use of an effective sunscreen is one of the most important steps. Visible effects of skin ageing are mainly due to ultraviolet light from sun. Conventional chemical sunscreens function primarily by binding to skin protein and sucking up ultraviolet B (UVB) photons (280–320 nm) and mostly possess para-aminobenzoic acid (or its derivatives), cinnamates, different salicylates and benzophenones, dibenzoylmethanes, anthraline derivatives, octocrylene, and homosalate. Avobenzone (Parsol-1789) is a benzophenone, which has the property of ultraviolet A (UVA) protection. Physical agents or sunblocks reflect or scatter radiation and act as barriers. Compounds containing metals such as zinc, iron, bismuth, and titanium function as direct physical blockers. Titanium dioxide and zinc oxide are highly reflective white powders. However, particles of submicron zinc oxide or titanium dioxide powder can transmit visible light, while preserving their UV-blocking characteristics, thus making the sunblock invisible on the skin. Benzophenone-8, Neo Heliopan MA and BB, Parsol MCX, and HS are some commercially available sunscreens. A patented sunscreen composition consists of activated platelet factor as a constituent in a cosmeceutically acceptable carrier. Such a preparation in the form of a shaving cream or foam, aftershave lotion, moisturizing cream, suntan lotion, and lipstick helps in restoring skin to its normal condition after damage due to cuts, abrasions, sun, and wind.

7.7.3 Moisturizers

Moisturizers play an important role in smoothing out age lines and aid in brightening the delicate skin. Emollients are usually incorporated in these moisturizers to smoothen the skin surface by scrapping the non-living outer layers of the skin, filling spaces between the layers and lubricating, and humectants to aid skin cells absorb and retain moisture in these layers. For women approaching their menopause, Healthy Remedies Balancing Lotion has been created, containing ingredients that reduce fine lines and wrinkles, uplift the neck area, and moisturize the sagging, dry skin. Black cohosh, soy extract, and vitamins A and E are some of those ingredients. A nourishing complex having hyaluronic acid and a revival complex possessing green tea leaf extract as well as glutathione augment the skin's natural moisture balance.

7.7.4 Bleaching Agents

Bleaching agents are applied for bleaching/fading various marks and blocking the formation of the skin pigment melanin. The bleaching agent most commonly used to reduce or remove brown marks, liver spots, and melasma is hydroquinone. Kojic acid extracted from mushrooms is a slightly less effective agent, which may be compounded with tretinoin or topical steroids, and hydroxy acids. Similar to any bleaching agent, aggressive exfoliation and sun protection are essential for good results. A synthetic detergent bar has been developed, which has hydroquinone as a skin-bleaching agent. The bar is maintained at a pH between 4 and 7. It contains a compressed mixture of a synthetic anionic detergent, hydroquinone, a stabilizer for hydroquinone, water, a buffer that maintains the pH of the bar, and excipients such as waxes, paraffin, dextrin, and starch. Likewise, a skin-bleaching combination consists of hydroquinone, tertiary butyl hydroquinone, and sometimes an additional stabilizer, and can additionally have a buffer to keep the pH between 3.5 and 7.5. Hydroquinone is not oxidized because of the maintenance of low pH and the presence of a stabilizer, and thus the product has an extended shelf life.

7.7.5 Hair Cosmeceuticals

Unlike all other land mammals, humans have direct control over the appearance of their hair. The length, colour, and style of hair can be modified according to one's wish. The physical appearance of people and their self-perception are greatly influenced by hair care, colour, and style. In ancient Egypt, the earliest forms of hair cosmetic procedures included setting hair with mud and hair colouring with henna. Countless ointments and tonics were recommended for beautifying hair as well as for treating scalp diseases in ancient Greece and Rome. Henry de Mondeville was the first to distinguish between medicinal therapies for treating diseases and cosmetic agents for beautification. With the development of cosmetics containing physiologically active ingredients, that is, cosmeceuticals, today delineation of cosmetics from pharmaceuticals has become more complex. By far the most frequent form of cosmetic hair treatment is shampooing. While shampoos were primarily aimed at cleaning hair and scalp, a number of variations in formulations have been adapted currently focusing on hair quality, hair care habit, and specific problems such as treatment of oily hairs, dandruff, and androgenic alopecia related to the superficial condition of the scalp.

Cosmetics utilized for treating hair are used topically to the scalp and hair. Although they cannot be used for therapy, they must be safe for the skin and scalp, hair and mucous membranes, and should not show any toxic effect in normal conditions of their use. A shampoo composition has been patented by Mausner (1979), which cleans the hair and scalp without damaging the fragile biological equilibrium of the scalp and hair. A hair-care cosmetic preparation consisting iodopropynyl butylcarbamate and/or a solution of zinc pyrithione in N-acyl ethylenediamine triacetate has been patented; this includes a suitable carrier and a non-allergenic dry extract of yarrow (*Achillea millefolium* L.), collected by oxidizing a water–alcohol solution extract of yarrow flower tops. The extract has less than 0.5% of polyphenolic derivatives (by weight) and is utilized for treating hair, particularly oily hair, based on yarrow extract. Buck (1997) patented a treatment method for androgenic alopecia in which liquor carbonic detergents are applied topically. A general perception is that genetic hair loss results when an inherited predisposition to circulating androgenic hormones activates.

Conditioning agents, special-care ingredients, and hair-growth stimulants are the general constituents of a hair cosmeceutical product.

Conditioning agents impart softness and gloss, reduce flyaway, and enhance disentangling facility. Many ingredients, including mostly fatty ingredients, quaternized cationic derivatives, hydrolysed proteins, cationic polymers, and silicons (Trueb 2001) might be utilized. Special-care ingredients aim to modify particular problems affecting the superficial scalp.

These shampoos are formulated by selecting specific ingredients and their clinical effectiveness in these conditions. Accordingly, the current antidandruff components are virtually all-effective antifungal agents—zinc pyrithione, octopirox, and ketoconazole. Hair-growth stimulants contained in shampoo are likely to be ineffective due to short contact time and water dilution. A cosmetic agent that acts as a topical hair-growth stimulant is a minoxidil-related compound (2,4-diamino-pyrimidine-3-oxide). It has been proposed to prevent inflammation and perifollicular fibrosis. In the prevention of seasonal alopecia, some degree of efficacy of 2,4-diamino-pyrimidine-3-oxide has been claimed. In the United States, the recent approval of two new products, Propecia and Rogaine Extra Strength (Minoxidil) 5%, have added a new dimension to the treatment options of androgenetic alopecia. These products are indicated to promote scalp hair growth in men.

7.7.6 Other Cosmeceuticals

Subcutaneous fat and oil glands are absent in the skin beneath the eye. Protection and plenty of moisture are required for this delicate skin to replenish and repair, which further aids in reducing the signs of premature ageing. When the skin ages, it becomes drier, rougher, and thinner. Overexposure to environmental pollution adversely affects the skin. Many topical skin-soothing products intervene in this process. Products specifically designed for this area need to be gentle and should contain ingredients that work from the inside out by acting on the cells under the skin's surface, without irritating the eyes. Numerous cosmeceutical eye creams are available for the nourishment of the under-eye skin with natural emollients and useful nutrients. Other active ingredients include butcher's broom, chamomile, antioxidants, vitamins A, C, and E, green tea, tiare flower, *Ginkgo biloba*, cucumber, calendula, and bisabolol. Yeast is a key ingredient in the eye-lifting moisturizing cream that provides medication against puffiness and irritation, and protects against future skin damage. The eye wrinkle cream generally contains wheat germ and corn oil, squalene, and carrot extract, which helps forestall the signs of ageing. Eye firming fluid contains aosain, an algae extract from seaweed, which helps in maintaining skin's elasticity. Table 7.5 provides common cosmeceutical contents.

7.7.7   Regulatory Aspects

Though the claims regarding drugs are subject to the FDA's review and approval process, cosmetics do not require mandatory FDA review. Regarding the status of 'cosmeceuticals', much confusion exists. There is no legal class called cosmeceuticals, but this term has found recognition in designating products at the borderline between cosmetics and pharmaceuticals. Cosmeceuticals do not require FDA review, and the Federal Food, Drug, and Cosmetic Act does not recognize the term itself. Unless the product has been approved by the FDA or an equivalent agency, it is also often difficult for consumers to validate the 'claims' about the actions or efficacies of cosmeceuticals. Some experts are demanding increased regulation of cosmeceuticals, which would necessitate proof of safety. In some countries, there are other classes of products that fall between the categories of cosmetics and drugs. For example, there are 'quasi-drugs' in Japan, 'controlled cosmetics' in Thailand, and 'cosmetic-type drugs' in Hong Kong. There is no harmonization of regulations on cosmeceuticals between the United States, and European, Asian, and other countries.

In the cosmetic industry, the global trend is towards producing 'medicinally' active cosmetics. In the pharmaceutical industry, the inclination is towards developing 'cosmetically' oriented medicinal products as part of the current 'life style' ideology. In the future, increasingly sophisticated formulations for cosmetics and skincare products will be available. Cosmetic companies are conducting research to deliver small-dose ingredients that do not need medical regulations and to include steroids and hormones into lip balms. They are trying to produce cosmeceuticals that could help in enhancing body mass, nail, and hair growth. As more chemicals with true biological activity are invented and tested, government regulatory agencies will face new challenges. To assess accurately the efficacy and safety issues, claim substantiation and premarketing testing should also evolve,

Table 7.5   Common cosmeceutical contents

Ingredient	Purported action	Source
Vitamins	Antioxidant	Vitamins A, C, and E
α-Hydroxy acids	Exfoliates and improves circulation	Fruit acids (glycolic acid, lactic acid, citric acid, tartaric acid, pyruvic acid, maleic acid, etc.)
β-Hydroxy acids	Antibacterial	Salicylic acid
Essential fatty acids	Smoothens, moisturizes, and protects	Linoleic, linolenic, and arachidonic acids
Coenzyme Q10 (ubiquinone)	Cellular antioxidant	Naturally occur in skin
Allatoin	Soothes	Comfrey
Aloe vera	Softens skin	*Aloe vera*
Arnica	Astringent and soothes	*Arnica montana*
Calendula	Soothes, softens, and promotes skin-cell formation	*Calendula officinalis*
β-Bisabolol	Anti-inflammatory, antibacterial, and calms irritated skin	Chamomile flower
Cucumber	Cools, refreshes, and tightens pores	Cucumber
Lupeol	Antioxidant and skin-conditioning agent	*Cratacva nurvula*
Ginkgo	Antioxidant that smoothens skin, rejuvenates, and promotes youthful appearance	*G. biloba*
Ivy	Stimulates circulation and helps other ingredients penetrate skin	*Hedera* spp. (ivy family)
Panthenol	Builds moisture and soothes irritation	Provitamin B5
Witch hazel	Tones	*Hamamelis virginiana*
Green tea extract	Antioxidant	Green teas
Neem oil limonoids	Antimicrobial	*Azadirachta indica*
Pycnogenol	Anti-ageing effect	Grape seed extract
α-Lipoic acids, resveratrol, polydatins	Potent free-radical scavengers and antioxidant	Fruits and vegetables
Furfuryladenine	Improves hydration and texture of skin	Plant-growth hormone
Kinetin	Free-radical scavenger and antioxidant	Plants and yeast
Sodium hyaluronate	Lubricates between skin tissues, and maintains natural moisture	Natural protein
β-Carotene	Minimizes lipid peroxidation and cellular antioxidant	Carrots and tomato

with important implications for total body health. New delivery systems and vehicles combined with established ingredients will change percutaneous absorption, which would require re-evaluation of substances with an assumed good safety profile. Biotechnology will face direct competition from pharmaceuticals and cosmetic industries. The links between internal health, beauty, and anti-ageing will be the most influential angle during the next 5 years. The next big beauty trends include skingestibles, which would enhance beauty from the inside out, borrowing pharmaceutical terms for cosmetic applications, amino peptides to increase skin elasticity, neuro-mediators (that would tell the brain to be happy), and the blurring of the boundaries between cosmetics and surgery. A better understanding of modern ingredients and assessment techniques will be essential for the trend towards therapeutic cosmetics.

7.8 BIOFUELS

To deal with energy crisis and find substitutes for petroleum-derived transportation, biofuels are drawing increasing attention worldwide. Biofuels are expected to help address energy cost, energy security, and global warming concerns related to liquid fossil fuels. The term biofuel here refers to any liquid fuel made from plant material, which can substitute petroleum-derived fuel. Ethanol made from sugarcane or diesel-like fuel made from soya bean oil, which are relatively familiar and dimethyl ether (DME) or Fischer–Tropsch liquids (FTL) made from lignocellulosic biomass can be included in biofuels.

According to a popular classification criterion, liquid biofuels can be categorized under 'first-generation' and 'second-generation' fuels. No strict technical definitions exist for these terms, but the feedstock used creates the main distinction between them. A first-generation fuel uses only a specific portion of the plant biomass that is often edible and is produced above the ground; sugars, grains, or seeds are its general constituents. Processing required is relatively simple. These are already being produced in significant commercial quantities in a number of countries.

Second-generation fuels are mostly made from inedible lignocellulosic biomass, either inedible residues of food crop production (such as corn stalks or rice husks) or inedible whole-plant biomass (such as grasses or trees grown particularly for energy). These fuels have not yet been commercially produced in any country.

In spark-ignition engines, alcohol fuels can be used as substitute for gasoline, while in compression ignition engines, the use of biodiesel, green diesel, and DME is most suitable. The Fischer–Tropsch process can develop a wide range of hydrocarbon fuels, the primary one being a diesel-like fuel for compression ignition engines.

While biofuels are attracting attention from the transport sector, use of biofuels for cooking is a potential application worldwide, particularly in rural areas of developing countries. In all cases, emissions from cooking with solid fuels will be greater than emissions of pollutants from combustion of biofuels. These fuels could play a critical role in health improvement because in developing countries, approximately 3 billion people cook with solid fuels and suffer severe health damages from the resulting indoor air pollution. It must be noted that the requirement of transportation fuel is more than cooking energy needs, so the scale of production of biofuels required is far smaller than the needs. According to an estimate, about 4–5 exajoules per year of clean cooking fuel would be enough to fulfil the basic cooking requirements of 3 billion people. This constitutes about 1% of the global commercial energy use today.

Expanded or new biofuel industries are being developed for the transport sector by many industrialized countries, and a growing interest is seen among many developing countries for similarly 'modernizing' the use of biomass and providing greater access to clean liquid fuels. Several reasons might be involved for their special interest in biofuels:

- Climates in many of these countries are favourable for growing biomass.
- Production of biomass is inherently rural and labour intensive and thus may provide an opportunity for new employment in regions where the majority of populations typically reside.
- Production of biomass energy to restore degraded lands may also be of interest in some areas.

The potential for generating rural income by producing high-value products (such as liquid fuels) is appealing. The possibility for export of fuels to industrialized countries may also be attractive. Moreover, the potential for decreasing emissions of greenhouse gases may offer the possibility for monetizing avoided emissions of carbon, for example, via Clean Development Mechanism credits.

The concerns raised by the expansion of biofuel production include diversion of land away from food production, preservation of biodiversity, or other important purposes.

At normal temperatures and pressures, DME is a colourless gas, with a slight ethereal odour. Much like propane, it liquefies under slight pressure. It is relatively inert, non-carcinogenic, non-corrosive, and almost non-toxic and does not develop peroxides upon prolonged exposure to air. It might act as a suitable substitute or a blending agent with liquefied petroleum gas (LPG, a mixture of propane and butane) due to its physical properties. If DME is blended (15%–25% by volume) with LPG, the mixture can be used with combustion equipment without any changes. DME is also an excellent diesel engine fuel, which does not produce soot during combustion and has high cetane number. As DME must be stored under mild pressure to maintain its liquid state, it is not feasible to blend DME with conventional diesel fuels in the current engines.

However, since combustion of DME is extremely clean in a properly designed compression ignition engine, it is an attractive option for compression ignition vehicles running in urban areas, where vehicular air pollution is severest. Fleet vehicles that are centrally maintained and fuelled (such as buses and delivery trucks) create an excellent potential market for DME because the equipment in vehicle refuelling stations differ from those in conventional refuelling stations that dispense petroleum-derived fuels, and so modified on-board fuelling systems are necessary. Since most of these vehicles run with petroleum fuels in urban areas, the extremely lower exhaust emissions from DME engines compared to diesel engines (particularly of harmful particles) are a source of strong public motivation for adopting DME fleets.

The main use of DME until recently was as an aerosol propellant in hair sprays and other personal-care products. The global production rate of DME is 150,000 tonnes/year, which is now increasing dramatically. In China, 265,000 tonnes/year of DME production capacity (110,000 tonnes of which are from natural gas and the rest from coal) was established from 2003 through 2006. Another 2.6 million tonnes/year of capacity addition (from coal) was expected there by 2009, and there were plans to develop a further 1 million tonnes/year capacity. In Iran, a gas-to-DME facility generating 800,000 tonnes/year came online in 2008. There are also plans for a facility to be built in Australia (with Japanese investment) to generate between 1 million and 2 million tonnes/year of DME from natural gas.

Essentially, all new DME produced will be utilized as an LPG substitute for domestic fuel. Some DME will also be used in buses in China, with its implementation in Shanghai initially and subsequently elsewhere. In China, commercial development of DME buses is under way. In Sweden, Volvo is developing heavy-duty vehicles (trucks and buses) fuelled with DME, which expects to have commercial vehicles soon. There are ongoing major efforts in Japan to commercialize heavy-duty DME road vehicles.

At present, the United States is drawing its attention towards the production of alcohol fuel via syngas processing. One such fuel is ethanol (or butanol); another fuel is a mixture of alcohols that includes a significant fraction of ethanol plus smaller fractions of several higher alcohols. Butanol and 'mixed-alcohol' fuels have the potential to be utilized in much the same way as ethanol is used today for blending with gasoline. The characterization is done on the basis of higher volumetric energy densities and lower vapour pressures than ethanol, making them more attractive as a fuel or blending agent. By using the process of catalytic synthesis, syngas can be converted into a mixture of alcohols. The process steps resemble those for making FT liquids. A mixture of alcohol molecules can be produced by passing a clean syngas over a catalyst. Various catalysts that were utilized for mixed alcohol production from syngas were patented in the late 1970s and early 1980s, but development efforts were stopped when oil prices fell in the mid-1980s. High oil prices have reignited interest, and recently a substantial grant has been awarded in support of one commercial-scale demonstration project by the US Department of Energy. Competing technologies are being developed by several start-up companies. Available published information apart from patents and patent applications concerning these private-sector activities is very less.

Pure ethanol or butanol can also be produced from syngas by using microorganisms that ferment the gas. If it can be made commercially viable, this combined thermo/biochemical route to a pure alcohol would enable the conversion of lignin in biomass feedstock (as well as hemicellulose and cellulose) to fuel, unlike the case for purely biochemical 'cellulosic ethanol' discussed earlier. BRI Energy, Inc., a private company, is actively seeking to commercialize the technology for fermenting syngas. It has announced its intention to build two commercial facilities near Oak Ridge, Tennessee, United States. While one facility would convert coal-derived syngas to ethanol, the other would convert municipal solid waste via gasification to ethanol. The US Department of Energy awarded the BRI a grant for a commercial-scale demonstration project. Little information is publicly available for independently evaluating BRI's technology.

7.9 SINGLE-CELL PROTEIN

Microorganisms that are used as source of protein in the purified and dried form are called single-cell proteins (SCPs). Historical evidences suggest that *Spirulina*, a filamentous alga, was extracted from Lake Chad in Africa and consumed as food in the early days. The Germans put *Candida utilis* in soups and sausages during the First World War. It was also extensively used during that Second World War. It was manufactured at the industrial scale in 1967.

The advantages of SCPs are as follows:

- These have high protein and low fat content.
- These are a good source of vitamins, particularly B-complex (e.g., mushrooms and yeasts).
- These can be produced throughout the year.
- Production of these proteins reduces environmental pollution as waste materials are utilized as substrate. This helps in recycling of materials.
- Large quantities of SCPs can be obtained in relatively small area and time as SCP organisms grow faster.
- These proteins are genetic engineering friendly as these are composed of amino acids, which can easily be selected with the help of genetic engineering.
- During the generation of SCP biomass, some organisms produce beneficial by-products such as organic acids and fats.

7.9.1 Sources of Single-cell Proteins

1. *Algae*: *Scenedesmus acutus*, *Spirulina maxima*, and microorganism comprised in genus *Chlorella* are grown for SCP. These have

about 60% protein with good amino acid composition, but less in sulphur-containing amino acids. *Chlorella* and *Spirulina* are used for commercial-scale production in Taiwan, Thailand, Japan, Israel, Mexico, and the United States. It is spray dried and sold as pills and powders. *Spirulina* grown on sewage water is free of pathogenic microorganisms.

The disadvantages are as follows:

- These are unsuitable for human consumption because these are rich in chlorophyll (except *Spirulina*)
- These have low density, that is, 1–2 g dry weight/L of substrate.
- During growth, there is a high risk of contamination.

2. **Yeasts and fungi**: For the production of SCPs, filamentous fungi such as *Chaetomium celluloliticum* growing on cellulose waste, *Fusarium graminearum* growing on starch, and *Paecilomyces varioti* growing on sulphur liquor are utilized. These have about 50%–55% protein. Yeasts such as *C. utilis* (torula yeast), *C. lipolytica* growing on ethanol, and *Saccharomyces cerviceae* growing on molasses are used for SCP production. Torula yeast is obtained as a food through fermentation with molasses as substrate. The protein–carbohydrate ratio in these products is higher than that of forages. It is rich in lysine but poor in methionine and cysteine. *Saccharomyces* has high protein with a good balance of amino acids and is rich in B- complex vitamins. It is more suitable as poultry feed. Several species of mushrooms are used as protein-rich food.

   The disadvantages are as follows:

   - These have rich composition of nucleic acids.
   - Growth rate of filamentous fungi is slower than that of yeasts and bacteria.
   - Risk of contamination is high.
   - Mycotoxins are produced by some strains and are thus required to be screened.

3. **Bacteria**: These have more than 80% protein. They are poor in sulphur-containing amino acids. Hydrocarbons and methanol act as substratum for *Brevibacterium* and *Methylophilus methylitropous*, respectively. These have has high nucleic acid content.

   The disadvantages are as follows:

- RNA content is high in these products.
- Risk of contamination during the production process is very high.
- Complicated procedure lies in the recovery of cells.
- Production of endotoxin should be tested carefully.

REFERENCES

Buck, C. J. 1997. Method for treatment for androenic alopecia. *US patent 5609858*

Dooley, T. P. 1997. Is there room for a moderate level of regulatory oversight? In *Drug Discovery Approaches for Developing Cosmeceauticals: advanced skin-care and cosmetic products,* Hori W (ed). Southbourough, MA: IBC

Grace, R. 2002. Cosmeceuticals: functional food for the skin. *Natural Food Merchandizer* XXIII: 92–9

Mausner, J. J. 1979. Protein shampoo. *US patent 4140759*

Peters, S. 2002. *Mycorrhiza 101.* Salinas, CA: Reforestation Technologies International

Trueb, R. M. 2001. The value of hair cosmetics and pharmaceauticals. *Dermatology* 202: 275–82

Important Issues in Plant Biotechnology

8.1 INTRODUCTION

As part of a wider scientific movement to modify organisms for economic, medical, and other human needs, development of the science and methods to generate transgenic crops began around 1983. Significant and complex ethical issues result following the modification of life. The development of transgenic crops has led to controversy among government organizations, businesses, researchers, and non-profit organizations. Civil society groups are particularly vocal about these issues. The quantifiable facts related to genetically modified (GM) foods appear to create fewer disputes than the increasing number of implications. These often take place in the form of ethical arguments, which supporters of transgenic crops dismiss as a defence of cultural artefacts. However, the new possibilities brought about by transgenic crops reveal an absence of research into their implications.

Biotechnology has no uniform definition. According to the broadest definition, the application of biological sciences to develop products and traditional plant and animal breeding techniques being under use since the dawn of civilization can be included under biotechnology. Biotechnology typically refers to new techniques used to develop products by modifying the genetic make-up of organisms and producing unique individuals or characteristics that conventional breeding techniques cannot easily produce. These products are usually termed bioengineered, transgenic, or GM because they possess the foreign genetic material. Agriculture is among the first industries radically impacted by this technology both at a production level and at a legal level. The products of transgenic engineering are mostly called GM organisms (GMOs). All these terms point to methods by which scientists splice genes of crop plants into DNA to transfer desired genetic traits. This technique is termed recombinant DNA (rDNA) technology. Genes are segments of DNA containing information that partly controls the function of an organism. Geneticists alter DNA by transferring genes from one species—an animal, a plant, a bacterium, or a virus—into another species, such as an agricultural crop plant. They also use an intermediate organism or virus to 'infect' the host DNA with the desired genetic material. A widely used technology to insert DNA from another species into plant cells is microparticle bombardment. The desired genetic material is precipitated onto micron-sized metal particles and put into any one of a number of devices meant to accelerate these 'microcarriers' to velocities needed to penetrate the plant cell wall. Transgenes are delivered into the cell's genome in this manner. Electroporation can also be employed to insert new DNA into a host cell. In this technique, a jolt of electricity is applied to cells to create openings in the plasma membrane surrounding a cell. To confirm the degree of effectiveness in delivering the foreign DNA, a marker gene is included in the package.

Gene stacking is gaining popularity since a whole array of traits can be added at once into the host. The entire process can be explained by its utilization in the development of transgenic rice, with the help of electroporation. When genetic engineering of plants emerged around 1983, it became likely that transgenic manipulation would benefit agriculture and even revolutionize it. The transfer of genetic characters across species held promises to resolve issues in managing agricultural crops, provide new opportunities to improve human and animal health, and offer a new revenue earning avenue for farmers through contract production of industrial crops. Possible environmental advantages included decreased use of toxic pesticides, enhanced weed control leading to less tillage and soil erosion, and conservation of water. Moreover, the new technology promised increased production.

8.2   COMMERCIAL STATUS AND PUBLIC ACCEPTANCE

On the one hand, GM crops have been hailed as the solution to world hunger and, on the other hand, they have been criticized as environmentally hazardous, 'playing god', and 'illegitimately crossing species boundaries' (Comstock 2008). There are valid and flawed arguments on both sides. However, when a normative argument is put forward to support or condemn something, it should take into account all related parties or else it becomes quite arbitrary. Hence, the arguments against GM crops should not be directly refuted; rather, reasons should be given supporting their use. Therefore, one would assert that the history of agriculture, social implications, and recent genetic advancements all highlight the ethical permissibility of GM technologies as aids to solve the problem of world hunger.

Genetically modified crops are engineered on a molecular level. Geneticists identify a particular gene from a foreign gene pool and insert it into an alternate genetic code. Foreign genes are taken from sources different from the organism's natural parents. The traditional methods of cross-breeding differ in a few ways with this type of gene selection (Comstock 2008). Conventionally, changes in plant genetics result from the natural reproductive process of two plants sharing their DNA. The exchanged DNA remains within the same species as a consequence of this reproductive process. Traditional techniques of breeding, such as the methods used by Gregor Mendel, did not choose genes singularly but instead transferred genes in groups. Traditional methods of gene selection do not differ much from the more specific selections through GM technologies. The history of agricultural and other sciences is inseparably linked to risk. Hence, risks of GM foods should be compared to their advantages for determining the ethical aspects of their use. Nearly every development involved some amount of risk since the beginning of civilization. Trial and error methods, and now the more sophisticated scientific methods, have posed danger in some form. Even though aircraft, medical research, exploration, and pesticides had their related risks, the anticipated benefits led scientists to follow their interests. Pesticides are linked to a variety of risks; however, we find pesticides useful against many pests that cause considerable damage (Ebenreck 1991). Similarly, one does not consider the risks of using an automobile even though a number of fatal accidents occur every year. A key element that decides whether risks outweigh benefits is 'need'. When the need persists, a community is willing to take an increasing amount of risk. The need of the Third World countries is more persistent than that of the developed countries.

Opponents of GM technologies have a common fear that the gene pool would be harmed by reducing genetic diversity (Resnick 2008). Basically, the fear is that if there remains only one type of crop because all the others have been replaced by a 'superior' GM type, one deadly disease or virus could wipe out all the food everywhere. This fear is unrealistic because the threats to a crop extend beyond insect predators and plant diseases. A crop's success also depends on the availability of water, climatic conditions, and the soil health, together with the economic forces of supply and demand. Inserting a gene that protects against cotton-devouring ladybugs does not protect against disease, drought, lack of nitrogen in soil, or a diminishing demand for genes. If a large quantity of crops gets destroyed in the country, it would worsen the problem of food supply. The history of agriculture points to the necessity to utilize GM technologies because need has always governed what agricultural pursuits were attempted and, in reality, the risk is still the same. Past cases of genetic selection can be weighed against prospects for the more selective biotechnology. Thus by contrasting various cases against one another, claims and interpretations can be more easily justified.

An example of genetic selection from the past is Norman E. Borlaug's experiments with wheat. Although he did not genetically engineer the wheat on the molecular level, he used similar methods. Borlaug collected hundreds of varieties of plants and employed 'shuttle breeding' to select for genetically preferable genes. In shuttle breeding, varieties of plants are grown in mass quantities in economically favourable locations. If the plants showed promising results, he shifted them to better climatic conditions. In fact, Borlaug crudely chose the preferred genes at a much faster rate than evolution would ever have allowed (Pollock 2008). This method suffers from some of the same risks as genetic engineering of plants at the molecular level. For example, both methods might lower genetic diversity and expose a crop to a supervirus. Sharing this risk is fundamental because it is one of the most practical arguments against GM crops. On the contrary, both have some similar benefits. Both techniques tend to increase production by protecting against diseases, viruses, and pests.

Concerns arose during 1939–1941 when a fungus called stem rust reduced Mexico's wheat production to half. The fungus was carried as an airborne spore and spread quickly, killing two-thirds of the durum wheat in the United States (Pollock 2008). It was feared that the stem rust fungus would destroy the entire wheat crop in North America. In 1970, Borlaug sent his wheat strains to India, where the wheat was experiencing similar problems with the fungus. Borlaug's wheat strains considerably improved wheat production in India, which reportedly fed 6.2 billion people. He was credited with feeding a billion people around the world and was awarded the Nobel Peace Prize. Feeding a starving nation made Borlaug win some of the greatest honours in recent history. This is an example of the benefit as more than 6 billion people were fed. On the flip side, the apprehension that a new strain of fungus would emerge became a reality when scientists reported a new form of stem rust called Ug99 from Uganda, Kenya, and Ethiopia in 2006. The problem was nearly as severe as the destruction of crops in the 1940s and 1950s. The suggested solution to the reappearance of stem rust is again to find resistant varieties (Pollock 2008). One of the main objectives of GM technologies is developing new resistances in plants, and it appears a feasible solution to use GM crops to that end. Similar results with GM crops can be expected if this success story has the same risks and benefits. There might be limitations to GM technology, but this case study highlights that agricultural risks can be overcome. Borlaug's research sheds light on the precautionary principle to argue against GM crops.

An interestingly different approach to the precautionary principle was taken by Engelhardt and Jotterand (2008), who commented that the precautionary principle dealt with the unfortunate consequences of not going for scientific innovations. They emphasized that the dangers of development must be dealt with technological innovation. The precautionary principle simply states that just as there is risk in action, there is risk in inaction. Borlaug's example of wheat shows that science can cope with the consequences of the GM technology, but science can hardly deal with the results of inaction. One cannot ignore the implications of GM foods for the society and potentialities they create across the globe. The problem of world hunger becomes increasingly noteworthy because it highlights the legitimacy of the moral basis of laws. Any country should uphold moral values through the legal system, laws, and regulations. Justice is one of the foundations of our civilization, as it is viewed as a defence of equality.

It is surprising to note that more than half of the child deaths each year results from malnutrition. If a child can be saved from starvation when there is a means, then all cases of child starvation should be prevented; otherwise, all children are not considered equal. About 60% of children below 5 years of age in Zambia had stunted growth due to the lack of calorie intake. Another reason might be that there was no means to feed these children. Now the question is whether GM crops can be a means to prevent the starvation of children. It is quite evident in this instance. The Zambian Government bought 'higher yielding' varieties of seeds from farmers, which provided families with sufficient food along with a little surplus to sell.[1] Since the ultimate benefit of GM foods is greater crop yields, they could provide a means to prevent the starvation of children. If large portions of our legal system are based on moral principles, then those moral portions must be applicable to all equal parties involved. Doing the opposite will debase the justifications for those laws. There is a social obligation towards innocent individuals, such as hungry people in the developing world, as well as towards the foundation of our own legal system.

That genetic technologies can be utilized to eliminate world hunger and agricultural shortages is justified by the history of risk and agricultural technologies similar to GM crops. First, need played an influential role in deciding the worth pursuing risks because benefit increases with need and desperation leads to a willingness to find a solution. Second, Borlaug and his experiments with wheat showed that problems caused by biotechnology can be solved by science. Moreover, it emphasizes that it is difficult to know whether naturally occurring problems can be solved without the help of biotechnology. Third, not allowing GM crops in other countries is a violation of the moral basis of laws. All these arguments are, perhaps, insufficient to validate risking the effects of GM technologies; however, some risks are worth taking considering the moral consequences alongside the history of agriculture.

8.3 BIOSAFETY GUIDELINES FOR RESEARCH INVOLVING THE BENEFITS AND RISKS OF GENETICALLY MODIFIED ORGANISMS

Although human health is not an area of expertise for the International Union for Conservation of Nature (IUCN), it is a major concern for IUCN members and opponents of GMOs. This is reflected in the IUCN GMO resolutions, as well as survey results from Europe and Japan in particular, where consumers fear potential health and environmental effects.

---

[1] Details available at <www.fao.org>

Some of the genes used in GMOs should not have been used in food crops since GM foods might pose a potential risk for human health. This risk can be analysed on the basis of the concept of 'substantial equivalence' supported by the Codex Alimentarius Commission established by the Food and Agriculture Organization and the World Health Organization (WHO). According to the concept, the observed difference between a natural product and the product produced through GM technologies is evaluated for risk. A GM crop is compared to a conventional counterpart on the basis of composition, and any extra components are comprehensively tested for their potential to cause allergies and toxicity. However, critics deem this concept insufficient since foods are complex and unexpected variations arising from genetic modification might be overlooked. According to the Food Standards Agency (FSA) of the United Kingdom, food from animals fed on these crops is as safe as food from animals fed on non-GM crops. The FAO has also come to the conclusion that risks to human and animal health from the use of GM crops and enzymes derived from GM microorganisms as animal feed are negligible (FAO 2004). An evidence-based study led the WHO to conclude that 'GM foods currently available on the international market have undergone risk assessments and are not likely, nor have shown, to present risks for human health any more than their conventional counterparts'. No known negative health effects have been reported to date from the consumption of GM foods. Objective evaluations have suggested that 70%–75% of processed food in the United States and Canada contains GM ingredients (Phillips and Corkindale 2003). About 300 million people have been consuming GM crops for more than a decade, and no health problem associated with eating GM crops has yet been reported. However, this study does not consider the fact that consumption habits differ across the world. One cannot assume that GM maize in one country will have the same apparently insignificant effect on consumers in another country where that crop is consumed in greater proportion. Scientists are also apprehensive about the long-term safety of consuming GM food. The WHO emphasizes on the need to continuously assess the safety of GM foods before they are marketed to thwart risks to human health. The main risks to human health are discussed in the following sections.

8.3.1 Allergenicity

One of the suspected adverse health impacts of GM crops is allergenicity, which came to light by research carried out by Hi-Bred in the mid-1990s. It found that soya bean plants engineered with a gene from Brazil nuts caused an allergic reaction in some people. Similar allergenic results of GM crops were not reported until 2005 when researchers from the Commonwealth Scientific and Industrial Research Organization (CSIRO) discovered that GM pest-resistant peas caused allergic lung damage in mice and abandoned the decade-long project as a result. The researchers had transferred into the pea a gene for the protein from the common bean that could kill pea weevil pests. The protein is not allergic when extracted from the bean, but it becomes structurally different from that in the original bean when expressed in the pea. This delicate change might lead to the unforeseen immune effects observed in mice, thus illustrating the unexpected impacts of gene transfer and the significance of using animals for testing the allergenic potential of GM foods. However, genetic modification has the potential of eliminating toxic elements from some food crops, which can prevent needless deaths. For example, genetic modification removed the allergenic traits of peanut and the cyanide characteristics of cassava. It is necessary that all GM foods should be clearly labelled for consumers so that they can avoid particular GM food types that they are found to be allergic.

Opponents of GMOs cite the risk of GM techniques resulting in changes that are toxic to human beings and animals. The most recent GM crop suspected to have toxicity is the GM maize line MON 863 (YieldGard rootworm corn), approved in the United States in 2003, which especially attacks the corn rootworm. MON 863 contains less Bt toxin than other Bt maize varieties, and it produces the toxin mostly in the roots, which is the site at which the western corn rootworm attacks (Alexander and van Mellor 2005). In the Philippines, concerns were raised in 2004 regarding the toxic effect of the pollen of Bt maize when about 100 people residing near GM maize fields reported symptoms of dizziness, headache, vomiting, extreme stomach pain, and allergies only when the pollen was airborne (Estabillo 2004). Although studies are still under way, a researcher reported at a Malaysian conference on GM food safety that an antibody response to the Bt toxin was developed in the blood, which confirmed that GM promoters could function in human cells. Hence, it can be suggested that if the promoters were to transfer out of GM food and integrate into human DNA, they could activate random genes inside humans and overproduce a toxin, allergen, or carcinogen (Smith 2006). No study has till date adequately proved a commercially produced GM crop to be toxic.

To enhance the success of GM crops, scientists have employed a method that involves antibiotic resistance genes in addition to the desired gene to identify plants that have successfully absorbed the introduced gene. Kanamycin is an antibiotic that is frequently used as a marker for plant modification. It is also used to treat many human infections. Since the genes have traditionally come from bacteria, human pathogens can increase their antibiotic resistance. These concerns prompted the FSA of the United Kingdom to carry out a number of research projects to examine the transfer and survival of DNA in the bacteria of the human gut. They opined that it is very unlikely that genes from GM food can end up in bacteria in the gut of people who consume them. In a recent study in human volunteers, it was found that no GM material survived the passage through the entire human digestive tract. Even though some DNA survived in the laboratory that simulated human or animal gastrointestinal tracts, researchers concluded that the possibility of bacteria in the human or animal gut taking up functioning DNA is very low but not impossible, given earlier cited evidence of gene transfer to bacteria in honeybees. There is no concrete proof that these drug-resistant strains will be more dangerous for humans. However, health authorities are still concerned about them because they are more difficult to treat in people. For example, the British Medical Association does not favour the use of antibiotic resistance markers in food. Considering the seriousness of the risk, scientists are adopting methods to remove the marker genes before a crop plant is developed for commercial use (Scutt, Zubko, and Meyer 2002). Of late, researchers have derived an alternative marker from tobacco instead of bacteria (Mentewab and Stewart 2005).

8.3.2 Potential Benefits

8.3.2.1 *Improved nutritional value*

Genetically modified crops and animals can directly affect human health by improving nutritional value, which WHO asserts 'can contribute directly to enhancing human health and development'. Since the 1940s, there has been the concept of adding specific nutrients to processed foods to improve their health benefits. Vitamins, iron, and calcium have been added to flour since long and, more recently, to dairy products and eggs. A dynamic and emerging segment of the food industry is the so-called functional food or nutraceutical market. Golden rice, which contains extra vitamin A, is a frequently cited example. Another example is a rice variety high in iron developed by the International Rice Research Institute. This 'biofortification' is meant to enhance the nutritional status of women who often suffer from anaemia. When non-anaemic religious sisters in the Philippines were involved in a nine-month feeding trial, it was found that women who ate biofortified rice had 20% higher iron content in blood than the women who ate traditional rice (Hass and Défago 2005).

Use of some Bt crops has also some indirect health benefits because it reduces the use of highly toxic chemical pesticides. Farm workers particularly benefit from this, especially in developing countries where manual labour is widespread and human–crop interactions are higher. For instance, pesticides used in farms harm more than 50,000 Chinese farmers annually, of which 400–500 die. It was found in trials before the commercial production that farmers growing GM rice used 80% less pesticides, which led to a decrease in pesticide-related health problems.

8.3.3 Ethical Considerations

The debate over GM food is more about conflicting views about democracy, capitalism, and global trade than about human health and environment (Charles 2001). Therefore, the WHO has called for widening the debate about GM foods to include ethical, social, and cultural considerations. Anti-GMO campaigners give more weightage to these issues than to ecological issues.

8.3.3.1 *Hunger and development*

Proponents of modern biotechnology cite its ability to increase food security as their principal argument for GM crops. Approximately one-third of adults in Sub-Saharan Africa are undernourished, and production of food has been decreasing in 31 of 53 African countries. (Studies revealed that the grain harvest in the world had fallen short of consumption 4 years in a row, which reduced the world grain stocks to the lowest level in 30 years. There has been a steady decline in the global sea fish catches since 1990. The pressure on food production systems has inevitably increased with the UN projection of a 40%–80% rise in the population during the coming four decades. This has, in turn, put increased pressure on the remaining wildlife habitats. There is a strong consensus that the world will be facing a food crisis by the middle of the next century.

The Millennium Development Goals aim at the commitment to tackle world hunger. It states that the objective of the signatory countries will be to 'reduce by half the proportion of people who suffer from hunger by 2015'. However, the current rate of progress indicates that this goal can be achieved only by 2150. A review in 2006 found that the nature of agricultural biotech partnerships in sub-Saharan Africa is often fragmented and supply driven and not end-user oriented. Hence, these have not been quite effective in reducing hunger by half (Ayele, Chataway, and Wield 2006).

Trewavas (1999) echoes the same view: 'The future will demand agriculture to be both flexible and diverse in technology, but efficient in land use. Farmers will have to be highly skilled at using technologies that must sustain farming for thousands of years. Increasingly, farm resources will need to be recycled; green manure and crop rotation will underpin soil fertility. Integrated pest management systems and zero tillage will be essential to minimise losses due to pests and weeds, and to limit soil erosion. Water will become an increasingly expensive commodity, and a premium will emerge on crops that use water efficiently without loss of yield. In all this future agriculture, genetic manipulation has a unique and intimate role'.

In 2005, FAO Director-General Jacques Diouf stated that GMOs are 'not the priority for reducing the number of hungry people by half by 2015. People in developing countries suffer from chronic hunger and malnutrition because they lack water, other inputs and credit to produce food, employment and income to access food. Lack of political will and financial resources are today's main obstacles to resolving the world's hunger problem'.

In addition, farmers who can take full advantage of the new technologies derive most of the benefits of GM crops (Conway 2003). There has been very little investment in developing new technologies for benefitting the rural poor, particularly in Africa. The Third World Network Africa conducted a study on the prospective impact of GM crops on the famine and poverty in Africa and found that GM technologies were ineffective and expensive and hence not appropriate for Africa. Also, the study found that GM technologies did not address the real causes of poverty and hunger (Degrassi, Alexandrova, and Ripandelli 2003). For example, the introduction of GM sweet potato in Kenya did not address weevils, which is a major problem of farmers. Yet supporters of GM technologies opine that biotechnology can enhance traditional breeding programmes and find solutions when other methods are not successful. With the help of genetic engineering, farmers can improve production on marginal lands, decrease the application of toxic pesticides, and enhance the nutritional content of food. According to the FAO, on suitably integrating GM technology with other technologies for the production of food, agricultural products, and services, biotechnology can considerably assist in fulfilling the needs of an expanding and increasingly urbanized population in the next millennium (FAO 2000). GM crops do not provide a magic remedy although they offer significant productivity gains and nutritional benefits. Farmers will benefit from this technology only when the new technology is applied in a way that enhances benefits to local communities, adjusts with local cultures, and acknowledges the need for complexity and diversity. The reports further recommended that 'there is an ethical obligation to explore these potential benefits responsibly, in order to improve food security, economically valuable agriculture and the protection of nature in developing countries'. The Consultative Group on International Agriculture Research (CGIAR) system promotes open access to research on GM crops for developing countries; however, their GM programmes are not large and inadequately funded because they give greater emphasis on traditional means of enhancing the productivity of poor farmers.

The opportunities that biotechnology presents for Africa to transform its economy were examined by a high-level African panel on modern biotechnology. It advised African countries to work together towards capacity building for applying biotechnology to enhance agricultural productivity and public health, increase industrial development and economic competitiveness, and consider the importance of encouraging the conservation and sustainable utilization of Africa's biodiversity

(High-Level African Panel on Modern Biotechnology of the African Union 2006). Some experts suggest the private sector to integrate commercialization and philanthropy to support poor and small-scale farmers to reap the potential benefits of GMOs, although a more systematic approach is required towards public–private partnerships. One of the basic principles is that farmers should have the right to choose for themselves and this right should be supported.

8.4 SOCIO-ECONOMIC IMPACTS AND ECOLOGICAL CONSIDERATION OF GENETICALLY MODIFIED ORGANISMS

For promoting GM crops, biotech companies are emphasizing that farmers can make huge savings by reducing the use of herbicides and pesticides. Brookes and Barfoot (2005) reported that the net financial benefits of GM crops at the farm level accrued to US$27 billion. The amount of pesticides utilized was decreased by 172 million kg, and the environmental footprint associated with the pesticide use reduced by 14%. On the contrary, Benbrook, McCullum, Knowles, *et al.* (2003) found that the use of herbicides increased on farms that were devoted to herbicide-resistant GM crops. However, many contradictory claims at the local level imply that the savings from the reduced pesticide use do not always completely compensate for higher prices of GM seeds, which can be more than thrice the price of traditional seeds (Qaim, Subramanian, Naik, *et al.* 2006). For instance, initial research in China suggested that the farmers who grew Bt cotton between 2001 and 2003 earned 36% more than conventional farmers by cutting their pesticide use by 70%. However, the profits did not last for long because there was an increase in the number of Bt-resistant pests. Also, in the United States, an analysis of the first decade of commercially released GM crops showed that Bt cotton and Bt corn were profitable when insect pressures were high, but when the pest pressure decreased, there were negative returns. The US finding also revealed that the majority of profits (40%) from herbicide-resistant soya beans were earned by seed firms, 28% by biotech firms, 20% by farmers, and 5% by consumers. A major challenge is to enhance benefits for farmers and consumers (Fernandez-Cornejo and Caswell 2006). Moreover, corporate control of agriculture is increasingly threatening poor farmers. The capacity of small farmers to produce enough food to survive is being undermined by subsidies, trade rules, and patents. In a review of the economic impact of prominent GM crops worldwide, conducted on behalf of the European Commission, highlighted that there was a variation in the net economic benefits for farmers across regions, but noted that both small and larger farmers had benefited. Detailed analyses 'show that increases in gross margin are comparatively larger for lower income farmers than for larger and higher income farmers'. They found that the adoption of Bt cotton led to a significant reduction in the use of insecticides in all cases studied, which provided a significant economic benefit. On the contrary, similar benefits were not observed in maize because the pests that Bt maize resists were generally not controlled by insecticide applications. As a result of adoption of herbicide tolerance, several herbicides have been displaced by one single product, which might be less toxic than the herbicides it replaces. However, the application of this herbicide, usually glyphosate, has increased; the prospective increase in cost has been related to decreased fuel consumption per hectare and the implementation of reduced soil tillage practices. Hence, it cannot be presumed that GM crops can automatically increase profits and reduce poverty. Besides the technology, differences in pesticide levels, irrigation intensities, and other farm parameters result in variations in productivity and, therefore, profits. Moreover, the price of GM products is influenced by global markets and consumer demand.

8.5 PRIVATE SECTOR, TRADE REGULATIONS, AND INTELLECTUAL PROPERTY RIGHTS

The major concern of civil society is that farmers will become vulnerable to the technological strategies of the powerful agribusiness companies that control the majority of research and GM markets. The opponents argue that the Trade Related Intellectual Property Rights (TRIPS) agreement under the World Trade Organization (WTO) cannot adequately protect traditional knowledge and biodiversity. They claim that the agreement also neglects basic human rights by allowing large biotechnology firms to buy and patent seeds, medicines, and traditional knowledge. This is the principal concern of many people who oppose GMOs, which is supported by the pre-existing discussions on 'farmer's rights' under the 1983 FAO International Undertaking on Plant Genetic Resources and International Union for the Protection of New Varieties of Plants (Wendt and Izquierdo 2001). The modern seed sector using genetic resources and traditional knowledge to develop new products should acknowledge and respect the rights of the traditional users who contributed to their conservation and improvement. The producers should also respect the rights of plant breeders. It is necessary to find a balance between respecting and protecting rights of traditional users and modern innovators. Further challenge to proper handling of GMO issues is posed by trade regulations, as they are not designed to ensure environmental safety and socio-economic benefits but serve the interests of influential corporations who enforce them. For instance, when a farmer in Canada grew Roundup Ready Canola, Monsanto successfully sued him because the trademarks and copyrights of the hybrid were owned by Monsanto. The court did not agree with the farmer's claim that pollen had blown on to his field from an adjacent farm and that he did not plant them himself (GMWatch.org 2003). This is a potential threat for small farmers whose livelihoods are compromised by circumstances not in their control if adequate biosafety regulations are not in place. Trade regulations should be designed incorporating the precautionary principles and peoples' rights to select what technology they want to use or not use at all.

8.6 GENETIC USE RESTRICTION TECHNOLOGY

A contentious issue that dominates debates on GM seeds is the so-called 'terminator technology'. This technology, officially known as the genetic use restriction technology (GURT), is employed to produce sterile seeds, thereby avoiding the escape of pollens into non-target fields and restricting farmers from harvesting seeds developed by biotechnology companies at high costs. Terminator seeds can become a weapon of mass destruction and assault our food sovereignty. Livelihoods and food security, as well as the health of the environment and people, are seriously threatened by GM technology. The argument of proponents is that the application of this technology could provide a mechanism for environmental containment and ensure that GM plants cannot cross-pollinate with naturally occurring varieties. On the contrary, opponents cite the impact on the conventional livelihoods of millions of farmers who depend on saved seeds but, in such cases, cannot collect and reuse seeds from GM plants. Moreover, the majority of these farmers cannot afford to buy new seeds every season. The opponents are also concerned with unintentional contamination of seeds and flow of genes from GM to traditional crops, along with the prospective ill effects on diversity. Considering these risks, besides the fact that GURTs mostly benefit the GM industry and provide few obvious benefits to needy farmers, the CGIAR has declined to conduct research on this technology. Huge opposition from people in the late 1990s made the CBD to accept a global de facto moratorium on the application of GURTs in 2000. The CBD's Subsidiary Body on Scientific, Technical and Technological Advice (SBSTTA) clearly stated: 'In the current absence of reliable data on GURTs, without which there is inadequate basis on which to assess their potential risks, and in accordance with the precautionary approach, products incorporating such technologies should not be approved by parties for field testing'. In March 2006, the eighth meeting of the Conference of the Parties (COP) to the CBD (Convention on Biological Diversity) in Curitiba, Brazil, reviewed and upheld its moratorium on GURTs. The Global Environment Facility (GEF), which is the financing body for the CBD, has provided considerable funding to implement the Biosafety Protocol. In the beginning, it offered support to 124 countries, which later increased to 140, to develop national biosafety frameworks (NBFs). An NBF combines administrative, legal, and technical policies that are developed to ensure an appropriate level of security to the safe handling, transfer, and use of living modified organisms generated from modern biotechnology that might have negative impacts on the preservation and sustainable use of biological diversity. It is prepared after considering risks to human health and especially focusing on transboundary movements, in accordance with the Cartagena Biosafety Protocol. Even though there is a difference in format from country to country (every country has its own way of promoting this objective), the common objective is to enhance the country's capacity to execute the Protocol.

8.7 BIOSAFETY

Even though the issues addressed so far are all significant, the primary objective of the Cartagena Protocol is not to discuss whether GM crops should or should not be consumed, imported, or cultivated since these activities are already taking place. Instead, it focuses on how countries should oversee these activities to safeguard both consumers and the environment. It also acts as a means to enlighten consumers and empower countries to reject the technology.

8.8   WEAKNESSES IN TESTING AND REGULATIONS

The private sector carries out most of the biotechnology research. Scientific assessment of the effects of GMOs by this sector is not possible because most of the existing scientific knowledge about GMOs is possessed by corporate and research organizations, which are often regarded as having financial motives in promoting GMOs to have positive contributions to human well-being. These apprehensions include the assertion that a high rate of failures is observed in GM projects, which are not clearly revealed or explained. Since the corporate protects trade secrets and the investments devoted to research and development, scientific examination of the debate is critically restrained because of the lack of access to this closely guarded information. The negative findings regarding the allergenic characteristics of GM crops among numerous positive examples highlight the necessity to examine new GM crops on an individual basis. Some analysts noted that because this research was conducted by public institutes, it was published in peer-reviewed journals and that private organizations do not usually make their negative findings public. However, there is no proof to support this claim, although there are indications of the scarcity of industry-derived publications in peer-reviewed journals. There are many reports highlighting that seed companies often make errors in field trials of GM crops by ignoring proper procedures such as not planting trees to act as a windbreak, failing to plant a buffer of hybrid maize to avoid spread of pollen, or planting maize very near to other crops.

In the United States, the Animal and Plant Health Inspection Service (APHIS) has supervised over 10,000 GMO field tests since 1987. However, the country was not aware of the locations of some of the field tests, which multiplied the risk that GMOs would 'inadvertently persist in the environment before they are deemed safe to grow without regulation'. The U.S. Department of Agriculture Office of the Inspector General did not find any proof of negative impacts on environment as a result of this ignorance, but it highlighted the significant loopholes in the procedures of testing and regulation. The threats of genetic contamination of regions intended for cultivating GM-free crops could result in devastating impacts if the GM crop travels from one country or region to another by accident. The recent and extensively reported example of contamination of long-grain rice with an illegal GM rice 'LLRICE601', field tested 5 years ago, has pointed out why it is necessary to formulate biosafety regulations and checks and expand them to all countries. Investments in biotechnology should concur with biosafety guidelines that protect the environment and human health, and this has been recognized by the Economic Community of West African States (ECOWAS). This need is reflected in the implementation of NBFs.

8.9 PUBLIC PARTICIPATION

Making an informed choice on whether to use or reject GM technology depends on an individual country, but for this it requires the correct information. Considering the controversy surrounding GM technology, there is a need for more transparent public policy decisions, such as regulatory frameworks, national policy dialogues, approval of individual GM products, and post-release monitoring. According to the CBD, governments have the responsibility to inform and consult the people before approving the introduction of GMOs. Other global instruments that require public participation include the Codex Principles on Risk Analysis and the Aarhus Convention on Access to Information, Public Participation in Decision-making, and Access to Justice in Environmental Matters. The convention acknowledges the legal right of people to take part in environmental decisions pertaining to the release of GMOs and their placement in the market (a right that the convention did not include previously). There is mixed opinion on biotechnology among people worldwide. Many people support genetic screening for hereditary diseases, using GMOs to reduce pollution, and employing genetic engineering to make medicines and vaccines. However, very few people favour the production of GM foods, cloning of human cells or tissue to cure patients, and cloning of animals for medical use. Customers need to know about the positive and negative traits of various GM crops and how to prevent the loss of local biodiversity. Their opinions of GM technology are predominantly influenced by knowledge about both benefits and risks of the product. It has been found that people who are aware of both benefits and risks will less likely consume GM food than those who are aware of only the benefits (Onyango and Nayga 2004). Many food producing companies have rejected GM ingredients after considering the importance of consumer choice. For instance, Frito-Lay Inc. declined to purchase GM maize for its maize-based chips and snacks. The risks of consumer rejection were more important to the company than the concern about health risks (Pollack and Shaffer 2000). Alongside the risk–benefit analysis, one should consider consumer's right to choose whether to eat food containing GM ingredients and products should be labelled to inform the consumer. Products that contain GMOs, and not products that are free from GM, have the legal onus. For example, in the European Union (EU), it is must to label a product containing GMOs or substances derived from GMOs unless the GM content is less than 0.9%. Labels declaring a product to be GM free are not yet included in the relevant EU directives, and so their legal status remains uncertain. However, to reassure customers, certain organic and health foods voluntarily carry a GM-free label. Attempts to encourage the safety of global trade in GMOs got a boost in 2004 when the labelling and documentation requirements were adopted. According to the new system approved by the 87 members of the Cartagena Protocol on Biosafety, all shipments of living organisms or GMOs meant for food, feed, or processing (such as soya beans and maize) should be labelled as 'may contain GMOs'.

About 10.3 million farmers in 22 countries are cultivating GM crops, and research on GM techniques is being conducted by at least 45 countries. There has been an increase in the commercial planting rates of GM crops by over 10% per year during the last decade. The fact that GMOs exist in the environment must be recognized, and they are unlikely to be removed in the near future. So far researchers have found no definite evidence of direct harmful impacts of commercially released GMOs on biodiversity. Moreover, there is currently no evidence to prove that GMOs are inherently dangerous or safe for human health. Laboratory tests have established the possible harm of GM applications through direct impacts on non-target species and gene flow, but they have also shown that biosafety processes in place are effective. Poor management generally results in contamination or indirect impacts through monoculture applications. This reiterates the necessity of strong governance structures, including the requirements of the Cartagena Protocol relating to use of the precautionary approach. This further affirms the notion that precaution should be taken for each new GM application in every ecosystem. Scientists should continue research on GM applications, but they should pursue only after assessing each GM application. There should be equal amount of funding for promoting and completely executing biosafety frameworks for each country that produces or imports GM products or that chooses not to use GM products. However, there are significant indirect impacts of GMOs currently in use, which deserve more attention.

The harm that monoculture techniques cause to ecosystems and biodiversity, whether genetic modification or conventional crops, is unquestionable and should be avoided. GM crops might induce the spread of agriculture into lands where conservation efforts are now under way. Biotech and agritech industries should apply sustainable farming methods while producing GM crops to prevent the direct impacts and lessen the indirect risks. The poor would be benefited from GMO technology only if the appropriate technology is applied in the right way and placed in the right hands. Government institutions should lead the way by carrying out ethical research for poor farmers, with poverty alleviation as the principal motive. This should be coupled with regular exchange of information between private organizations and the general public who pay the costs. Robust safeguards for protecting the environment and health should be based on strengthening institutions and regulatory frameworks. Most countries require help to develop relevant public institutions and adopt adequate biosafety laws. Absence of people's involvement, mostly in developing countries, shows that the society has not been able to fully participate in a debate that may potentially significantly impact the environment and livelihoods. To take informed decisions, both consumers and producers have the right to know whether GM ingredients are present in the food or seeds that they purchase.

8.10 GENE FLOW, IPR, AND IPP

'Property' is tangible, which can be owned and which benefits the owner. However, 'intellectual property' is something intangible, which constitutes a bundle of rights associated with the works of intellects that are granted provisional legal protection against unauthorized uses. There are two major types of intellectual property—copyright and industrial intellectual property rights:

Intellectual property right and protection

Intellectual property right and protection

The country's rank in the international market can be assessed by the technology it has, how many Intellectual property rights (IPR) it own or other intellectual things. The IPR have been defined as ideas, inventions, and creative expressions based on which there is a public compliance to bequeath the status of property. They provide certain rights that benefit the owner and provide social benefit to them too. Each company establishes its own research and development department so that they could get some patent of new or novel things and this provide the economic edge over the others. The World Trade Organization's agreement on trade related aspects of Intellectual Property Rights is also providing the protection to the world on global scale.

- Copyright shields creative 'works' such as literary works (novels, poems, newspapers, scientific articles, and so on), paintings, music, performances, sculptures, and architectures. Copyright safeguards the owners' right by enabling them to permit or forbid distribution, reproduction, communication to the people and reutilization of these works. Hence, to reproduce or publish the work, a third party has to obtain permission from the right-holders. Copyright applies to original works, even though the concept of originality considerably varies in the EU from one jurisdiction to another because of the non-harmonization of the notion of copyright at the EU level. The copyrighted work should be available in some tangible form (e.g., on computer screen, paper, canvas, or stone), whereas a mere idea cannot be copyrighted as such. Hence, copyright laws protect the concrete expression embodied into a creative work without extending the protection to the ideas underlying the work. Copyright protection expires after 70 years of the author's death. Contrary to other intellectual property rights conferred by trademarks, patents, and industrial designs, it operates automatically since the author does not have to register the work at a public office to obtain protection. Since the authors own the right to allow or disallow the use of their work, they are referred to as 'rights owners'. For instance, copyright laws entitle musicians to claim paternity over their music by giving them the moral right of attribution on their work and to protect the financial benefits deriving from commercial usage (e.g., broadcasting, web-casting, and recording activities).

- The protection of ideas that can be marketed in new, innovative products and processes falls under industrial property rights. Patent is the most important example of industrial property. It grants the inventor exclusive rights to use and apply an idea for a fixed period of time. Trademarks, registered designs, and appellations of origin are related concepts.

  The common characteristic of the two aforementioned categories of intellectual property is that such a property can be freely acquired, sold, and assigned. It is possible to separate the intellectual property rights existing on a given item from the item itself because of the clear distinction between intellectual and real (i.e., tangible) property regimes. For instance, one can donate a painting to a gallery without conferring to the gallery the right to copy and sell the reproductions.

Although law often deals with copyright and industrial property rights under the same head, copyright is most applicable to traditional heritage institutions involved in digitization projects. Industrial intellectual property is surveyed. Intellectual property laws provide a framework for rewarding creative people. In the absence of these laws, inventors would not benefit from new ideas, artists would not earn from their work, and the investments made in works such as books, films, and softwares would never be recouped. New creativity is stimulated on being rewarded, which has benefits for the entire society. Industrial intellectual property safeguards the consumer. Consumers get confidence from branding and trademarks, and they are assured that the product will meet their expectations.

For a cultural heritage institution involved in digitization projects, the enforcement of intellectual property laws implies that the institution can safely place information online. This stimulates interest on its holdings without risking the published material to be reused without permission. Nevertheless, the institution is also responsible to ensure that it has copyright clearance from the right-holders and to take suitable measures to protect its intellectual property rights. Gene flow between cultivars and between wild and weedy relatives has always been observed in many important crops in the world. It is difficult to evaluate factors influencing this gene flow, such as the system of mating, mode of pollination and seed dispersal, and the particular features of the habitat. Hence, the quantification of gene flow is not easy. In relation to the ecological risks associated with the commercial release of transgenic plants, transgene flow from engineered crops to other cultivars or to their wild and weedy relatives is one of the major threats. It is necessary to quantify this gene flow with transgenic crops and try to formulate strategies to control or minimize it, after considering the potential ecological effects of the newly introduced genes, whether advantageous or disadvantageous. It has been proved that the use of transgenic plants is effective in quantifying the gene flow to other cultivars of the same species or to wild and weedy relatives in all crops investigated. After reviewing the major case studies in this area, it can be concluded that the possible risk of gene flow should be assessed on a case-by-case basis and general conclusions should be made with caution.

Agricultural practices have greatly improved crops by selection and breeding since their inception. Most of the developed crop cultivars are compatible among themselves. However, they are also compatible with their wild or weedy relatives, which suggests that gene flow has always taken place.

There was commercial production of 52.6 million ha of GM crops in 10 countries in 2001 (James 2003). The problem of gene flow has acquired special significance with transgenic plants. The difference between conventional plant breeding techniques and those involving transgenic plants is that only genes coming from compatible species can be introduced in the first case, whereas there is introduction of genes from non-compatible organisms in the second case. The potential for pollen-based gene flow varies with the geographical spread of the crop and that of its wild or weedy relatives. In Europe, wild relatives of plants such as rapeseed or rice grow in similar locations and this creates some potential for mating. On the contrary, maize, cotton, and potato do not have wild or weedy relatives in the region, which makes the gene transfer unlikely. Many researchers have investigated problems associated with traditionally improved crops to infer potential risks of transgenic crops. Breeding between wild relatives and cultivars in many breeding programmes has often been very important to obtain new varieties. However, the flow of genes from wild relatives to cultivars can be beneficial for cultivation in the field. For the investigation of transgenic flow, the opposite direction of gene flow from cultivars to wild relatives has often been regarded as the possible risk. Introgression and hybridization have always occurred between cultivated and wild and weedy relatives

(Harlan 1965). Many plants have long-distance pollen and seed dispersal mechanisms. Regarding the potential of gene escape, the extent of gene flow is influenced by factors such as the closeness of compatible wild relatives, fitness of hybrids, system of mating and mode of pollination, and the method of seed dispersal. These factors have been extensively reviewed (Ellstrand, Prentice, and Hancock 1999). It can be concluded from these studies that the most significant crops in the world have compatible wild relatives and that natural hybrids have been reported for most of them, generally without ecological impacts.

In the case of gene flow between crop species and their wild and weedy relatives, it is essential to probe whether the produced hybrids can survive in the field and to what extent the ecosystem is affected by undesirable consequences. The fitness of crop/congener hybrids was generally assumed to be low under field conditions. However, research conducted to measure spontaneous hybridization between wild radish (*Raphanus sativus*) and cultivated radish and that between *Sorghum bicolor* and *S. halepense* (a widespread weed) (Arriola and Ellstrand 1997) demonstrated that spontaneous hybridization takes place in reality. Moreover, the fitness of the hybrids was equal to that of their non-hybrid relatives. It has been established that flow of genes from crops to weeds has generated new or more difficult weeds.

Loss of biodiversity is another problem resulting from the crop-to-weed gene flow. Under suitable conditions, hybridization between a common species and a rare one can result in extinction of the rare species within a few generations. Crops planted in new locations are most likely to face the risk of extinction by gene flow because of being in the vicinity of wild relatives, which increases the hybridization rate (Ellstrand 2001). There are many cases in which hybridization between a crop and its wild relatives has enhanced the extinction risk for the wild taxon. For instance, it has been implicated that natural hybridization with cultivated rice has resulted in the near extinction of the endemic Taiwanese taxon *Oryza rufipogon* ssp. *formosana* (Kiang, Antonovics, and Wu 1979). In traditional agriculture, gene flow has clearly taken place and has often had undesirable consequences.

Natural and engineered characteristics are likely to have similar patterns of gene dispersal. As a result, problems arising from cross-pollination with wild relatives in transgenic crops will be similar to those observed in traditional breeding programmes.

For plants cultivated for non-GM, GM-free, or 'organic' niche markets, cross-pollination of GM varieties with traditional varieties and germination of volunteer GM seeds (seeds dropped, blown, or unintentionally planted) have been of great concern for the corresponding producers. This chapter will deal with studies carried out with transgenic plants to quantify gene flow and evaluate strategies to control or minimize it.

8.10.1 Gene Flow Studies Using Transgenic Plants

Many countries have produced transgenic plants from different crops, and the number of crops is expected to increase in the coming years. Worldwide, the major GM crops include soya bean (grown in 33.3 million ha in 2001, constituting 63% of the global area), GM corn (9.8 million ha; 19% of the global area), transgenic cotton (6.8 million ha; 13% of the global area), and canola (2.7 million ha; 5% of the global area) (James 2003). There is still limited knowledge of the ecological consequences of gene flow from transgenic crops to wild relatives, weeds, or other non-transgenic varieties. The movement of genes across species boundaries creates many opportunities for expected and unexpected risks. Besides food safety, there are other concerns such as ecological risks (e.g., new or increased resistance to insecticides and herbicides because of excessive selection pressure or hybridization), alterations in the ecological competitiveness of crops, and the potential loss of genetic diversity in regions of crop origin (Amand, Skinner, and Peaden 2000).

Attempts have been made by many researchers to understand better and quantify factors affecting the flow of genes. This information can be utilized to design suitable confinement measures for release and monitoring mechanisms to ensure the effectiveness of the confinement (Staniland, McVetty, Friesen, *et al.* 2000). Simulation models might also be useful to evaluate the results of introducing new genes into the environment (Lavigne, Klein, Vallée, *et al.* 1998; Colbach and Sache 2001 a, b).

Ellstrand, Prentice, and Hancock (1999) have compiled data on spontaneous hybridization between important food crops and their wild relatives. They obtained the data basically by measuring the spontaneous hybridization rate by conventional genetic plant-breeding methods or through the application of molecular markers. In some of these crops, it has been proved that transgenic plants with a marker gene can be a potent tool to detect, quantify, and assess the extent and ecological risks of gene flow. Numerous wild relatives of rapeseed (*Brassica napus* L., AACC, $2n = 38$), which is a partially allogamous species, are present in cultivated areas. The genes can be dispersed through volunteers from rapeseeds lost before or during the harvest or via pollen. Rapeseed pollen can travel over large distances, and a transgenic rapeseed cultivar can pollinate crops, as well as related weedy species, growing in adjacent fields. An important component of risk assessment and management is the existence of feral plants outside of fields because these plants can act as reservoirs for genetic pollution between GM and non-GM crops. A transgenic crop in which risk assessment of gene flow has been extensively studied using herbicide-tolerant varieties is oilseed rape because this characteristic is easy to analyse in large populations. All commercial GM rapeseed crops are tolerant to herbicides (James 2003). For sustainable agricultural systems, the dispersal of herbicide-resistant genes might be a serious problem. It can make the weed and rapeseed volunteer populations tolerant to one or a variety of herbicides, and thus weed control becomes very difficult. If transgenic rapeseed volunteers exist in a conventional rapeseed crop, they would contaminate the traditional harvest and prevent farmers from obtaining a 'non-GM' label for their product.

Many researchers have utilized herbicide-resistant GM rapeseed to study the impact of gene flow to weedy relatives. Hence, under field conditions, fertile interspecific hybrids of GM rapeseed (*B. napus*) and *B. campestris* (*B. rapa*) have been produced. Some *B. campestris*-like crops having a transgene from the rapeseed were generated after two generations of hybridization and backcrossing. This suggested the possible transfer of genes from oilseed rape to the weedy relative *B. campestris*. Baranger, Chevre, Eber, *et al.* (1995) used five male sterile F$_1$ hybrid rapeseed genotypes, heterozygous for the *bar* gene, and five parental male sterile lines as controls in a field experiment to assess their spontaneous outcrossing with wild radish. They recovered interspecific hybrids of the five rapeseed genotypes and wild radish. But various rapeseed genotypes greatly differ in their capacity to generate these hybrids. It appears that the production of these transgenic interspecific hybrids is easy under natural conditions, largely because of the reduction in interspecific barriers owing to a male sterile recipient and high pollen pressure from the wild species.

Scientists used these F_1 oilseed rape–wild radish interspecific hybrids to study the traits of consecutive backcross generations cultivated under field conditions using wild radish as a pollinator (Chèvre, Eber, Baranger, *et al.* 1997, 1998).

The steady introgression of transgenes within natural populations is dependent on the genomic structure, fertility, and transmission of oilseed rape genes within consecutive generations. It has been demonstrated from cytogenetic, isozyme, and RAPD (random amplified polymorphic DNA) analyses that intergeneric gene flow might happen mainly by transgene introgression; however, it occurs at a slow rate and low probability under natural optimal conditions. Four generations were required to obtain herbicide-resistant plants with a chromosome number and morphology close to that of the wild radish. Hence, when the wild radish is a female parent, introgression would be rare under normal agricultural conditions. To measure the rate of hybridization between oilseed rape and wild radish within normal agronomic conditions, Chèvre, Eber, Darmency, *et al.* (2000) carried out field experiments in France in three plots, each of size 1 ha. In each case, they transplanted wild radish plants at different densities in the middle, at the border, or near the margin of the herbicide-resistant oilseed rape field. For the total harvest of the wild radish, they assessed the frequency of interspecific hybrids in the range of 10–3310. Wild radish plants growing at the border of the field gave the highest frequency. For the total harvest of the oilseed rape seedlings, it was found that the frequency of interspecific hybrids ranged from 2310 to 5310. Hence, under normal agronomic conditions, interspecific hybridization in both directions occurs between oilseed rape and wild radish, but at a very low rate. Gene flow is also greatly influenced by the frequency and distance of pollen dispersal. Scheffler, Parkinson, and Dale (1993) designed a field trial to observe the pollen movement. In the trial, they planted transgenic Westar plants with the *bar* gene in a 9 m diameter circle at the centre of a 1.1 ha field surrounded by non-transgenic plants with a distance of at least 47 m in all directions. At the centre of the transgenic area, non-transgenic plants were sown within a circle of 1 m diameter to estimate the level of pollen dispersal when plants were in close contact. To enhance cross-pollination, beehives were put at the trial site. The researchers analysed the frequency of pollen dispersal in seed samples collected from non-transgenic plants planted at different distances and sprayed with glufosinate. The percentage of pollen dispersal in the non-transgenic circle was estimated to be 4.8%. The estimated frequency was 1.5% at a distance of 1 m, which reduced to 0.00033% at 47 m. No obvious directional effects that could be ascribed to wind or insect activity were detected. Scheffler, Parkinson, and Dale (1995) carried out another study to assess the usefulness of isolation distances for field plots of transgenic oilseed rape resistant to herbicides. They evaluated the effectiveness of 200 m and 400 m isolation distances for small-scale trials. These distances were used because they are commonly adopted for the production of basic and certified breeders' seeds. They found the frequency of hybrid formation to be 0.156% at 200 m and 0.0038% at 400 m. These estimates are within the limits set for the production of basic seeds (0.1%). It is generally assumed that hybridization occurs between transgenic and non-transgenic crops, and also with some wild relatives, efforts were made to formulate strategies to reduce this gene flow.

Staniland, McVetty, Friesen, *et al.* (2000) analysed the effectiveness of border areas (15–30 m wide) in limiting the spread of transgenic herbicide-resistant *B. napus* pollen. When seed samples harvested from transgenic plants from the border area at distances from 0 m to 30 m were screened, it was found that pollen-mediated gene flow in *B. napus* decreased rapidly with an increase in distance from the pollen source, provided there was a continuous border/pollen trap. The currently stipulated 10 m wide continuous borders (established by some government safety regulations) effectively control the majority of pollen-mediated gene flow, but they do not completely eliminate gene escape under field conditions. Some efforts have been made in the case of oilseed rape to establish a model for predicting the dispersal of pollen from transgenic plants (Lavigne, Klein, Vallée *et al.* 1998). Colbach and Sache (2001) and Colbach, Clermont-Dauphin, and Meynard (2001) have prepared a model that shows the influence of cropping system on gene escape from herbicide-tolerant winter rapeseed cultivars to rapeseed volunteers. In their model, factors such as evolution of genotype proportions over time, temporal evolution of a population of rapeseed volunteers in field, and impact of genotype on herbicide efficiency, seeds, and pollen production have been considered. The authors expected that this model (called GeneSys) should be easily adaptable to herbicide-tolerant spring rapeseed cultivars and also to other genes, such as those coding for fatty acids, whether introduced by classic selection or genetic engineering.

8.10.1.1 Maize (Zea mays L.)

According to the Directorate-General for Agriculture and Rural Development of the European Commission (2001), the second most important GM crop in the world is corn. Two-thirds of this corn is resistant to insects (Bt corn), half of the rest is tolerant to herbicides, and the remaining half contains both modifications. Pollination in maize primarily occurs through wind, and maize pollen remains alive for a short period. Most pollen falls within 5 m of the field's edge, and 98% of the pollen remains mostly within a 25–50 m radius of the maize field. It appears that the dispersal and flow of pollen between GM and non-GM maize crops depend on the scale of pollen release and the distance between sources and recipients. As maize and its progenitor are pollinated through wind and have the ability of outcrossing, the ultimate introgression of transgenes from commercial hybrids into landraces and wild relatives is possible if they are grown in close proximity (Christou 2002). Teosinte is a grass recognized as the feral progenitor species of corn that is cultivated in Mexico and some Central American countries. Studies on distances travelled by corn pollen revealed that the possibility of gene flow between corn and teosinte is very high, whether both plants are grown in the same field, or whether teosinte is grown along the borders or in dense patches outside the corn field. It is largely presumed that a gene incorporated in maize by genetic engineering could get transferred to teosinte and spread through some teosinte populations in some circumstances (Doebley 1990; Castillo-Gonzalez and Goodman 1995). On the contrary, in spite of the huge potential for gene flow between modern corn and teosinte, there is considerable uncertainty in the literature on the origin of maize and the potential of introgression with teosinte that introgression occurs from cultivated corn to teosintes (Doebley 1984) or that introgression can become fixed in the absence of selection pressure (Martinez-Soriano and Leal-Klevezas 2000). The eventual introgression of transgenes from commercial GMOs into landraces should also be considered. All native maize landraces grow in Oaxaca (Mexico) are regarded as the international centre of corn biodiversity. The conventional agronomic system prevalent in this region, described by Louette and Smale (2000), could help in introducing a new genetic material. Small-scale farmers select their seeds for desirable traits and replant them. In addition, they exchange seeds from local varieties to enhance productivity (and genetic diversity) of their corn. When they cultivate varieties in which there occurs open pollination, there is a possibility of cross-pollination (i.e., gene flow) with compatible varieties or non-local varieties. Although transgenic maize is not allowed in Mexico, some transgenic Bt corn could be inadvertently sown. However, there is no proof of introgression of transgenes in these landraces.

Flow of genes throughout cross-pollination might occur between the conventional maize crops and GMOs. Conventional plant breeding programmes recommend the separation distances needed to maintain cross-pollination in fields (having an area of 2 ha or more) to be 200 m for 99% grain purity and 300 m for 99.5% grain purity (Ingram, Boisnard-Lorig, Dumas, *et al.* 2000). However, corn pollen can be found at distances more than 800 m from a field (Eastham and Sweet 2002). The potential risk of pollen significantly increases with the area and number of fields planted (Treu and Emberlin 2000). In addition, Jones and Brooks (1950) reported that there is a relation between the percentage of outcrossing within a field and the depth of the field in the direction of the source of contamination. The authors also observed that the percentage of outcrosses taking place in successive rows at varying distances of isolation shows that the first five rows near the contamination source act as a barrier to the dispersal of contaminating pollen. In 1999, Simpson (unpublished data) calculated the gene flow from a 36,312 m plot of glufosinate-tolerant maize to a neighbouring barrier crop of traditional maize. Samples were collected from various distances along three transects starting at 1 m from the GM maize plot up to 51 m away. Results indicated a sharp decrease in the levels of cross-pollination to below 1% at 18 m from the source of pollen. Therefore, the percentage of cross-breeding with other adjacent maize crops will be dependent on the separation distance, local barriers to pollen dispersal, and local climate and topography (Eastham and Sweet 2002).

8.10.1.2 Rice

Cultivated rice (*O. sativa* L.) is an autogamous plant. One of the main factors that control self-pollination in rice is the floral architecture. The time of day when the flower opens and the period of flowering in rice depend on the cultivar used and on humidity and temperature. It is believed that dissemination of rice pollen is low because anther dehiscence generally takes place just before, or at the same time as, the lemma and palea open. As a result, most pollen grains fall on the stigma. In addition, the viability of pollen is low (Yoshida 1981). There is limited horizontal movement of pollen. In hybrid seed production fields, an isolation distance of 10 m is considered sufficient to prevent contamination with pollen from neighbouring fields (Khush 1993).

However, cross-pollination in rice is possible to some extent, and the amount depends largely on the differences in climate and variety. Ellstrand, Prentice, and Hancock (1999) and Messeguer, Fogher, Guiderdoni, *et al.* (2001) have recently studied the flow of genes between cultivated rice and its wild and weedy relatives. Transgenic herbicide-resistant rice crops were used to quantify the gene flow between transgenic and non-transgenic plants of the same cultivars. In a field trial in Amposta (Spain), the frequency of gene flow from transgenic to non-transgenic plants of the Mediterranean japonica rice variety Thaibonnet was estimated (Messeguer, Fogher, Guiderdoni, *et al.* 2001). Transgenic crops (having the *pat* gene, which confers resistance to the herbicide ammonium glufosinate) were planted in the centre, and non-transgenic crops were planted in concentric circles around transgenic plants, at the distance of 1 m or 5 m. All panicles of the non-transgenic plants were individually harvested, and their location (relative to the geographic orientation in the field) was recorded. In a greenhouse, samples of seeds from each plant were sown, and a commercial herbicide was sprayed on seedlings at three-leaf to four-leaf stages. Using PCR (polymerase chain reaction) and Southern analysis, the integration of the *pat* gene was confirmed. On placing the plants at a distance of 1 m from the transgenic central nucleus, a gene flow slightly lower than 0.1% was observed in the circular plot. A significant asymmetric distribution of gene flow was noticed in this circle, where the largest values (0.53%) were recorded in the case where the direction of the dominant wind was followed. As expected according to the characteristic features of rice pollen, a considerably lower value (0.01%) was detected in the other circle (5 m from the transgenic plants). Using a circular design, a second field trial was carried out in the year 2000 to estimate the frequency of gene flow from transgenic plants (Senia, Mediterranean variety with *bar* and *gus* genes) to non-transgenic plants planted at varying distances. The influence of wind on the dispersal of pollen and cross-pollination was also analysed. Preliminary results of this field trial indicated that the detected gene flow rate had the same magnitude (about 0.1%) as that obtained with the variety Thaibonnet. As the distance between transgenic and non-transgenic plants increased, the gene flow rate decreased, and it was established that the dominant wind of the region strongly influenced the gene flow rate between transgenic and non-transgenic plants.

The aforementioned results were contrary to those reported else where. They analysed the gene flow from transgenic to non-transgenic purple rice by using transgenic rice with a *Xa* gene for bacterial blight resistance. On the basis of a pathogenicity test, they failed to find any proof of gene flow from transgenic to non-transgenic crops. Rice cultivars easily cross with the weedy relative (red rice) found in directly seeded paddy fields and produce viable and fertile hybrids (Langevin, Clay, and Grace 1990). Owing to the inheritance of the dominant wild traits (Langevin, Clay, and Grace 1990; Noldin, Chandler, and McCauley, *et al.* 1999), cross-pollination and introgression might convert transgenic rice to a hybrid transgenic red rice weed. To study the segregation of the *bar* gene and evaluation of traits, researchers have recently analysed samples from controlled crosses between red rice and transgenic rice lines tolerant to ammonium glufosinate. Their study showed that the transfer of the *bar* gene did not change the fitness values for traits such as dormancy and seed production, which are related to the reproductive success of the hybrids (Oard, Cohn, Linscombe, *et al.* 2000). However, the ecological effect of this 'reverse' gene flow can be insignificant because hybrid seeds will be harvested during the cultivation of crop in the field. Wheeler, Chappey, Lash, *et al.* (2000) carried out field and greenhouse experiments to find whether glufosinate-resistant genes were transferred to red rice as a result of outcrossing with transgenic white rice and at what rate the outcrossing took place. With the help of three transgenic white rice cultivars tolerant to glufosinate and eight red rice ecotypes, they established that red rice seedlings expressed the phenotype of resistance to the herbicide (survival) in all combinations of red rice and white rice. The average level of resistance across biotypes and cultivars ranged from 0.8% to 19.7%. Therefore, red rice seedlings tolerant to herbicide treatment are generated in large numbers due to outcrossing. Molecular evidence is still required to establish the transfer of genes to red rice. Recently, gene flow from transgenic rice to red rice has been analysed by employing a circular design. In their study, transgenic Senia *bar* tolerant plants were planted in two circles and surrounded by red rice. The gene flow rate indicated in the primary results was lower than that reported by other authors, and the gene flow was influenced by the wind direction ranging from 0.002% to 0.06%. Gene flow rates measured using rice transgenic plants undoubtedly demonstrated that genes will escape, to some extent, from transgenic to non-transgenic plants or to red rice. A mathematical model can be designed by taking into account the features of this species, such as self-pollination, relatively high pollen weight, low pollen viability, decrease in the average gene flow with distance between transgenic and non-transgenic plants, and the strong influence of wind. The model can be commercially applied to predict the rate of gene flow and make strategies to minimize it.

8.10.1.3   *Potato* (Solanum tuberosum L.)

For most crops, isolation distances have become well established because of many years of experience in the production of seeds from conventional plant cultivars, and these distances are strictly observed to maintain purity of seeds to defined legal specifications (Frankel and Galun 1977). On the other hand, true seed production under field conditions is not important for asexually propagated clonal crops; therefore, there is relatively less experience and information. This is also applicable in the case of potatoes, and there is very less correct information on effective pollination distances for this reason. Many field trials have been conducted to assess pollen dispersal and estimate the appropriate isolation distances between transgenic and non-transgenic potatoes. When transgenic potato lines with *npt*II and *gus* genes (Dale, Howes, Price, *et al.* 1992) or with *npt*II, *gus*, and *als* genes were used (Tynan, Williams, and Conner 1990; the frequency of transgenic progeny in pollen trap cultivars steeply declined with distance. In conclusion, it can be said that an isolation distance of 20 m is enough for the preliminary field assessment of transgenic potatoes.

8.10.1.4   *Sugar beet* (Beta vulgaris L.)

Beets belonging to the *B. vulgaris* species are present in crop, wild and weedy forms, and all of them are interfertile (Desplanque, Boudry, Broomberg, *et al.* 1999). Sugar beet (*B. vulgaris* L. subsp. *vulgaris*) is a biennial plant in which overwintering is typically necessary to become generative and produce pollen and seeds. The biennial wild relative (*B. vulgaris* subsp. *maritima*—*B. maritima*) grows in coastal regions (e.g., Northern Europe) with a mild winter climate (Rothmaler 1990). For analysing the potential of cross-pollination between sugar beet and wild beet, Pohl-Orf, Brand, Drieben, *et al.* (1999) studied the capacity of transgenic lines to survive in central Europe during winter compared to non-transgenic lines. Survival of both transgenic and non-transgenic lines is possible to some extent. Thus, these overwintering sugar beets can be a source of pollen dispersal, and there is a possibility of cross-pollination with wild relatives. Flow of genes between transgenic sugar beet and wild beet was also examined at the farm scale. It was found in this study that gene flow occurred at a low density between bolters of the crop in a 1 ha field and weed beet growing in an adjacent field. In another study, Saeglitz, Pohl, and Bartsch (2000) investigated the efficacy of a containment strategy based on the use of a strip of bait crops such as hemp (*Cannabis sativa* L.). The containment strategy failed because transgenic offspring were detected more than 200 m behind the hemp containment. Pohl-Orf, Brand, Schuphan, *et al.* (1998) assumed that gene flow takes place and they studied the ecological implications of gene flow and transfer of transgenic characteristics to intraspecific relatives, for example, Swiss chard and wild beet.

Transgenic plants tolerant to rhizomania, caused by the soil-borne beet necrotic yellow vein furovirus, were analysed in this study. Although there is high probability of outcrossing of transgenic virus resistance into wild beet varieties, only insignificant ecological impacts can be expected because of the absence of any ecological advantage in coastal ecosystems where Swiss chard and wild beet grow.

8.10.1.5   *Cotton* (Gossypium hirsutum L.)

Australia is home to the wild *Gossypium* species that are distant diploid relatives of the commercial tetraploid cottons *G. barbadense* and *G. hirsutum* (Brubaker, Paterson, and Wendel 1999). The effectiveness of the wild Australian *Gossypium* species in cotton breeding is dependent on the ability to produce fertile hybrids. To the extent possible under glasshouse conditions, it presumes that fertile hybrids between transgenic cotton and spatially related wild species may arise. Brubaker, Paterson, and Wendel (1999) showed that the probability of wild species serving as recipients of transgenes was practically zero, particularly in the presence of the profound prezygotic barriers that isolate the cultivated tetraploid cottons from their wild Australian relatives. Several studies have been performed about the extent of natural crossing in cotton by using transgenic plants. Llewellyn and Fitt (1996) assessed the usefulness of containment techniques under Australian environmental conditions by measuring the dispersal of pollen away from a test plot of transgenic cotton (Bt cotton with *npt*II gene conferring kanamycin resistance) into a surrounding buffer field of non-transgenic cotton plants that acted as a sink for pollen carried by insects. Dispersal was calculated by measuring the frequency of *npt*II gene in the progeny of buffer plants. Outcrossing declined with distance, and a 20 m buffer zone served to prevent dispersal of transgenic pollen in small-scale field tests. Zhang, Kumar, and Nei (1997) used a selectable marker gene (*tfdA*) for tolerance to the herbicide 2,4-D to evaluate the pollen dispersal frequencies. Seeds of non-transgenic cotton plants cultivated around a 436 m plot of transgenic cotton were collected at different distances in eight directions and sprayed with the herbicide. The estimated pollen dispersal frequency was 11.2% at a distance of 1 m, which decreased to 0.03% at 50 m. These results agreed with those obtained by Zhang, Timakov, Stankiewicz *et al.* (2000). In this study, Bt cotton NewCott 33B and transgenic *tfdA* cotton were selected for assessing pollen dispersal frequency. The findings indicated that transgenic cotton could be dispersed into the environment. The largest dispersal frequency was 10.48%. The longest distance of pollen dispersal from transgenic cotton in these tests was 50 m.

For important crops in the world, gene flow and introgression have always occurred between cultivars and wild and weedy relatives (Ellstrand, Prentice, and Hancock 1999). Factors influencing this gene flow, such as the system of mating, mode of pollination, method of dispersal of seeds, and specific features of the habitat where the crops grow, are difficult to assess and, consequently, it is not easy to quantify the gene flow. In traditional agriculture, segregation of specific crops is a major issue. Several segregation systems are in place, although most of them are not related to GM crops. For instance, industrial oils can be produced from oilseed rape varieties having high erucic acid. However, since erucic acid is an anti-nutrient, it is necessary that these crops do not contaminate oilseed rape varieties grown for human consumption, which have low concentration of both erucic acid and glucosinolate. In Europe and the United States, erucic acid levels in oil should be less than 2% for human consumption (Bock 2002). Hence, there are well-established isolation distances to make sure that cross-pollination is below stipulated limits (Ingram 2000). Currently, the chances of mixing of GM crops with organic or conventional crops pose serious threat. The EU has fixed a threshold of 1% GMO content at the marketing stage for traditional food products and ingredients. Food containing more than 1% GMOs related to the different ingredients must be labelled. However, utilization and presence of GMOs have been excluded in the case of organic farming. For the quantification of gene flow to other cultivars of the same species or to wild and weedy relatives, transgenic plants can be an effective tool. Plant breeders predict that gene flow occurs in both transgenic and non-transgenic plants because it depends on the traits of the species and not on the specific characteristic of the introduced transgene. Gene flow has been adequately characterized in allogamous species such as sugar beet and rapeseed, and efforts have been made to establish containment policies. However, these policies have proved to be efficient only in reducing the chances of gene flow and not in fully controlling gene escape.

A mathematical model for predicting pollen dispersal and the influence of the cropping system on gene escape has been designed for rapeseed. Bock (2002) has recently employed this model (GeneSys) for evaluating cropping practices or the value of certain varietal traits for reducing the presence of GM plants in the field. They concluded in the study that volunteers are a prominent source of adventitious GM crops in rapeseed farms. In maize, factors such as barriers to pollen movement, separation distance, and local climate and topography influence the introgression between GM and non-GM varieties. However, volunteers are unlikely to survive because of their inability to shed seeds naturally. The probability of gene flow between cultivars in self-pollinating species such as rice can be reduced by increasing the security distance and considering the dominant winds that enhance cross-pollination and pollen dispersal. But this cannot be noticed in red rice weed. In this case, crop management should be altered to avoid gene flow. If transgenic crops are introduced, the environment will probably get modified, which has happened for centuries with the traditional breeding crops. Since plants are immobile, they have to depend on dispersal of pollen. In the absence of gene flow from dispersed pollen, plants might face genetic bottlenecks from inbreeding and, consequently, might exhibit decreased genetic diversity and possible fitness. As such, plants pollinated by wind and insects are naturally more promiscuous for their own benefit. In addition, food crops are not ancient species without any modification; humans have genetically manipulated them by educated trial and error coupled with intense selective pressure. Continuous improvement in agronomic characteristics was possible due to the fact that genes could be easily exchanged between food crops and their ancestors.

Molecular biology techniques have resulted in quicker improvement of a particular trait in crops. The ecological influence of hybridization between crops and wild relatives is dependent on the introduced trait and not on how the characters were bred into the crop. According to the National Research Council of the United States, the method of crop breeding, whether by laborious hand selection, crossing over several years, or through molecular biology, is not relevant to evaluating ecological risks. The focus should be on the traits produced by the technique, and these traits should be evaluated in the relevant environments. Since only the gene of interest and no marker genes will be present in the plant genome in the second-generation transgenic plants, the ecological advantage or disadvantage that the introduced traits produce will determine the ecological impact of both traditional and transgenic crops. For example, let a characteristic that confers resistance against insects be introgressed into a wild relative. Likewise, if a herbicide-tolerant gene is introgressed into a wild relative, it will not be of use unless herbicides have been sprayed in areas where the plants are grown. A new trait produced by a known single gene can be tested to find whether it increases fitness. For instance, a decade long field research in the United Kingdom (Crawley, Brown, Hails, *et al.* 2001) showed that GM herbicide-resistant maize, oilseed rape, sugar beet, and potato expressing either Bt toxin or pea lectin did not persist outside the field and never adopted weedy character. In these experiments, GM traits (resistance to herbicides or insects) not expected to increase fitness of plants in natural habitats were involved. The environmental impact of plants with GM traits such as pest resistance or drought tolerance, which might enhance performance under field conditions, should be assessed during the development of such plants. A case-by-case approach is required to focus on gene flow and biodiversity in specific crop species and ecological situations. For example, in the case of concerns about the loss of teosinte (the wild ancestor of maize cultivars) on the introduction of transgenic maize in Mexico, it should be noted that teosinte has remained distinct for a long time with its known diversity of subspecies despite being compatible with maize landraces and that crops have been cultivated nearby for many hundred years. This observation suggests no impact on biodiversity owing to corn–teosinte interactions. However, this does not reduce the importance of careful testing under field conditions.

It can be concluded from case histories presented in this chapter and from the work of Bock (2002) and Eastham and Sweet (2002) that some extent of gene flow and introgression will occur. Although containment policies can greatly decrease the possibility of gene flow through cross-pollination in some crops, it is almost impossible to control the emergence of transgenic varieties due to the seeds blown, dropped, or accidentally planted during harvest and traditional cultivation practices. Many transgenic crops have been introduced into the environment and some of them (such as maize) have been planted for more than 8 years. However, no consistent evidence is available for that period to suggest that their release has been more dangerous to the environment than traditional plant breeding crops. In fact, the environmental benefits of these plants, including their impact on decreased use of pesticides, are now being documented. Phipps and Park (2002) believe that GM technology has minimized the use of pesticides on a global scale and that the amount of reduction varies between crops and the introduced trait. According to an estimate, the use of GM oilseed rape, soya bean, cotton, and maize varieties decreased pesticide use by about 22.3 million kg of the formulated product in 2000. Moreover, diesel consumption was reduced, which resulted in a decrease in the release of carbon dioxide. Hence, the possible risk of gene flow cannot invalidate the efficiency of genetic transformation. Traditional agricultural practices should also account for the environmental conservation. These practices have frequently tended towards degradation instead of conservation, perhaps because of the absence of adequate scientific basis for correct resource management. The knowledge gathered in the recent decades must be applied to both transgenic and traditional breeding for augmenting food supply, but after taking care of the environment.

REFERENCES

Alexander, C. E. and T. Van Mellor. 2006. Determinants of corn rootworm resistant corn adoption in Indiana. *The Journal of Agrobiotechnology Management and Economics* 8(4): 197–204

Amand, P. S., D. Z. Skinner, and R. N. Peaden. 2000. Risk of alfalfa transgene dissemination and scale-dependent effects. *Theoretical and Applied Genetics* 101(1-2): 107-114

Arriola, P. E. and N. C. Ellstrand. 1997. Fitness of interspecific hybrids in the genus Sorghum: persistence of crop genes in wild populations. *Ecological Applications* 7(2): 512-518

Ayele, S., J. Chataway, and D. Wield. 2006. Partnerships in African crop biotech. *Nature Biotechnology* 24(6): 619-621

Baranger, A., A. M. Chevre, F. Eber, and M. Renard. 1995. Effect of oilseed rape genotype on the spontaneous hybridization rate with a weedy species: an assessment of transgene dispersal. *Theoretical and Applied Genetics* 91(6-7): 956–963

Benbrook, C., C. McCullum, L. Knowles, S. Roberts, and T. Schryver. 2003. Application of modern biotechnology to food and agriculture: Food systems perspective. *Journal of Nutrition Education and Behavior* 35(6): 319-332

Bock, R. 2002. Transgenic plastids in basic research and plant biotechnology. *Journal of Molecular Biology* 312(3): 425–438

Brookes, G. and P. Barfoot. 2005. GM crops: the global economic and environmental impact-the first nine years 1996-2004. *The Journal of Agrobiotechnology Management and Economics* 8(2-3): 187-196

Brubaker, C. L., A. H. Paterson, and J. F. Wendel. 1999. Comparative genetic mapping of allotetraploid cotton and its diploid progenitors. *Genome* 42(2): 184–203

Charles, D. 2001. Lords of the Harvest: Biotech, Big Money, and the Future of Food. Cambridge, MA: Basic Books

Chèvre, A. M., F. Eber, A. Baranger, and M. Renard. 1997. Gene flow from transgenic crops. *Nature* 389(6654): 924-924

Chèvre, A. M., F. Eber, A. Baranger, , G. Hureau, P. Barret, H. Picault, and M. Renard. 1998. Characterization of backcross generations obtained under field conditions from oilseed rape-wild radish F1 interspecific hybrids: an assessment of transgene dispersal. *Theoretical and Applied Genetics* 97(1–2): 90–98

Chèvre, A. M., F. Eber, H. Darmency, A. Fleury, H. Picault, J.C. Letanneur, and M. Renard. 2000. Assessment of interspecific hybridization between transgenic oilseed rape and wild radish under normal agronomic conditions. *Theoretical and Applied Genetics* 100(8): 1233–1239

Christou, P. 2002. No credible scientific evidence is presented to support claims that transgenic DNA was introgressed into traditional maize landraces in Oaxaca, Mexico. *Transgenic Research* 11(1): 3–5

Colbach, N. and I. Sache. 2001. Blackgrass (*Alopecurus myosuroides Huds.*) seed dispersal from a single plant and its consequences on weed infestation. *Ecological Modelling* 139(2): 201–219

Colbach, N., C. Clermont-Dauphin, and J. M. Meynard. 2001. GeneSys: a model of the influence of cropping system on gene escape from herbicide tolerant rapeseed crops to rape volunteers: I. Temporal evolution of a population of rapeseed volunteers in a field. *Agriculture, Ecosystems & Environment* 83(3): 235–253

Comstock, G. 2008. Ethics and genetically modified foods. In *Contemporary Issues in Bioethics*, edited by T. L. Beauchamp, LeRoy Walters, J. Kahn, and A. Mastroianni, pp. 752–759. California: Wadsworth Pub Company

Conner, J.M, A. Paterson, and J. R. Piggott. 1994. Interactions between ethyl esters and aroma compounds in model spirit solutions. *Journal of Agricultural and Food Chemistry* 42(10): 2231–2234

Conway, G. 2003. From the green revolution to the biotechnology revolution: food for poor people in the 21st century. Paper presented at the Woodrow Wilson International Center for Scholars Director's Forum, 12 March 2003. Details available at <www.unc.edu/courses/2006spring/geog/021/001/Conway.pdf

Crawley, M. J., S. L. Brown, R. S. Hails, D. D. Kohn, and M. Rees. 2001. Biotechnology: Transgenic crops in natural habitats. *Nature* 409(6821): 682–683

Dale, L., G Howes, B. Price, B. and J. C. Smith. 1992. Bone morphogenetic protein 4: a ventralizing factor in early Xenopus development. *Development* 115(2): 573–585

Degrassi, G., N. Alexandrova, and D. Ripandelli. 2003. Databases on biotechnology and biosafety of GMOs. *Environmental Biosafety Research* 2(3): 145-160

Desplanque, B., P. Boudry, , K. Broomberg, P. Saumitou-Laprade, J. Cuguen, and H. Van Dijk. 1999. Genetic diversity and gene flow between wild, cultivated and weedy forms of Beta vulgaris L.(Chenopodiaceae), assessed by RFLP and microsatellite markers. *Theoretical and Applied Genetics* 98(8): 1194–1201

Doebley, J. F. 1984. Maize introgression into teosinte-a reappraisal. *Annals of the Missouri Botanical Garden* 1100–1113

Doebley, J. 1990. Molecular evidence and the evolution of maize. *Economic Botany* 44(3): 6–27

Eastham, K. and J. Sweet. 2002. Genetically Modified Organisms (GMOs): The Significance of Gene Flow through Pollen Transfer (pp. 1–74). Copenhagen, Denmark: European Environment Agency

Ebenreck, S. 1991. Pest control for a whole earth. In *Ethics and Agriculture*, edited by Charles V. Blatz. Moscow: University of Idaho Press

Ellstrand, N. C. 2001. When transgenes wander, should we worry? *Plant physiology* 125(4): 1543–1545

Ellstrand, N.C., H. C. Prentice, and J. F. Hancock. 1999. Gene flow and introgression from domesticated plants into their wild relatives. Annual Review of Ecology and Systematics 30: 539–563

Engelhardt H. T. and F. Jotterand. 2008. In *Contemporary Issues in Bioethics*, edited by T. L. Beauchamp, L. Walters, J. Kahn, and A. Mastroianni, pp. 760–763. California: Wadsworth Pub Company

Estabillo, A. V. 2004. Farmer's group urges ban on planting Bt corn; says it could be cause of illnesses, *Mindanews*, 19 October 2004. Details available at <www.mindanews.com/2004/10/19nws-btcorn.html>

FAO. 2000. Details available at http://WWW.Faostat.com

FAO. 2004. Details available at http://WWW.Faostat.com

Harlan, J.R. 1965. The possible role of weed races in the evolution of cultivated plants. *Euphytica* 14(2): 173–176

Fernandez-Cornejo, J. and M. F. Caswell. 2006. The first decade of genetically engineered crops in the United States. *USDA-ERS Economic Information Bulletin* 11(1): 1–36

Frankel, R. and E. Galun. 1977. Allogamy. In *Pollination Mechanisms, Reproduction and Plant Breeding*, pp. 79–234. Berlin Heidelberg: Springer

Haas, D. and G. Défago. 2005. Biological control of soil-borne pathogens by fluorescent pseudomonads. *Nature Reviews Microbiology* 3(4): 307-319

Ingram, G. C., C. Boisnard Lorig, C. Dumas, and P. M. Rogowsky. 2000. Expression patterns of genes encoding HD ZipIV homeo domain proteins define specific domains in maize embryos and meristems. *The Plant Journal* 22(5): 401–414

James, C. 2003. Global review of commercialized transgenic crops: 2002 feature: Bt maize. ISAAA Briefs No. 26. ISAAA: Ithaca, NY

Jones, M. D. and J. S. Brooks. 1950. Effectiveness of distance and border rows in preventing out-crossing in corn. Technical Bulletin. *Oklahoma Agricultural Experiment Station* 38: 18

Kiang, Y. T., J. Antonovics, and L. Wu. 1979. The extinction of wild rice (*Oryza perennis formosana*) in Taiwan. *Journal of Asian Ecology* 1(1): 9

Khush, G .S. 1993. Floral structure, pollination biology, breeding behaviour, transfer distance and isolation considerations. *Biotechnology Series* (1)

Langevin, S. A., K. Clay, and J. B. Grace. 1990. The incidence and effects of hybridization between cultivated rice and its related weed red rice (*Oryza sativa* L.). *Evolution* 44(4): 1000–1008

Lavigne, C., E. K. Klein, P. Vallée, J. Pierre, B. Godelle, and M. Renard. 1998. A pollen-dispersal experiment with transgenic oilseed rape. Estimation of the average pollen dispersal of an individual plant within a field. *Theoretical and Applied Genetics* 96(6–7): 886–896

Llewellyn, D., and G. Fitt. 1996. Pollen dispersal from two field trials of transgenic cotton in the Namoi Valley, Australia. *Molecular Breeding* 2(2): 157–166

Louette, D. and M. Smale. 2000. Farmers' seed selection practices and traditional maize varieties in Cuzalapa, Mexico. *Euphytica* 113(1): 25–41

Martínez-Soriano, J. P. R. and D. S. Leal-Kievezas. 2000. Transgenic maize in Mexico: No need for concern. *Science-International Edition-AAAS* 287(5457): 1399

Mentewab, A. and C. N. Stewart, Jr. 2005. Overexpression of an *Arabidopsis thaliana* ABC transporter confers kanamycin resistance to transgenic plants. *Nature Biotechnology* 23: 1177–1180. Details available at <http://plantsciences.utk.edu/pdf/nbt1134.pdf>

Messeguer, J., C. Fogher, E. Guiderdoni, V. Marfa, M. M. Catala, G. Baldi, and E. Mele. 2001. Field assessments of gene flow from transgenic to cultivated rice (*Oryza sativa* L.) using a herbicide resistance gene as tracer marker. *Theoretical and Applied Genetics* 103(8): 1151–1159

Noldin, J. A, J. M. Chandler, and G. N. McCauley. 1999. Red rice (*Oryza sativa*) biology. I. Characterization of red rice ecotypes. *Weed Technology* 13(1): 12–18

Oard, J., M. A. Cohn, S. Linscombe, D. Gealy, and K. Gravois. 2000. Field evaluation of seed production, shattering, and dormancy in hybrid populations of transgenic rice (Oryza sativa) and the weed, red rice (Oryza sativa). *Plant Science* 157(1): 13–22

Onyango, B. M. and R. M. Nayga Jr. 2004. Consumer acceptance of nutritionally enhanced genetically modified food: relevance of gene transfer technology. *Journal of Agricultural and Resource Economics* 567–583

Phillips, P. W. B. and D. Corkindale. 2003. Marketing GM foods: the way forward. *AgBioForum* 5(3): 113–121. Details available at www.agbioforum.missouri.edu/v5n3/v5n3a06-phillips.htm

Phipps, R. H. and J. R. Park. 2002. Environmental benefits of genetically modified crops: global and European perspectives on their ability to reduce pesticide use. *Journal of Animal and Feed Sciences* 11(1): 1–18

Pohl-Orf, M., U. Brand, I. Schuphan, and D. Bartsch. 1998. What makes a transgenic plant an invasive alien? Genetically modified sugar beet and their potential impact on populations of the wild beet Beta vulgaris subspec. maritima Arcang. *Plant Invasions: Ecological Mechanisms and Human Responses* 235–243

Pohl-Orf, M., U. Brand, S. Drieben, P. R. , Hesse, M. , Lehnen, C. Morak, and D. Bartsch. 1999. Overwintering of genetically modified sugar beet, *Beta vulgaris* L. subsp. vulgaris, as a source for dispersal of transgenic pollen. *Euphytica* 108(3): 181–186

Pollock, J. 2008. Green revolutionary. *Technology Review* 111: 74–76

Pollack, M. A. and G. C. Shaffer. 2000. Biotechnology: the next transatlantic trade war? Washington Quarterly 23(4): 41–54

Qaim, M., A. Subramanian, , G. Naik, and D. Zilberman. 2006. Adoption of Bt cotton and impact variability: Insights from India. *Applied Economic Perspectives and Policy* 28(1): 48–58

Resnick, D. 2008. Human genetic engineering. In *Contemporary Issues in Bioethics*, edited by T. L. Beauchamp, L. Walters, J. Kahn, and A. Mastroianni, pp. 266–272. California: Wadsworth Pub Company

Saeglitz, C., M. Pohl, and D. Bartsch. 2000. Monitoring gene flow from transgenic sugar beet using cytoplasmic male sterile bait plants. *Molecular Ecology* 9(12): 2035–2040

Scheffler, J. A., R. Parkinson, and P. J. Dale. 1993. Frequency and distance of pollen dispersal from transgenic oilseed rape (*Brassica napus*). *Transgenic Research* 2(6): 356–364

Scheffler, J. A., Parkinson, and P. J. Dale. 1995. Evaluating the effectiveness of isolation distances for field plots of oilseed rape (Brassica napus) using a herbicide resistance transgene as a selectable marker. *Plant Breeding* 114(4): 317–321

Scutt, C. P., E. I. Zubko, and P. Meyer. 2002. Techniques for the removal of marker genes from transgenic plants. *Biochimie* 84:1119–1126

Smith, J. M. 2006. Debate on GM foods. *Food consumer*. Details available at www.foodconsumer.org/777/8/printer_Part_3_Debate_on_GM_foods.shtml

Staniland, B. K., P. B. McVetty, L. F. Friesen, S. Yarrow, G. Freyssinet, and M. Freyssinet. 2000. Effectiveness of border areas in confining the spread of transgenic *Brassica napus* pollen. *Canadian Journal of Plant Science* 80(3): 521–526

Stierle, Andrea. Personal communication. 9 May 2006

Treu, R. and J. Emberlin. 2000. Pollen Dispersal in the Crops Maize (*Zea mays*), Oil Seed Rape (*Brassica napus* ssp *oleifera*), Potatoes (*Solanum tuberosum*), Sugar Beet (*Beta vulgaris* ssp. *vulgaris*) and Wheat (*Triticum aestivum*). A Report for the Soil Association from the National Pollen Research Unit, University College

Trewavas, A. 1999. Much food, many problems. *Nature* 402 (6759): 231-232

Tynan, J. L., M. K. Williams, and A. J. Conner. 1990. Low frequency of pollen dispersal from a field trial of transgenic potatoes. *Journal of Genetics and Breeding* 44(4): 303–305

Wendt, J. and J. Izquierdo. 2001. Biotechnology and development: A balance between IPR protection and benefit-sharing. *Electronic Journal of Biotechnology* 4(3): 15–16

Wheeler, D. L., C. Chappey, A. E. Lash, D. D. Leipe, T. L. Madden, G. D. Schuler, and B. A. Rapp. 2000. Database resources of the national center for biotechnology information. Nucleic Acids Research 28(1): 10–14

Yoshida, S. 1981. *Fundamentals of Rice Crop Science*. Manila, Philippines: International Rice Research Institute

Zhang, J., S. Kumar and M. Nei. 1997. Small-sample tests of episodic adaptive evolution: a case study of primate lysozymes. Molecular Biology and Evolution 14:1335–1338

Zhang, P., B. Timakov, R. L. Stankiewicz, and I. Y. Turgut. 2000. A trans-activator on the Drosophila Y chromosome regulates gene expression in the male germ line. *Genetica* 109(1-2): 141–150

Index

About the Author

S Umesha is Associate Professor at the Department of Biotechnology in University of Mysore. He did his doctorate in Applied Botany from University of Mysore and postdoctorate from Virginia Tech University, USA. His fields of specialization include plant biotechnology, molecular diagnostics of plant/animal pathogens, and host–pathogen interactions. He was visiting researcher in Copenhagen University, Denmark (2002), Darmstadt University, Germany (2005), and Chinese Academy of Agriculture Research, Wuhan, People's Republic of China (2011).

Dr Umesha has published 90 research/review articles in reputed international/national journals, presented 50 research papers in national/international conferences/seminars, and contributed four book chapters. He has completed six research projects as Principal Investigator for Department of Science and Technology, Department of Biotechnology, Ministry of Science and Technology, and University Grants Commission, Government of India. Under his supervision, 8 students have been awarded PhD degrees and 22 students have been awarded MPhil degrees.

He has been awarded Professor K S Jagdish Chandra Young Scientist award from University of Mysore, Young Scientist award from Department of Science and Technology, Government of India under FTP programme, Overseas Fellowship award from Department of Biotechnology, Government of India, and Best Research Paper awards from professional societies, such as SBC, AMI, IBS, ISMPP, and IPS. Dr Umesha has been conferred with fellowship from National Academy of Biological Sciences, Chennai and Indian Society of Mycology and Plant Pathology, Udaipur.

Printed and bound by CPI Group (UK) Ltd, Croydon, CR0 4YY

17/10/2024

01775681-0014